T0189945

Handbook of Fractional Calculus for Engineering and Science

Advances in Applied Mathematics

Series Editors

Daniel Zwillinger, H. T. Banks

Introduction to Quantum Control and Dynamics
Domenico D'Alessandro

Handbook of Radar Signal Analysis
Bassem R. Mahafza, Scott C. Winton, Atef Z. Elsherbeni

Separation of Variables and Exact Solutions to Nonlinear PDEs
Andrei D. Polyanin, Alexei I. Zhurov

Boundary Value Problems on Time Scales, Volume I
Svetlin Georgiev, Khaled Zennir

Boundary Value Problems on Time Scales, Volume II
Svetlin Georgiev, Khaled Zennir

Observability and Mathematics. Fluid Mechanics, Solutions of Navier-Stokes Equations, and Modeling
Boris Khots

Handbook of Differential Equations, 4th Edition
Daniel Zwillinger, Vladimir Dobrushkin

Experimental Statistics and Data Analysis for Mechanical and Aerospace Engineers
James Middleton

Advanced Engineering Mathematics with MATLAB
Dean G. Duffy

Handbook of Fractional Calculus for Engineering and Science
Harendra Singh, H. M. Srivastava, and Juan J Nieto

https://www.routledge.com/Advances-in-Applied-Mathematics/book-series/CRCADVAPPMTH?pd=published,forthcoming

Handbook of Fractional Calculus for Engineering and Science

Edited by

Harendra Singh

H. M. Srivastava

Juan J. Nieto

CRC Press is an imprint of the
Taylor & Francis Group, an informa business
A CHAPMAN & HALL BOOK

First edition published 2022
by CRC Press
6000 Broken Sound Parkway NW, Suite 300, Boca Raton, FL 33487-2742

and by CRC Press
2 Park Square, Milton Park, Abingdon, Oxon, OX14 4RN

Library of Congress Cataloging in Publication Data
A catalog record has been requested for this book

ISBN: 978-1-032-04779-9 (hbk)
ISBN: 978-1-032-20430-7 (pbk)
ISBN: 978-1-003-26351-7 (ebk)

DOI: 10.1201/9781003263517

Typeset in Palatino
by SPi Technologies India Pvt Ltd (Straive)

Contents

Preface

This book includes several topics in the area of fractional calculus. It offers a collection of research-article chapters on mathematical models formulated by fractional differential equations describing engineering and science problems. These chapters include useful methods for solving various types of such models, as well as their significance and relevance in other areas of scientific research and study. The book consists of twelve chapters and is organized as follows.

Chapter 1 presents a solution to the vibration model with a space Liouville derivative and a time Caputo derivative when the order of the fractional derivative is in the range (1, 2). It suggests an analytical solution based on the Fourier method. An implicit difference method for obtaining the numerical solution to fractional vibration equations (FVEs) is constructed. First, the FVEs are equivalently transformed by the integral operator into their integro partial differential problem. Second, the classical central difference approximation and a midpoint formula are used for the second- and first-order derivatives, respectively. Meanwhile, to discretize Caputo and Liouville derivatives, a Crank–Nicholson technique based on the shifted and weighted Grunwald difference scheme is utilized. This model is used to describe the mechanical oscillation process in a viscoelastic medium. The analysis of stability and convergence is rigorously discussed. Finally, the numerical experiments show that the suggested methods are very effective.

Chapter 2 aims to find the solution for systems of nonlinear fractional-order differential equations arising in a helium-burning network using the q-homotopy analysis transform method (q-HATM). It considers a model which illustrate the chemical reactions involved in the helium-burning network. The Atangana–Baleanu (AB) operator with fractional order and algorithm is an elegant consolidation of Laplace transform with q-HAM. The existence, uniqueness, and competence of this model are illustrated. The chapter concludes that the results achieved confirm that the fractional differential system is very effective, highly methodical, accurate, and easily applicable to a range of scientific disciplines.

Chapter 3 presents a reproducing kernel Hilbert space method (RKHSM) for studying time-fractional PDEs. Specifically, it applied the RKHSM to time-fractional heat-like and Navier–Stokes equations when the equations' coefficients vary between constants and variables. These equations have significant applications in applied science and engineering. The current approach utilizes some important binary reproducing kernel Hilbert (RK) spaces with appropriate RK functions. Error estimations and convergence analysis of the proposed method are discussed. The assessment of the

RKHSM is made by testing illustrative applications which have exact solutions. The results suggest that the RKHSM is a very effective and highly convenient solution method.

Chapter 4 is a systematic study of the spectral collocation of models of fractional differential equations with the help of such special functions as Chebyshev and Legendre polynomials. The models are converted into those involving a set of ordinary differential equations, which are solved using known methods such as the finite difference method. The Newton–Raphson method (NRM) is used to find the numerical solutions for a set of nonlinear algebraic equations. These results are verified in the classical case by comparison with an exact solution. For non-integer cases, the accuracy of the solution is verified by computing the residual error function (REF). In all cases, the results are found to be fairly accurate and, in all calculations, the software program Mathematica is used.

Chapter 5 applies the powerful analytical approach of the sine Gordon equation method to construct new complex dark–bright solitons and other solutions to the conformable (2+1)-dimensional generalized Bogoyavlensky–Konopelchanko equation. It plots 3D and 2D graphs to illustrate the characteristics of the results under suitable conditions.

Chapter 6 presents the nonlinear propagation of dust-ion-acoustic (DIA) waves in a dusty plasma with bi-Maxwellian electrons by deriving the Gardner equation using the reductive perturbation technique. The time-fractional Gardner's equation is obtained by the Euler–Lagrange equation. The basic features (amplitude, width, etc.) of the hump (positive potential)- and dip (negative potential)-shaped DIA solitons are found to exist beyond the Korteweg de Vries (KdV) limit. These solitons are qualitatively different from the KdV and modified KdV solitons. The time-fractional damped Gardner equation is solved analytically using the extended G0/G-expansion method.

Chapter 7 deals with the construction of a numerical scheme to obtain the numerical solution of fractional differential equations (FDEs). The construction of the scheme is based on the generalized Taylor's series and standard Runge–Kutta method. FDEs have wide areas of applications in science and engineering such as convection diffusion problems, plasma physics, mechanics, polymer physics and rheology, diffraction theory, and regular variations in thermodynamics. The chapter seeks to develop the most appropriate RK-type method to find the numerical solution of IVP in fractional-order differential equations. Numerical experiments governing the applications of FDEs in physical problems are given to ensure the efficiency and utility of the proposed scheme.

Chapter 8 studies the fractional kinetic equation. The Laplace transform approach is employed to obtain the solution for a generalized fractional kinetic problem, in terms of polynomial weighted incomplete $\bar{I}$-functions, incomplete I-functions, incomplete H-functions and incomplete $\bar{H}$-functions.

Fractional kinetic equations are relevant to a variety of problems related to science and engineering. The analytic solutions for numerous forms of kinetic equations are examined using specific examples.

Chapter 9 discusses the slippage of oscillations exhibited by fractional second-grade fluid according to its viscosity. The mathematical model is fractionalized and then tackled by means of Laplace transform. The optimal analytical solutions for velocity and shear stress are investigated by invoking the Wright generalized hypergeometric function. Newtonian and non-Newtonian solutions are employed to demonstrate the effective influence of slipping and non-slipping oscillations of fractional and non-fractional second-grade fluid. The sub-solutions of fractional as well as non-fractional second-grade fluid are compared for slippage effects on the oscillations of fluid. Variable viscosity under the assumption of slip condition on boundary condition produces strong similarities and differences between non-integer and integer mathematical models on the basis of the material and embedded parameters.

Chapter 10 deals with a new fractional chaotic system described by the Caputo derivative. The Lyapunov exponents give the nature of the behaviors – chaotic and hyper chaotic – of the solutions of the fractional system under consideration. Phase portraits are obtained after implementation of the numerical scheme, which includes the numerical discretization of the Riemann–Liouville fractional integral. The stability of the equilibrium points is illustrated using the Matignon criterion. The chapter also discusses the impacts of the initial conditions on the behaviors of solutions. In this chapter, the bifurcation diagrams and the Lyapunov exponents play a crucial role in the nature of the dynamics. The circuit schematic should be implemented for real-world science and engineering applications.

Chapter 11 focuses on travelling wave solutions to the conformable Gross–Pitaevskii equation. Two powerful methods, the sine-Gordon expansion method and the modified exponential function method, are used. Many entirely new travelling wave solutions to the governing model are extracted. Various simulations are also plotted with the help of a software package.

Chapter 12 establishes new fractional calculus results for the generalized Mathieu-type and alternating Mathieu-type series, by extensive use of Marichev–Saigo–Maeda operator tools. The results are expressed in terms of the generalized H-function. From the application point of view, all the results are also deduced in terms of Fox's H-function.

Editors

Dr. Harendra Singh is an Assistant Professor in the Department of Mathematics, Post-Graduate College, Ghazipur, Uttar Pradesh, India. He holds a PhD in mathematics from the Indian Institute of Technology (BHU), Varanasi, India. He has qualified GATE, JRF and NBHM in Mathematics. He has also been awarded a post-doctoral fellowship in mathematics at the National Institute of Science Education and Research (NISER), Bhubaneswar Odisha, India. His research is widely published. He has edited *Methods of Mathematical Modelling Fractional Differential Equations* (CRC Press, 2019) and *Advanced Numerical Methods for Differential Equations: Applications in Science and Engineering* (CRC Press, 2021). His 45 research papers have been published in various journals of repute with h-index of 19 and i10 index 26.

Dr. H. M. Srivastava is a Professor Emeritus, Department of Mathematics and Statistics, University of Victoria, British Columbia, Canada. He holds a PhD from Jai Narain Vyas University of Jodhpur in India. He has held numerous visiting, honorary and chair professorships at many universities and research institutes in different parts of the world. He is also actively associated with numerous international journals. Professor Srivastava's research interests include several areas of pure and applied mathematical sciences. He has published 36 books and more than 1,350 peer-reviewed journal articles.

Dr. Juan J. Nieto is a Professor at the University of Santiago de Compostela, Spain. Professor Nieto's research interests include several areas of pure and applied mathematical sciences. He has published many books, monographs, and edited volumes, and more than 650 peer-reviewed international scientific research journal articles. Professor Nieto has held numerous visiting and honorary professorships. He is also actively associated editorially with numerous journals.

Contributors

Kashif Ali Abro
Institute of Ground Water Studies, Faculty of Natural and Agricultural Sciences
University of the Free State
Bloemfontein, South Africa

Ali Akgül
Art and Science Faculty, Department of Mathematics
Siirt University
Siirt, Turkey

Lanre Akinyemi
Department of Mathematics
Lafayette College
Easton, PA, USA

Temirkhan S. Aleroev
Department of Applied Mathematics
Moscow State University of Civil Engineering
Moscow, Russia

Abdon Atangana
Institute of Ground Water Studies, Faculty of Natural and Agricultural Sciences
University of the Free State
Bloemfontein, South Africa

Nourhane Attia
Dynamic of Engines and Vibroacoustic Laboratory
University M'hamed Bougara of Boumerdes
Boumerdes, Algeria

Haci Mehmet Baskonus
Faculty of Education
Harran University
Sanliurfa, Turkey

Sanjay Bhatter
Department of Mathematics
Malaviya National Institute of Technology
Jaipur, India

Hasan Bulut
Department of Mathematics
Firat University
Elazig, Turkey

Naresh M. Chadha
School of Physical Sciences
DIT University
Uttarakhand, India

Fayık Değirmenci
Department of Mathematics
Harran University
Sanliurfa, Turkey

Asmaa M. Elsayed
Department of Applied Mathematics
Moscow State University of Civil Engineering
Moscow, Russia

Wei Gao
School of Information Science and Technology
Yunnan Normal University
Yunnan, China

Kamlesh Jangid
Department of HEAS (Mathematics)
Rajasthan Technical University
Kota, Rajasthan, India

M. M. Khader
Department of Mathematics and Statistics, College of Science
Al Imam Mohammad Ibn Saud Islamic University (IMSIU)
Riyadh, Kingdom of Saudi Arabia

Ajay Kumar
Bakhtiyarpur College of Engineering (Department of Science and Technology, Govt. of Bihar)
Patna, India

Sapna Meena
Department of Mathematics
Malaviya National Institute of Technology
Jaipur, India

Kajal Mondal
Department of Mathematics
Cooch Behar Panchanan Barma University
India

Ram K. Pandey
Department of Mathematics and Statistics
Dr. H.S. Gour University
Sagar, India

Rakesh K. Parmar
Department of HEAS (Mathematics)
University College of Engineering and Technology
Bikaner, Rajasthan, India

S. D. Purohit
Department of HEAS (Mathematics)
Rajasthan Technical University
Kota, India

Arjun K. Rathie
Department of Mathematics
Vedant College of Engineering and Technology (Rajasthan Technical University)
Bundi, India

Santanu Raut
Department of Mathematics
Mathabhanga College
Cooch Behar, India

Khaled M. Saad
Department of Mathematics, College of Arts and Sciences
Najran University
Najran, Kingdom of Saudi Arabia

Bilgin Senel
Fethiye Faculty of Business Administration
Mugla Sitki Kocman University
Mugla, Turkey

Mine Senel
Fethiye Faculty of Business Administration
Mugla Sitki Kocman University
Mugla, Turkey

Ndolane Sene
Laboratoire Lmdan, Département de Mathématiques de la Décision, Faculté des Sciences Economiques et Gestion
Université Cheikh Anta Diop de Dakar Dakar Fann, Senegal

Harendra Singh
Department of Mathematics
Post-Graduate College
Ghazipur, India

Ambreen Siyal
Department of Basic Sciences and Related Studies
Mehran University of Engineering and Technology
Jamshoro, Pakistan

H. M. Srivastava
Department of Mathematics and Statistics
University of Victoria
Victoria, British Columbia, Canada

Neelam Tiwari
Department of Physics
Govt. P.M.R.S. College
GPM (C. G.), India

Shruti Tomar
School of Physical Sciences
DIT University
Uttarakhand, India

P. Veeresha
Department of Mathematics
CHRIST (Deemed to be University)
Bengaluru, India

Gulnur Yel
Final International University
Kyrenia, Turkey

1

Analytical and Numerical Methods to Solve the Fractional Model of the Vibration Equation

Temirkhan S. Aleroev
Moscow State University of Civil Engineering, Moscow

Asmaa M. Elsayed
Moscow State University of Civil Engineering, Moscow
Zagazig University, Zagazig, Arab Republic of Egypt/Egypt

CONTENTS

1.1 Introduction

Fractional partial differential equations (FPDEs) are found in many different branches of science and engineering, such as hydrodynamics, electroanalytical chemistry, quantum science, viscoelastic mechanics, signal image processing, chain-breaking of polymer materials, molecular spectrum, and

DOI: 10.1201/9781003263517-1

anomalous diffusion process of ions in nerve cells [1–4]. Also, PDEs with a fractional order have been used to simulate the flow and filtration of a fluid in a porous fractal medium. The use of fractional derivatives (FDs) for modeling real physical processes or environments leads to the appearance of equations containing derivatives and integrals of fractional order in addition to the classical ones. Researchers have focused their efforts on fractional-order physical models [5, 6] because of the material's dynamic and viscoelastic behavior [7]. As a result, the model of fractional order is widely employed to model the frequency apportionment of structural damping mechanisms [8, 9]. An intrinsic multiscale existence of these operators is an interesting feature. As a consequence, memory effects (i.e., a system's response is a function of its previous history) are enabled by time-fractional operators, while non-local and scale effects are enabled by space-fractional operators. Fractional analysis is used in many areas of science, including nonlinear biological processes, solid-state mechanics, field theory, control theory, friction, and fluid dynamics [10]. For the study of fractionally damped viscoelastic material, Josefson and Enelund [11] employed the finite element scheme. In problems of linear viscoelasticity, fractional calculus has been widely used [12–14]. In the design of buildings and other facilities, concrete structures are used, because they are sturdy, reliable, and robust. At the same time, the surface of concrete structures is susceptible to major damage. Therefore, a composite with enhanced operating characteristics is currently being developed, based on a concrete-blend, polymer concrete, which is characterized by greater tolerance to moisture, chemical compounds, low temperatures, and toughness relative to concrete. It is possible to simulate polymer concrete as a collection of solid filler granules contained in a viscoelastic medium [15–18]. The fractional oscillator equation describes transverse movement under the control of the force of gravity or the exterior force of a filler granule. Thus, the substitution of concrete for polymer concrete equates to the substitution of second-order differential equations with FDEs. Special attention is given to the use of fractional calculus to establish better mathematical models of many real-world issues. Many scholars have described the theoretical evolution and implementation of fractional calculus. FDEs are more pragmatic to display natural phenomena and have been utilized in many branches of applied mathematics [19–23]. In some significant and groundbreaking books [12, 24–27], the FDs method has been applied to a wide range of physical problems in science and engineering and in physical models. The dynamic formulation of the problem with viscoelastic damping and its application in science and engineering are examined in the next section.

The chapter is organized as follows. In Section 1.2, we discuss the model's construction and its application in science and engineering. FVEs are presented in Section 1.3. In Section 1.4, we derive some simple definitions and essential lemmas. In Section 1.5, the Fourier method for solving FVEs is discussed. We also describe the ADI scheme for time–space FVEs in Section 1.5. Section 1.6 considers numerical solutions using the finite difference scheme.

The suggested method in Section 1.7 shows the proposed scheme's stability and convergence. To validate our theoretical findings, numerical experiments are performed in Section 1.8. Finally, a brief conclusion is presented in the last section.

1.2 The Problem Formulation for FVEs with Viscoelastic Damping and its Application in Science and Engineering

In the description of vibration models, fractional differentiation operators are commonly utilized. It is well known that FD equations accurately describe the motion of vibrations with elastic and viscoelastic components [28, 29]. Damped vibrations with fractional damping are also defined by these equations; where damping is characterized as the restraint of oscillatory or vibration motion, it decreases, restricts, and prevents an oscillatory mechanism from oscillating. When a damping force becomes viscoelastic, it incorporates viscous and elastic properties to suppress or damp the system's vibration. Likewise, the damping force is known as viscous-viscoelastic when the model reaches viscous rubbing at low speed and reaches pure viscoelastic friction at high speed, in particular the vibrations of an aircraft wing in a supersonic gas flow, vibrations of nanoscale sensors, etc. [30]. Figure 1.1 depicts free oscillation with viscoelastic damping, the model under discussion in this chapter.

The viscoelastic vibration mathematical model is described as:

$$mw'' + bD^{\gamma}w + aw = F, \tag{1.1}$$

where m is the mass of the system, γ, b are the fractional order viscoelastic vibration and coefficient of the damping, respectively. a is a natural frequency and D^{γ}, F are the FD operator and external force respectively. The findings of [31] demonstrate that the outcome of solving problems can be used to simulate alteration in the deformation-strength properties of polymer concrete under the effect of gravity force (2.1). The researchers looked at samples

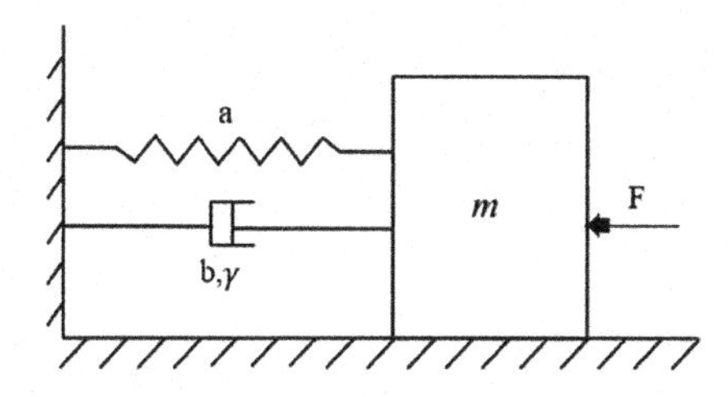

FIGURE 1.1
Viscoelastic damping of the vibration model.

of polyester resin-based polymer concrete (chloride-1, Diane, 1-dichloro-2, diacyl, and 2-diethylene).

FDs are frequently employed to characterize viscoelastic characteristics of sophisticated materials, as well as dissipative forces in structural dynamics [32]. In [33], the analytical solution for linear fractional vibration dependent on various fractional calculus constitutive problems is given using the Laplace transform technique. In the fractional calculus model of viscoelastic conduct, fractional-order derivatives are used to link stress and strain fields in viscoelastic materials. The evolution of such models has previously been discussed, and it has been proved that FD-constitutive equations have been related to molecular theories explaining the macroscopic behavior of viscoelastic media. The improvement of new damping mechanisms in technology and engineering focused on a continuum of damping factors spread uninterruptedly during the relaxation or creep periods instead of a single method of damping elements has reignited interest in viscoelastic models and their application to complex problems [34]. Previous efforts to explain the mechanical characteristics of viscoelastic solids have failed because the mathematical models that describe the action of these materials have not been precisely related to the underlying physical concepts. To explain the mechanical properties of these components, engineers were forced to use phenomenological (empirical) methods. Also on an electronic machine, the order of the extended formulas for systems of engineering importance could be very high, and the scale of the matrices prohibitive to control. FD strain-stress fundamental relations for viscoelastic solids do not just explain the mechanical characteristics of certain components but also contribute to closed-form formulations of numerical simulation motion equations for viscoelastically damped systems. In the study of vibration damped vibrations, numerical techniques are often used. In 1983, Bagley and Torvik were among the first to use numerical approaches to solve dynamic models involving viscoelasticity damped models.

1.3 Mathematical Model of FVEs

First, we note that FDs in space can be utilized to simulate irregular diffusion or scattering, and FDs in time can be utilized to stimulate certain processes with memory. Let us study the following vibration string equation in the domain $D = \{0 < q < I, 0 < p < \aleph\}$:

$$\frac{\partial^2 w(p,q)}{\partial q^2} = \frac{\partial^2 w(p,q)}{\partial p^2} + a\,{}^{C}D_q^{\beta} w(p,q) + b\,{}^{R}D_p^{\alpha} w(p,q), \tag{1.2}$$

with boundary conditions

$$w(0,q) = w(\aleph,q) = 0, \tag{1.3}$$

and initial conditions

$$\begin{aligned} w(p,0) &= \varphi(p), \\ w_q(p,0) &= \psi(p), \end{aligned} \tag{1.4}$$

where D_q^β -denotes the Caputo's temporal derivative with respect to the variable q of order $\beta(1 < \beta < 2)$, D_p^α is the Liouville's spatial derivative with respect to the variable p of order $\alpha(1 < \alpha < 2)$, i.e.,

$$D_q^\beta w(p,q) = \frac{1}{\Gamma(2-\beta)} \int_0^q (q-\tau)^{1-\beta} w''(p,\tau) d\tau,$$

$$D_p^\alpha w(p,q) = \frac{1}{\Gamma(2-\alpha)} \frac{d^2}{dp^2} \int_0^p (p-\xi)^{1-\alpha} w(\xi,q) d\xi.$$

The connection between derivatives of fractional order for both Caputo and Liouville operators can be expressed as

$${}^{Rl}D_q^\gamma g(q) = {}^C D_q^\gamma g(q) + \sum_{k=0}^{n-1} \frac{g^{(k)}(0) q^{k-\gamma}}{\Gamma(k+1-\gamma)}. \tag{1.5}$$

Analytical methods have the advantage of explaining the fundamentals of mechanical engineering problems and their physical connotations, making it possible to analyze a variety of physical and mechanical engineering problems and taking less time than the numerical method. However, scholars have found that obtaining exact solutions to PDEs is extremely complicated. Thus, many numerical methods such as the finite difference method, homotopy perturbation methods, finite element method, Galerkin method, Adomian decomposition method, and spectral method, have been developed to obtain the numerical solutions for FPDEs; see [18, 20, 21, 35–38] for examples. Our main goal in this work is to create a high-order numerical method for Eq. (1.2) and perform the corresponding numerical analysis for the method suggested, which is extended in [3, 18, 36, 39, 40]. Up to now, for FPDEs with nonlinear terms, many linearized schemes have been constructed. For diffusion problem in the time-derivative term, Li and Xu [41] established a time–space spectral system. By combining the compact difference method for spatial discretization and L_1 approximation for temporal discretization, a finite difference scheme was derived in [42]. In 2020, Guo et al. [43] developed a novel linearized finite difference/spectral-Galerkin scheme in order to solve the three-dimensional distributed-order space–time fractional nonlinear reaction-diffusion-wave equation, and proved the stability and convergence of the suggested scheme with numerical results. In 2019, Huang et al. [44] considered a two-dimensional nonlinear super-diffusion problem in the time-derivative term and proposed two conservative linearized ADI schemes to get the approximate solution of the model. Furthermore, a proposed scheme was proved that is uniquely solvable and convergent with $\mathcal{O}\left(\tau + h_x^2 + h_y^2 + \tau^\alpha\right)$ order in L_2 norm with mesh size h and time step τ. For the space–time fractional telegraph equation, Zhao and Li [45] proposed a linearized fractional difference/finite element approximation, and the suggested scheme proved to be unconditionally stable and accurate in both time and space using the energy method and mathematical induction. In [46], Sun and Wu introduced a finite difference method by adding two additional parameters to turn the original equation into a low-order equation system enabling error analysis. In [47], the fractional-order delay model was analyzed by Harendra Singh using the Chebyshev polynomials method. He and Pan [48] proposed an ADI scheme for Ginzburg-Landau FD equations, and the proposed method was proved to be unconditionally stable using the energy method and mathematical induction. To construct a compact difference method for fractional-order diffusion-wave equations, the equivalent integrodifferential equations and product trapezoidal law were used by Chen and Li [49]. Harendra Singh [50] proposed a computational method for solving the fractional-order advection-dispersion problem using the Jacobi collocation method. In 2016, Wang et al. [51] studied finite difference methods for both temporal and spatial FDs for differential equations. To improve the efficiency, they also proposed a preconditioner for the implementation of the scheme. They also developed a compact ADI finite difference method for studying the two-dimensional time fractional diffusion-wave equation in [52]. A reliable algorithm for giving the numerical solution of the Lane-Emden nonlinear equations and analyzing the error of the proposed scheme was studied in [53].

1.4 Notations and Preliminaries

This section describes several definitions and lemmas that are used in subsequent sections of this chapter.

Lemma 1.4.1

[54] If $w(q) \in C^2([0,I])$, then

$$ {}_0J_q w(q_{j+\frac{1}{2}}) = \frac{1}{2}\left[{}_0J_q w(q_{j+1}) + {}_0J_q w(q_j)\right] + \mathcal{O}(\tau^2), $$

where ${}_0J_q$ is the integral operator of the first order, and $q_{j+\frac{1}{2}} = q_j + \frac{1}{2}\tau$ with a size step τ; $q_j = j\tau$, moreover if $w(q) \in C^3([0,I])$; $0 < \theta < 1$,

$$ w'(q_{j+\frac{1}{2}}) = \frac{w(q_{j+1}) - w(q_j)}{\tau} + \mathcal{O}(\tau^2), $$

and

$$ {}_0^{Rl}D_q^{\theta} w(q_{j+\frac{1}{2}}) = \frac{1}{2}\left[{}_0^{Rl}D_q^{\theta} w(q_{j+1}) + {}_0^{Rl}D_q^{\theta} h(q_j)\right] + \mathcal{O}(\tau^2). $$

Lemma 1.4.2

(see [55]) assume $w(p) \in C^4([p_{i-1}, p_{i+1}])$, let $\zeta(r) = w^{(4)}(p_i + rh) + w^{(4)}(p_i - rh)$, then

$$ \delta_p^2 w(p_i) = \frac{w(p_{i+1}) - 2w(p_i) + w(p_{i-1})}{h^2}. $$

Lemma 1.4.3

(see [56]) Assume ϖ_j is the weights of the generating function $(3/2 - 2z + z^2/2)^{-1}$, that is expressed as, $\varpi_j = 1 - 3^{-(j+1)}$. If $w(q) \in C^2([0,I])$ and $w(0) = w'(0) = 0$, for the first-order integral, we get the second-order approximation as follows

$$ \left|{}_0J_q w(q_{j+1}) - \tau\sum_{j=0}^{m+1} \varpi_{m+1-j} w(q_j)\right| \le C \max_{0 \le q \le q_{j+1}} |w''(q)|\tau^2, \tag{1.6} $$

where C represents a general constant the value of which varies from one line to another.

Lemma 1.4.4

(see [57]) Suppose $w(q) \in L^1(R)$, the approximation of the Riemann-Liouville operator using the shifted and weighted Grünwald difference formula holds

$$ {}_0^{Rl}D_q^{\theta} w(q_{j+1}) = \tau^{-\theta} \sum_{j=0}^{m+1} \lambda_j^{(\theta)} w(q_{m+1-j}) + \mathcal{O}(\tau^2), \quad 0 < \theta < 1, \tag{1.7} $$

where

$$\lambda_0^{(\theta)} = \frac{2+\theta}{2}\rho_0^{(\theta)}, \quad \lambda_j^{(\theta)} = \frac{2+\theta}{2}\rho_j^{(\theta)} - \frac{\theta}{2}\rho_{j-1}^{(\theta)} \quad ; j \geq 1,$$

$$\rho_j^{(\theta)} = (-1)^j \begin{pmatrix} \theta \\ j \end{pmatrix} \text{for } j \geq 0.$$

Theorem 1.4.5

Suppose $\left\{\lambda_s^{\theta}\right\}_{s=0}^{\infty}$ *and* $\left\{\varpi_s\right\}_{s=0}^{\infty}$ *are the weights mentioned above in Lemma 1.4.3 and 1.4.4, respectively. So, for each integer value k as well as any real vector* $(W_1, W_2, \ldots, Ws)^T \in R^s$*, the inequalities*

$$\sum_{m=0}^{s-1}\left(\sum_{k=0}^{m}\lambda_k^{\theta}W_{m+1-k}\right)W_{m+1} < 0,$$

and

$$\sum_{m=0}^{s-1}\left(\sum_{k=0}^{m}\varpi_k W_{m+1-k}\right)W_{m+1} < 0,$$

hold.

Proof. The proof of the second inequality can be found in [58]. Therefore, we prove here only the first inequality. To prove that the above quadratic form is negative, we need only prove that the following G symmetric Toeplitz matrix is negatively defined

$$G = \begin{pmatrix} \lambda_1^{(\theta)} & \lambda_0^{(\theta)} & & & \\ \lambda_2^{(\theta)} & \lambda_1^{(\theta)} & \lambda_0^{(\theta)} & & \\ \vdots & \lambda_2^{(\theta)} & \lambda_1^{(\theta)} & \ddots & \\ \lambda_{n-2}^{(\theta)} & \cdots & \ddots & \ddots & \lambda_0^{(\theta)} \\ \lambda_{n-1}^{(\theta)} & \lambda_{n-2}^{(\theta)} & \cdots & \lambda_2^{(\theta)} & \lambda_1^{(\theta)} \end{pmatrix},$$

let $S = \dfrac{G+G^T}{2}$ be the symmetric part of the matrix G, where the generating functions of G and G^T, can be expressed respectively in the following form

$$g_G(p) = \sum_{s=0}^{\infty}\lambda_s^{(\theta)}e^{i(s-1)p}, \quad g_{G^T}(p) = \sum_{s=0}^{\infty}\lambda_s^{(\theta)}e^{-i(s-1)p},$$

therefore, the generating function of S is written as $g(\theta,p)=\frac{g_G(p)+g_{G^T}(p)}{2}$, which is a continuous and periodic real-value function in the interval $[-\pi,\pi]$, such that $g_G(p)$ and $g_G{}^T(p)$ are conjugated, and using the coefficients of $\lambda_s^{(\theta)}$ given by Lemma 1.4.4, then

$$\begin{aligned} g(\theta,p) &= \frac{1}{2}\left(\sum_{s=0}^{\infty}\lambda_s^{(\theta)}e^{i(s-1)p}+\sum_{s=0}^{\infty}\lambda_s^{(\theta)}e^{-i(s-1)p}\right) \\ &= \frac{1}{2}\left(\frac{2+\theta}{2}e^{-ip}\sum_{s=0}^{\infty}\rho_s^{(\theta)}e^{isp}-\frac{\theta}{2}\sum_{s=0}^{\infty}\rho_s^{(\theta)}e^{isp}+\frac{2+\theta}{2}e^{ip}\sum_{s=0}^{\infty}\rho_s^{(\theta)}e^{-isp}-\frac{\theta}{2}\sum_{s=0}^{\infty}\rho_s^{(\theta)}e^{-isp}\right) \\ &= \frac{2+\theta}{4}\left(e^{-ip}\left(1-e^{ip}\right)^{\theta}+e^{ip}\left(1-e^{-ip}\right)^{\theta}\right)-\frac{\theta}{4}\left(\left(1-e^{ip}\right)^{\theta}+\left(1-e^{-ip}\right)^{\theta}\right). \end{aligned}$$

Now, we will prove for $1<\alpha<2$ the function $g(\theta,p)\leq 0$, since $g(\theta,p)$ is a continuous real-value function and even, hence, we only assume its principal value in $[0,\pi]$, which leads to

$$e^{i\gamma}-e^{i\eta}=2i\sin\left(\frac{\gamma-\eta}{2}\right)e^{i\frac{(\gamma+\eta)}{2}},$$

then

$$g(\theta,p)=\left(2\sin\left(\frac{p}{2}\right)\right)^{\theta}\left(\frac{2+\theta}{2}\cos\left(\frac{\theta}{2}(p-\pi)-p\right)-\frac{\theta}{2}\cos\left(\frac{\theta}{2}(p-\pi)\right)\right),$$

so, $g(\theta,p)$ decreases with respect to θ, that's mean that $(g(\theta,p)\leq 0)$.

1.5 Solving a FVEs by Fourier Method

Let us show how to solve Eq. (1.2) using the Fourier method. To apply the Fourier method, assume that

$$w(p,q)=\Upsilon(p)\Psi(q), \tag{1.8}$$

substitute from Eq. (1.8) in Eq. (1.2)

$$\frac{\Psi''(q)}{\Psi(q)}-\frac{a}{\Psi(q)}\cdot D_q^{\beta}\Psi(q)=\frac{\Upsilon''(p)}{\Upsilon(p)}+\frac{b}{\Upsilon(p)}\cdot D_p^{\alpha}\Upsilon(p)=-\lambda. \tag{1.9}$$

For the unknown function $\Upsilon(p)$, we obtain the ordinary linear differential equation

$$\begin{aligned}\Upsilon''(p)+bD_p^{\alpha}\Upsilon(p)+\lambda\Upsilon(p)=0;\\ \Upsilon(0)=0,\Upsilon(\aleph)=0,\end{aligned} \tag{1.10}$$

the negative number λ is defined as the eigenvalue of Eq. (1.10) if and only if λ satisfies the function

$$\omega(\lambda)=\sum_{m=0}^{\infty}(-1)^m\sum_{j=0}^{m}\frac{\binom{m}{j}b^j\lambda^{m-j}}{\Gamma(2m+2-j\alpha)}. \tag{1.11}$$

In order to discuss the solution to the problem (1.10), let us integrate both sides of Eq. (1.10) for q from 0 to p,

$$\Upsilon'(p)+\frac{b}{\Gamma(1-\alpha)}\int_0^p\Upsilon(q)(p-q)^{-\alpha}dq=\lambda\int_0^p\Upsilon(q)dq+\Upsilon'(0), \tag{1.12}$$

for q from 0 to p, we can again integrate both sides of Eq. (1.12)

$$\Upsilon(p)+\int_0^p\frac{b}{\Gamma(2-\alpha)}(p-q)^{1-\alpha}-\lambda(p-q)\Upsilon(p)dq=p.\Upsilon'(0)+\Upsilon(0). \tag{1.13}$$

Table 1.1 lists the first seven eigenvalues that are found numerically using the computing language Wolfram Mathematica (Figures 1.2 and 1.3).

Then the eigenfunctions $\Upsilon_m(p)$ of the Eq.(5.3), have the form

$$\Upsilon_k(p)=A\left[p+\sum_{m=1}^{\infty}(-1)^m\sum_{j=0}^{m}\frac{\binom{m}{j}b^j\lambda_k^{m-j}}{\Gamma(2m+2-j\alpha)}p^{2m+1-j\alpha}\right];k=1,2,\ldots,7. \tag{1.14}$$

TABLE 1.1

The First Seven Eigenvalues of the Boundary Value Problem (1.10) for $\aleph=2$; $\alpha=1.47$, $b=1.8$

λ_1	λ_2	λ_3	λ_4	λ_5	λ_6	λ_7
4.983	17.479	35.949	60.421	90.511	126.105	175.501

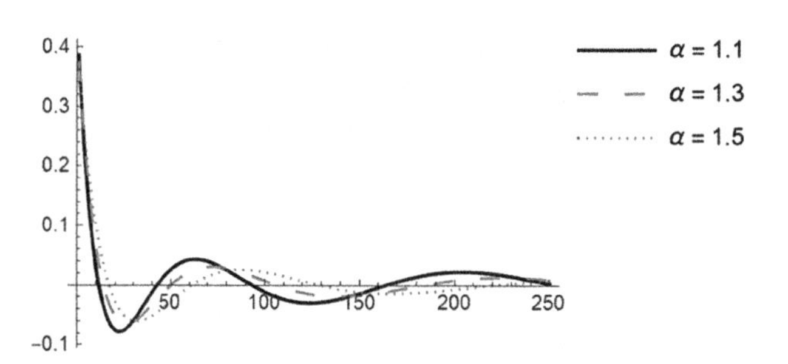

FIGURE 1.2
Eigenfunction $\omega(\lambda)$ of Eq. (1.11) at $b = 1.8$, corresponding to $\alpha = 1.1, 1.3, 1.5$.

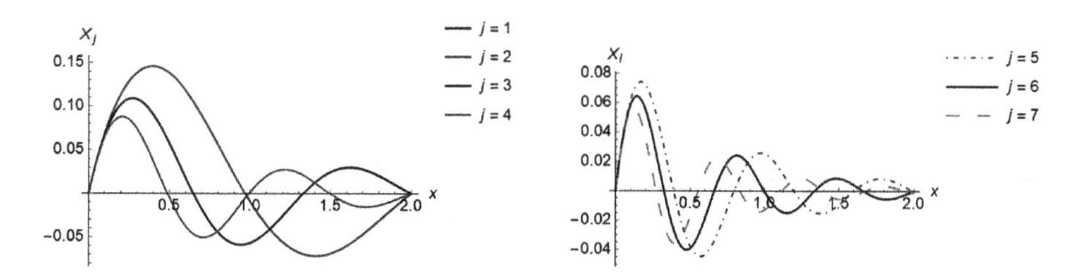

FIGURE 1.3
Eigenfunctions $\Upsilon j(p)$; $j = 1, \ldots, 7$, of Eq. (1.14) corresponding to $b = 1.8$, $\alpha = 1.47$.

The eigenfunction system (1.10) is complete [59], but it is not orthogonal. As a consequence, therefore, we propose (according to [3]) the following conjugate system to obtain that the system is biorthogonal to Eq. (1.10)

$$\begin{aligned} &\widetilde{\Upsilon}''(p) + bD_p^\alpha \widetilde{\Upsilon}(p) + \lambda \widetilde{\Upsilon}(p) = 0, \\ &\widetilde{\Upsilon}(0) = \widetilde{\Upsilon}(\aleph) = 0; \qquad \alpha \in (1,2), \end{aligned} \tag{1.15}$$

where, $\widetilde{\Upsilon}_k(p)$ is the eigenfunctions of Eq. (1.15), then

$$\widetilde{\Upsilon}_k(p) = \Upsilon_k(1-p), \tag{1.16}$$

then,

$$\widetilde{\Upsilon}_k(p) = \sum_{m=0}^{\infty}(-1)^m \sum_{j=0}^{m} \frac{\binom{m}{j} b^j \lambda_k^{m-j}}{\Gamma(2m+2-j\alpha)}(1-p)^{2m+1-j\alpha}; k = 1,2,\ldots,7. \tag{1.17}$$

TABLE 1.2

The Inner Product of $\langle \Upsilon_k, \tilde{\Upsilon}_m \rangle$ at $b = 1.8$ and $\alpha = 1.47$

$\langle \Upsilon_k, \tilde{\Upsilon}_m \rangle$	Υ_1	Υ_2	Υ_3	Υ_4	Υ_5	Υ_6	Υ_7
$\tilde{\Upsilon}_1$	0.05106	0	0	0	0	0	0
$\tilde{\Upsilon}_2$	0	−0.0095	0	0	0	0	0
$\tilde{\Upsilon}_3$	0	0	0.0031	0	0	0	0
$\tilde{\Upsilon}_4$	0	0	0	−0.0013	0	0	0
$\tilde{\Upsilon}_5$	0	0	0	0	0.0006	0	0
$\tilde{\Upsilon}_6$	0	0	0	0	0	−0.0003	0
$\tilde{\Upsilon}_7$	0	0	0	0	0	0	0.0002

The inner product $\langle \Upsilon_k(p), \tilde{\Upsilon}_m(p) \rangle$, for the case $\aleph = 2$; $\alpha = 1.47$, $b = 1.8$ can be calculated approximately, that are given in Table 1.2

Next, we seek the general solution of the following Equation

$$\Psi''(q) - aD_q^{\beta}\Psi(q) + \lambda\Psi(q) = 0, \tag{1.18}$$

Let us take

$$\Psi''(q) = Y(q), \tag{1.19}$$

integrate Eq. (1.19) more than once from 0 to q

$$\Psi(q) = \int_0^q (q-s)Y(s)ds + \Psi'(0)q + \Psi(0), \tag{1.20}$$

solving Eq. (1.20), we obtain

$$Y(q) + \int_0^q \left[\frac{-a}{\Gamma(2-\beta)}(q-\zeta)^{1-\beta} + \lambda(q-\zeta) \right] Y(\zeta)d\zeta = -\lambda \mathcal{A}q - \lambda \mathcal{B}, \tag{1.21}$$

such that $\mathcal{A} = \Psi'(0), \mathcal{B} = \Psi(0)$. Then the solution of Eq. (1.21) takes the form

$$\Psi(q)=\mathcal{A}\left[q-\frac{\lambda q^3}{6}+\sum_{m=1}^{\infty}(-1)^{m+1}\sum_{k=0}^{m}\frac{\binom{m}{k}(-a)^k\lambda^{m+1-k}}{\Gamma(2m+4-k\beta)}q^{2m+3-k\beta}\right];$$
$$+\mathcal{B}\left[1-\frac{\lambda q^2}{2}+\sum_{m=1}^{\infty}(-1)^{m+1}\sum_{k=0}^{m}\frac{\binom{m}{k}(-a)^k\lambda^{m+1-k}}{\Gamma(2m+3-k\beta)}q^{2m+2-k\beta}\right], \tag{1.22}$$

For simplicity, we can put

$$T(q)=q-\frac{\lambda q^3}{6}+\sum_{m=1}^{\infty}(-1)^{m+1}\sum_{k=0}^{m}\frac{\binom{m}{k}(-a)^k\lambda^{m+1-k}}{\Gamma(2m+4-k\beta)}q^{2m+3-k\beta}, \tag{1.23}$$

and

$$\widetilde{T}(q)=1-\frac{\lambda q^2}{2}+\sum_{m=1}^{\infty}(-1)^{m+1}\sum_{k=0}^{m}\frac{\binom{m}{k}(-a)^k\lambda^{m+1-k}}{\Gamma(2m+3-k\beta)}q^{2m+2-k\beta}, \tag{1.24}$$

then the solution of Eq. (1.18) can be rewritten as

$$\Psi(q)=\mathcal{A}.T(q)+\mathcal{B}.\widetilde{T}(q). \tag{1.25}$$

Hence, the solution of Eq. (1.2) is written out in a standard form

$$w(p,q)=\sum_{k=1}^{\infty}\Upsilon_k(p)\Psi_k(q)=\sum_{k=1}^{\infty}\Upsilon_k(p)\left[\mathcal{A}_k.T_k(q)+\mathcal{B}_k.\widetilde{T}_k(q)\right]. \tag{1.26}$$

We will use at $q=0$, the following form of the initial condition (1.4) and (1.26)

$$w(p,0)=\sum_{k=1}^{\infty}\Upsilon_k(p)\left[\mathcal{A}_k.T_k(0)+\mathcal{B}_k.\widetilde{T}_k(0)\right]=\varphi(p),$$

Further, to use the initial condition (1.4), we differentiate both sides of (1.16) with respect to q at $q = 0$:

$$w_q(p,0) = \sum_{k=1}^{\infty} \Upsilon_k(p)\left[\mathcal{A}_k.T_k'(0) + \mathcal{B}_k.\widetilde{T}'_k(0)\right] = \psi(p). \tag{1.27}$$

Multiply both sides of each equation of Eq. (1.27) by $\widetilde{\Upsilon}_k$, therefore, from the orthogonality of the functions Υ_k and $\widetilde{\Upsilon}_k$ we obtain

$$\begin{aligned} \mathcal{A} &= \frac{\langle \varphi(p), \widetilde{\Upsilon}_k(p)\rangle}{\langle \Upsilon_k(p), \widetilde{\Upsilon}_k(p)\rangle}; \\ \mathcal{B}_k &= \frac{\langle \psi(p), \widetilde{\Upsilon}_k(p)\rangle}{\langle \Upsilon_k(p), \widetilde{\Upsilon}_k(p)\rangle}, \end{aligned} \tag{1.28}$$

which allows us to write out the solution of problem Eq. (1.2) in the following formula

$$w(p,q) = \sum_{k=1}^{\infty} \frac{\Upsilon_k(p)}{\langle \Upsilon_k(p), \widetilde{\Upsilon}_k(p)\rangle}\left[T_k(q)\langle \psi(p), \widetilde{\Upsilon}_k(p)\rangle + \widetilde{T}_k(q)\langle \varphi(p), \widetilde{\Upsilon}_k(p)\rangle\right]. \tag{1.29}$$

1.6 Solving FVEs Numerically by Finite Difference Scheme

Considering Eq. (1.2) with condition (1.3), we find that if we assume for Eq. (1.2) an equivalent form, the precision of the discrete approximations can be improved. First, let us take the fractional integral operator ${}_0J_q$ in Eq. (1.2), then

$$\frac{\partial w(p,q)}{\partial q} - aD_q^{\theta} w(p,q) = J_q \frac{\partial^2 w(p,q)}{\partial p^2} + bJ_q D_p^{\alpha} w(p,q), \tag{1.30}$$

where $0 < \theta = \beta - 1 < 1$ and $J_q D_q^{\beta} w(p,q) = D_q^{\beta-1} w(p,q)$.

In order to discretize Eq. (1.30), we introduce the temporal step size $\tau = I/\mathbb{N}$ with a non-zero integer $\mathbb{N}$, and $q_n = n\tau$; $n = 0, 1, \ldots, \mathbb{N}$, and define a grid function time $\Omega_\tau = \{q_n \mid n \geq 0\}$. For a spatial discretization, let $h = \aleph/\mathbb{M}$ with a non-zero integer $\mathbb{M}$, and $p_i = ih$; $0 \leq i \leq \aleph$, and define a grid function space $\Omega_h = \{p_i \mid \ 0 \leq i \leq \mathbb{M}\}$. Suppose that on $\Omega_h \times \Omega_\tau$, there exists a grid function $W = \{w_i^n \mid \ 0 \leq i \leq \mathbb{M}, n \geq 0\}$, such that for any $w, g \in W$,

$$w_i^{n+\frac{1}{2}} = \frac{1}{2}\left[w_i^{n+1} + w_i^n\right], \qquad \delta_q w_i^{n+\frac{1}{2}} = \frac{1}{\tau}\left[w_i^{n+1} - w_i^n\right],$$
$$\langle w^n, g^n\rangle = h\sum_{i=1}^{M-1} w_i g_i, \qquad \| w^n \|^2 = \langle w, w\rangle,$$
$$\| w^n \|_\infty = \max_{0\le i\le M} | w_i^n |, \qquad \langle \delta_p^2 w, g\rangle = -\langle \delta_p w, \delta_p g\rangle.$$

Then using Eq. (1.5) and Lemma 1.4.1, a weighted Crank–Nicolson method for Eq. (1.30) at the point $(p_i, q_{n+\frac{1}{2}})$ formed as

$$\frac{w(p_i,q_{n+1}) - w(p_i,q_n)}{\tau} - \frac{a}{2} D_q^\theta \left(w(p_i,q_{n+1}) + w(p_i,q_n) - 2w(p_i,0)\right) =$$
$$\frac{1}{2}\left(J_q \frac{\partial^2 w(p_i,q_{n+1})}{\partial p^2} + J_q \frac{\partial^2 w(p_i,q_n)}{\partial p^2}\right) + \frac{b}{2}\left(J_q D_p^\alpha w(p_i,q_{n+1}) + J_q D_p^\alpha w(p_i,q_n)\right) + \mathcal{O}(\tau^2), \tag{1.31}$$

assume that $w(p,q) \in C_{p,q}^{5,3}([0,\aleph]\times[0,\Im])$, let $w_i^n = w(p_i,q_n)$, so we suggest the following Crank-Nicolson method, which based on Lemma 1.4.4 and Theorem 1.4.5. to Eq. (1.31), then

$$\frac{w_i^{n+1} - w_i^n}{\tau} - \frac{a}{2}\tau^{-\theta}\left(\sum_{k=0}^{n+1} \lambda_k^{(\theta)} w_i^{n+1-k} + \sum_{k=0}^{n} \lambda_k^{(\theta)} w_i^{n-k} - 2\lambda_0^{(\theta)} \varphi_i\right) =$$
$$\frac{\tau}{2}\left(\sum_{k=0}^{n+1} \omega_k \delta_p^2 w_i^{n+1-k} + \sum_{k=0}^{n} \omega_k \delta_p^2 w_i^{n-k}\right) + \frac{b}{2}\tau\left(\sum_{k=0}^{n+1} \omega_k \delta_p^\alpha w_i^{n+1-k} + \sum_{k=0}^{n} \omega_k \delta_p^\alpha w_i^{n-k}\right) + \mathcal{O}(\tau^2 + h^2), \tag{1.32}$$

since $w(i,0) = \varphi_i(p)$, and multiply τ in Eq. (1.32) then

$$w_i^{n+1} - \frac{a}{2}\tau^{1-\theta}\left(\sum_{k=0}^{n+1} \lambda_k^{(\theta)} w_i^{n+1-k} + \sum_{k=0}^{n} \lambda_k^{(\theta)} w_i^{n-k} - 2\lambda_0^{(\theta)} \varphi\right)$$
$$= w_i^n + \frac{\tau^2}{2}\left(\sum_{k=0}^{n+1} \omega_k \delta_p^2 w_i^{n+1-k} + \sum_{k=0}^{n} \omega_k \delta_p^2 w_i^{n-k}\right)$$
$$+ \frac{b}{2}\tau^2\left(\sum_{k=0}^{n+1} \omega_k \delta_p^\alpha w_i^{n+1-k} + \sum_{k=0}^{n} \omega_k \delta_p^\alpha w_i^{n-k}\right) + \mathcal{O}(\tau^3 + \tau h^2), \tag{1.33}$$

ignoring the truncation error term $\mathcal{O}\left(\tau^3 + \tau h^2\right)$ from Eq. (1.33) and replacing w_i^n by its numerical solution W_i^n, we obtain the following scheme for Eq. (1.33)

$$W_i^{n+1} - \frac{a}{2}\tau^{1-\theta}\left(\sum_{k=0}^{n+1}\lambda_k^{(\theta)}W_i^{n+1-k} + \sum_{k=0}^{n}\lambda_k^{(\theta)}W_i^{n-k} - 2\lambda_0^{(\theta)}\varphi\right) =$$
$$W_i^n + \frac{\tau^2}{2}\left(\sum_{k=0}^{n+1}\omega_k\delta_p^2 W_i^{n+1-k} + \sum_{k=0}^{n}\omega_k\delta_p^2 W_i^{n-k}\right) + \frac{b}{2}\tau^2\left(\sum_{k=0}^{n+1}\omega_k\delta_p^\alpha W_i^{n+1-k} + \sum_{k=0}^{n}\omega_k\delta_p^\alpha W_i^{n-k}\right). \tag{1.34}$$

1.7 Stability and Convergence Analysis

This section discusses the stability and convergence of the suggested ADI method (1.33). Let a grid function $W = \{w_i^n \mid 0 \le i \le \mathbb{M}, n \ge 0; w_0 = w_{\mathbb{M}} = 0\}$ is a grid function on $\Omega_h \times \Omega_\tau$.

Lemma 1.7.1

(see[55]) For any $w, u \in W$, there exist linear difference operator to the operator δ_p^α that is denoted by $\delta_p^{\alpha/2}$, where

$$\langle \delta_p^\alpha w, u\rangle = \langle \delta_p^{\alpha/2} w, \delta_p^{\alpha/2} u\rangle, \qquad \langle \delta_p^2 w, u\rangle = -\langle \delta_p w, \delta_p u\rangle.$$

Theorem 1.7.2

Let $w(p,q) \in c_{p,q}^{6,3}([0,\aleph]\times[0,\Im])$ and suppose $w(p,q)$ is the analytical solution of Eq. (1.2) and the numerical solution of scheme (1.33)–(1.34) is $W(p,q)$, which is defined as $\{W_i^n \mid 0 \le i \le \mathbb{M}, 0 \le j \le \mathbb{N}\}$. Then for every $n\tau \le T$, it achieves that

$$\| W^j - w^j \| \le \tilde{c}(\tau^2 + h^2).$$

Proof. Subtracting Eq. (1.34) from (1.33) and denoting $e_i^j = w_i^j - W_i^j$, then we have

$$e_i^{m+1} - e_i^m - \frac{a}{2}\tau^{1-\theta}\left(\sum_{j=0}^{m+1}\lambda_j^{(\theta)}e_i^{m+1-j} + \sum_{j=0}^{m}\lambda_j^{(\theta)}e_i^{m-j} - 2\lambda_0^{(\theta)}e_i^0\right)$$
$$= \frac{\tau^2}{2}\left(\sum_{j=0}^{m+1}\omega_j\delta_p^2 e_i^{m+1-j} + \sum_{j=0}^{m}\omega_j\delta_p^2 e_i^{m-j}\right) + \frac{b}{2}\tau^2\left(\sum_{j=0}^{m+1}\omega_j\delta_p^\alpha e_i^{m+1-j} + \sum_{j=0}^{m}\omega_j\delta_p^\alpha e_i^{m-j}\right)$$
$$+\mathcal{O}(\tau^3 + \tau h^2). \tag{1.35}$$

Multiplying Eq. (1.35) by $h\left(e_i^{m+1} + e_i^m\right)$ summing over i from 0 to $\mathbb{M}-1$, we obtain that

$$\| e^{m+1} \|^2 - \| e^m \|^2 = \frac{a}{2}\tau^{1-\theta}\left(\sum_{j=0}^{m}\lambda_j^{(\theta)}\langle e^{m+1-j} + e^{m-j} - e^0, e^{m+1} + e^m\rangle + \lambda_{m+1}^{(\theta)}\langle e^0, e^{m+1} + e^m\rangle\right)$$
$$+\frac{\tau^2}{2}\left(\sum_{j=0}^{m}\omega_j\langle\delta_p^2(e^{m+1-j} + e^{m-j}), e^{m+1} + e^m\rangle + \omega_{m+1}\langle\delta_p^2 e^0, e^{m+1} + e^m\rangle\right)$$
$$+\frac{b}{2}\tau^2\left(\sum_{j=0}^{m+1}\omega_j\langle\delta_p^\alpha(e^{m+1-j} + e^{m-j}), e^{m+1} + e^m\rangle + \omega_{m+1}\langle\delta_p^\alpha e^0, e^{m+1} + e^m\rangle\right)$$
$$+\langle\mathcal{O}(\tau^3 + \tau h^2), e^{m+1} + e^m\rangle. \tag{1.36}$$

Since $e_i^0 = 0$ for $0 \le i \le \mathbb{M}$, and summing over m from 0 to $R - 1$, we obtain that

$$\begin{aligned} \| e^R \|^2 = {} & \frac{a}{2}\tau^{1-\theta}\sum_{m=0}^{R-1}\sum_{j=0}^{m}\lambda_j^{(\theta)}\langle e^{m+1-j} + e^{m-j}, e^{m+1} + e^m\rangle \\ & + \frac{\tau^2}{2}\sum_{m=0}^{R-1}\sum_{j=0}^{m}\omega_j\langle\delta_p^2\left(e^{m+1-j} + e^{m-j}\right), e^{m+1} + e^m\rangle \\ & + \frac{b}{2}\tau^2\sum_{m=0}^{R-1}\sum_{j=0}^{m}\omega_j\langle\delta_p^\alpha\left(e^{m+1-j} + e^{m-j}\right), e^{m+1} \\ & + e^m\rangle + \sum_{m=0}^{R-1}\langle\mathcal{O}(\tau^3 + \tau h^2), e^{m+1} + e^m\rangle. \end{aligned} \tag{1.37}$$

The use of Theorem 1.4.5 yields the following inequality, from which it can be inferred that the first three terms on the whole of the right side of Eq. (1.37) are negative.

$$\| e^R \|^2 \le \sum_{m=0}^{R-1}\langle\tilde{c}(\tau^3 + \tau h^2), e^{m+1} + e^m\rangle. \tag{1.38}$$

Suppose,

$$\| e^S \| = \max_{0 \le R \le N} \| e^R \|,$$

then we obtain

$$\| e^S \| \le \tilde{c}(\tau^2 + h^2). \tag{1.39}$$

Thus, the proof is complete.

Theorem 1.7.3

Consider the numerical solution of scheme Eqs (1.33)–(1.34) is stable in a grid function $\{W_i^n \mid, 0 \le i \le \mathbb{M} \mid, 0 \le n \le \mathbb{N}\}$, *which holds*

$$\| W^J \| \le \tilde{c}.$$

Proof. Multiplying Eq. (1.34) by $h\left(W_i^{m+1} + U_i^m\right)$ and summing from $1 \le i \le \mathbb{M} - 1$,

$$\begin{aligned}
&\| W^{m+1} \|^2 - \| W^m \|^2 \\
&= \frac{a}{2}\tau^{1-\theta}\left[\sum_{j=0}^{m}\lambda_j^{(\theta)}\langle W^{m+1-j} + W^{m-j}, W^{m+1} + W^m\rangle + \lambda_{m+1}^{(\theta)}\langle \varphi, W^{m+1} + W^m\rangle\right. \\
&\left.-2\lambda_0^{(\theta)}\langle \varphi, W^{m+1} + W^m\rangle\right] + \frac{\tau^2}{2}\sum_{j=0}^{m}\omega_j\langle \delta_p^2(W^{m+1-j} + W^{m-j}), W^{m+1} + W^m\rangle \\
&+\frac{\tau^2}{2}\omega_{m+1}\langle \delta_p^2\varphi, W^{m+1} + W^m\rangle + \frac{b}{2}\tau^2\sum_{j=0}^{m+1}\omega_j\langle \delta_p^\alpha(W^{m+1-j} + W^{m-j}), W^{m+1} + W^m\rangle \\
&+\frac{b}{2}\tau^2\omega_{m+1}\langle \delta_p^\alpha\varphi, W^{m+1} + W^m\rangle.
\end{aligned} \tag{1.40}$$

Without loss of generality, we take a homogeneous initial condition Eq. (1.4); $w_i^0 = 0$ and summing Eq. (1.40) over m from 0 to $J - 1$, we get

$$\begin{aligned}
\| W^J \|^2 - \| W^0 \|^2 &= \frac{a}{2}\tau^{1-\theta}\sum_{j=0}^{m}\lambda_j^{(\theta)}\langle W^{m+1-j} + W^{m-j}, W^{m+1} + W^m\rangle \\
&+\frac{\tau^2}{2}\sum_{j=0}^{m}\omega_j\langle \delta_p^2(W^{m+1-j} + W^{m-j}), W^{m+1} + W^m\rangle \\
&+\frac{b}{2}\tau^2\sum_{j=0}^{m+1}\omega_j\langle \delta_p^\alpha(W^{m+1-j} + W^{m-j}), W^{m+1} + W^m\rangle,
\end{aligned} \tag{1.41}$$

using Theorem 1.4.5 and Lemma 1.7.1, and by arguments similar in the proof of Theorem 1.7.2, we get

$$\| W^J \| \le \tilde{c}.$$

Theorem 1.7.4

Consider the numerical solution of scheme Eqs (1.33)–(1.34) is stable in a grid function $\{W_i^n|,0\le i\le \mathbb{M}|,0\le n\le \mathbb{N}\}$; $W_i^0\neq 0$, *which holds*

$$\|W^J\|\le\tilde{c}\left(\|\varphi\|^2-\frac{a}{4}T^{1-\theta}\|\varphi\|^2+\frac{T^2}{4}\|\delta_p^2\varphi\|^2+\frac{b}{4}T^2\|\delta_p^\alpha\varphi\|^2\right).$$

Proof. If we take initial conditions that are not equal to zero in (1.34) summation over m from 0 to $J-1$, and use Cauchy Schwartz inequality, Theorem 1.4.5, then we get

$$\begin{aligned}\|W^J\|^2\le\|W^0\|^2&+\frac{a}{2}\tau^{1-\theta}\left[\sum_{m=0}^{J-1}\lambda_{m+1}^{(\theta)}\|\varphi\|\|W^{m+1}+W^m\|-2\sum_{m=0}^{J-1}\lambda_0^{(\theta)}\|\varphi\|\|W^{m+1}+W^m\|\right]\\&+\frac{\tau^2}{2}\sum_{m=0}^{J-1}\omega_{m+1}\|\delta_p^2\|\|W^{m+1}+W^m)\|+\frac{b}{2}\tau^2\sum_{m=0}^{K-1}\omega_{m+1}\|\delta_p^\alpha\|\|W^{m+1}+W^m)\|,\end{aligned}\tag{1.42}$$

Using Young inequality, Eq. (1.42) yields

$$\begin{aligned}\|W^J\|^2\le\|\varphi\|^2&+\frac{a}{4}\tau^{1-\theta}\sum_{m=0}^{J-1}\lambda_{m+1}^{(\theta)}\|\varphi\|^2+\|\frac{a}{4}\tau^{1-\theta}\sum_{m=0}^{J-1}W^{m+1}+W^m\|^2-\frac{a}{2}\tau^{1-\theta}\sum_{m=0}^{J-1}\lambda_0^{(\theta)}\|\varphi\|^2\\&-\frac{a}{2}\tau^{1-\theta}\sum_{m=0}^{J-1}W^{m+1}+W^m\|^2+\frac{\tau^2}{4}\sum_{m=0}^{J-1}\omega_{m+1}\|\delta_p^2\varphi\|^2+\frac{\tau^2}{4}\sum_{m=0}^{J-1}\|W^{m+1}+W^m\|^2\\&+\frac{b}{4}\tau^2\sum_{m=0}^{J-1}\omega_{m+1}\|\delta_p^2\varphi\|^2+\frac{b}{4}\tau^2\sum_{m=0}^{J-1}\|W^{m+1}+W^m\|^2.\end{aligned}\tag{1.43}$$

Since for $\tau^{1-\theta}\sum_{m=0}^{J-1}\omega_{m+1}$ and $\tau^{1-\theta}\sum_{m=0}^{J-1}\lambda_{m+1}^{(\theta)}$ limited (bounded), and we can use Gronwall's inequality $\|W^{m+1}+W^m\|\to 0$, then Eq. (1.43) is written as

$$\begin{aligned}\|W^J\|^2\le\tilde{c}\|\varphi\|^2&+\frac{a}{4}\tau^{1-\theta}\sum_{m=0}^{J-1}\lambda_{m+1}^{(\theta)}\|\varphi\|^2-\frac{a}{2}\tau^{1-\theta}\sum_{m=0}^{J-1}\lambda_0^{(\theta)}\|\varphi\|^2\\&+\frac{\tau^2}{4}\sum_{m=0}^{J-1}\omega_{m+1}\|\delta_p^2\varphi\|^2+\frac{b}{4}\tau^2\sum_{m=0}^{J-1}\omega_{m+1}\|\delta_p^\alpha\varphi\|^2.\end{aligned}\tag{1.44}$$

Let $\tau \sum_{m=0}^{N-1} = T$, thus

$$\| W^J \|^2 \leq \left(\| \varphi \|^2 - \frac{a}{4} T^{1-\theta} \| \varphi \|^2 + \frac{T^2}{4} \| \delta_p^2 \varphi \|^2 + \frac{b}{4} T^2 \| \delta_p^\alpha \varphi \|^2 \right). \tag{1.45}$$

1.8 Numerical Examples

This section presents numerical examples to demonstrate the computational performance and theoretical findings of our proposed methods.

Example 1.8.1

$$\begin{aligned} \frac{\partial^2 w(p,q)}{\partial q^2} &= \frac{\partial^2 w(p,q)}{\partial p^2} + aD_q^\beta w(p,q) + bD_p^\alpha w(p,q), \quad 0 < p < \aleph, 0 < q < \Im, \\ w(0,q) &= w(\aleph,q) = 0, \\ w(p,0) &= p(1-p), \\ w_q(p,0) &= p^3(p-1), \end{aligned} \tag{1.46}$$

when $a = -1.8$, $b = 0.5$, $I = 1$ and $\aleph = 1$ corresponding to $\alpha = 1.5$, $\beta = 1.47$. First, we note that the analytical solution $w(p,q)$ of Eq. (1.46) fulfill all the smoothness conditions needed by schemes (1.29) and (1.34). Then in Figure 1.4, let us take the step size $\tau = h = 1/20$ for plotting a curve at $I = 1$ of the analytical solutions and the numerical solutions for $\alpha = 1.5$, $\beta = 1.47$.

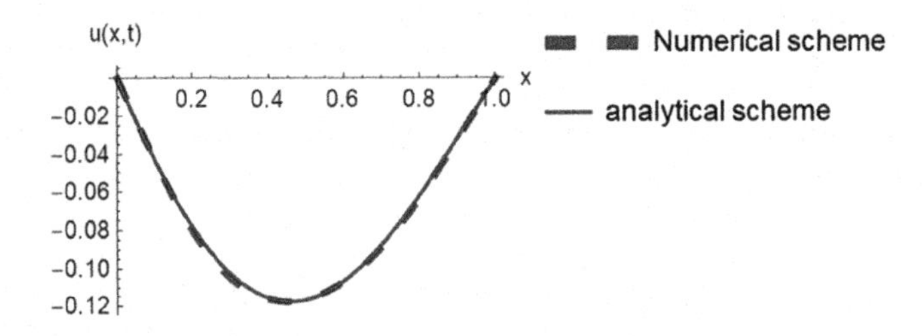

FIGURE 1.4
Analytical solution (1.29) and numerical solution (1.34) of ADI scheme for $\alpha = 1.5$, $\beta = 1.47$.

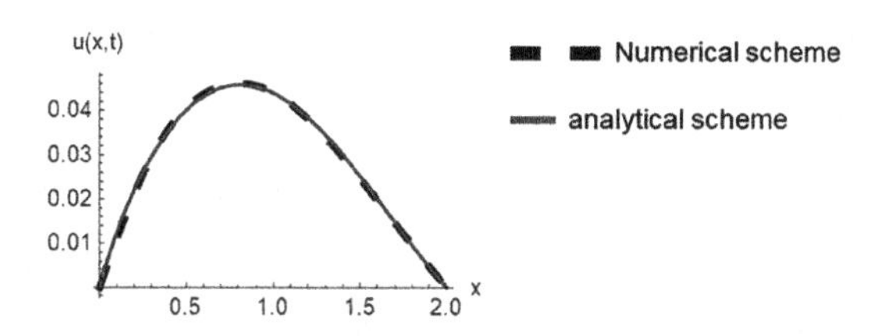

FIGURE 1.5
Analytical solution Eq. (1.29) and numerical solution Eq. (1.34) of ADI scheme for $\alpha = 1.47, \beta = 1.47$.

Example 1.8.2

$$\begin{aligned}&\frac{\partial^2 w(p,q)}{\partial q^2} = \frac{\partial^2 w(p,q)}{\partial p^2} + aD_q^\beta w(p,q) + bD_p^\alpha w(p,q), \quad 0 < p < \aleph, 0 < q < \Im,\\ &w(0,q) = w(\aleph,q) = 0,\\ &w(p,0) = 0,\\ &w_q(p,0) = p(p-2),\end{aligned} \tag{1.47}$$

when $a = -0.5$, $b = 1.8$, $I = 2$ and $\aleph = 2$ corresponding to $\alpha = 1.47, \beta = 1.47$.

In Figure 1.5, let us take step size $\tau = h = 2/16$ for plotting the curves at $I = 2$ of the approximate solution Eq. (1.34) and the analytical solution of the scheme Eq. (1.29) corresponding to $\beta = 1.47$ and $\alpha = 1.47$,

From Figures 1.4 and 1.5, we find that the analytical solution agrees well with our numerical solution.

1.9 Conclusion

This chapter has described and demonstrated FVEs with viscoelastic damping. Using the Fourier method, we deduced the analytical solution of the problem. The numerical scheme was constructed. The suggested ADI scheme demonstrated stability and convergence with second-order accuracy in space and time. We found that the analytical solution is in perfect agreement with our numerical solution. Both the analytical and numerical methods showed that the suggested methods are efficient for solving one-dimensional time–space FVEs and are also applicable to other FPDEs. For future work, the proposed computation technology will be compared to other compact methods for homogeneous and non-homogeneous models, and the model's applicability in dynamic systems and mechanical engineering will be studied.

Author Contributions

In this research paper, all authors contributed equally. The published version of the paper has been read and approved by all authors.

Conflicts of Interest

The authors declare no conflict of interest.

Acknowledgment

The researcher (A. Elsayed) is funded by a partial scholarship from the Ministry of Higher Education of Egypt.

References

[1] Tarasov V.E., Review of some promising fractional physical models, *International Journal of Modern Physics B* 2013, *27*, 1330005.
[2] Hilfer R., *Applications of fractional calculus in physics*, World Scientific, Singapore, 2000.
[3] Aleroev T.S., Elsayed A.M., Mahmoud E.I., Mathematical model of the polymer concrete by fractional calculus with respect to a spatial variable, *IOP Conference Series: Materials Science and Engineering* 2021, *1129*(1), 012031.
[4] Harendra S., Srivastava H.M., Numerical investigation of the fractional-order Liénard and duffing equations arising in oscillating circuit theory, *Frontiers in Physics* 2020, *8*, 120.
[5] Miller K. S., Ross B., *An introduction to the fractional calculus and fractional differential equations*, John Wiley and Sons, New York, 1993.
[6] Podlubny I., *Fractional differential equations*, Academic Press, San Diego, 1991.
[7] Shen K.L., Soong T.T., Modeling of viscoelastic dampers for structural applications, *Journal of Engineering Mechanics* 1995, *121*, 694–701.
[8] Ingman D., Shzdalnitsky J., Iteration method for equation of viscoelastic motion with fractional differential operator of damping, *Computer Methods in Applied Mechanics and Engineering* 2001, *190*, 5027–5036.
[9] Das S., *Functional fractional calculus*, Springer, New York, 2011.

[10] Rami El-N.A., Cosmology with fractional action principle, *Romanian Reports in Physics* 2007, *3*(59), 763–771.
[11] Enelund M., Josefson B.L., Time-domain finite element analysis of viscoelastic structures with fractional derivative constitutive relations, *American Institute of Aeronautics and Astronautics Journal* 1997, *35*,1630–1637.
[12] Kilbas A.A., Srivastava H.M., Trujillo J.J., *Theory and applications of fractional differential equations*, Elsevier, Amsterdam, 2006, *523*.
[13] Mahmoud E.I., Orlov V.N., Numerical solution of two dimensional time-space fractional Fokker Planck equation with variable coefficients, *Mathematics* 2021, 9, 1260.
[14] Torvik J.P., Determination of system damping from the response amplitude at resonance, *Journal of Vibration and Acoustics* 2016, *138*(1), 011008.
[15] Aleroev T.S., The analysis of the polymer concrete characteristics by fractional calculus, *Journal of Physics: Conference Series* 2020, *1425*(1), 012112.
[16] Aleroev T.S., Aleroeva T.H., Huang F.J., Nie M.N., Tang F.Y., Zhang Y.S., Features of seepage of a liquid to a chink in the cracked deformable layer, *International Journal of Modeling, Simulation, and Scientific Computing* 2010, *1*, 333–347.
[17] Aleroev S.T., Kekharsaeva R.E., Boundary value problems for differential equations with fractional derivatives, *Integral Transforms and Special Functions* 2017, *28*, 900–908.
[18] Kirianova L., Modeling of strength characteristics of polymer concrete via the wave equation with a fractional derivative, *Mathematics* 2020, *8*(10), 1843.
[19] Harendra S., Analysis of drug treatment of the fractional HIV infection model of CD4+ T-cells, *Chaos Solitons and Fractals* 2021, *146*, 110868.
[20] Zhao-Peng H., Wanrong C., Shengyue L., Numerical correction of finite difference solution for two-dimensional space-fractional diffusion equations with boundary singularity, *Numerical Algorithms* 2021, *86*(3), 1071–1087.
[21] Harendra S., AbdulMajid W., Computational method for reaction diffusion model arising in a spherical catalyst, *International Journal of Applied and Computational Mathematics* 2021, *7*(3), 65.
[22] Odibat Z., Momani S., The variational iteration method: An efficient scheme for handling fractional partial differential equations in fluid mechanics, *Computers and Mathematics with Applications* 2009, *58*, 2199–2208.
[23] Harendra S., Analysis for fractional dynamics of Ebola virus model, *Chaos Solitons and Fractals* 2020, *138*, 109992.
[24] Podlubny I., *Fractional differential equations: An introduction to fractional derivatives, fractional differential equations, to methods of their solution and some of their applications*, Academic Press, Cambridge, MA, 1998, 198.
[25] Harendra S., Jagdev S., Sunil D.P., Kumar D., *Advanced numerical methods for differential equations: Applications in science and engineering*, CRC Press Taylor and Francis, Boca Raton, 2021.
[26] Rudolf H., *Applications of fractional calculus in physics*, World Scientific, Singapore, 2000.
[27] Harendra S., Devendra K., Dumitru B., *Methods of mathematical modelling: Fractional differential equations*, CRC Press Taylor and Francis, Boca Raton, 2019.
[28] Ingman D., Suzdalnitsky J., Control of damping oscillations by fractional differential operator with time-dependent order, *Computer Methods in Applied Mechanics and Engineering* 2004, *193*, 5585–5595.

[29] Ray S.S., Sahoo S., Das S., Formulation and solutions of fractional continuously variable order mass spring damper systems controlled by viscoelastic and viscous-viscoelastic dampers, *Advances in Mechanical Engineering* 2016, *8*(5), 1–13.
[30] Draganescu G.E., Cofan N., Rujan D.L., Nonlinear vibrations of a nano-sized sensor with fractional damping, *The Journal of Optoelectronics and Advanced Materials* 2005, *7*(2), 877–884.
[31] Aleroev T.S., Erokhin S., Kekharsaeva E., Modeling of deformation-strength characteristics of polymer concrete using fractional calculus, *IOP: Materials Science and Engineering* 2018, *365*(3), 032004.
[32] Xu M., Tan W., Intermediate processes and critical phenomena: Theory, method and progress of fractional operators and their applications to modern mechanics, *Science in China Ser G: Physics, Mechanics and Astronomy* 2006, *49*, 257–272.
[33] Chen J., Liu F., Anh V., Shen S., Liu Q., Liao C., The analytical solution and numerical solution of the fractional diffusion-wave equation with damping, *Applied Mathematics and Computation* 2012, *219*, 1737–1748.
[34] Sarver J.J., Robinson P.S., Elliott D.M., Methods for quasi-linear viscoelastic modeling of soft tissue: application to incremental stress-relaxation experiments, *The Journal of Biomechanical Engineering* 2003, *125*(5), 754–758.
[35] Bhrawya H.A., Zaky A.M., A method based on the Jacobi tau approximation for solving multi-term time-space fractional partial differential equations, *Journal of Computational Physics* 2015, *281*, 876–895.
[36] Aleroev T.S., Elsayed A.S., Mahmoud E.I., Solving one dimensional time-space fractional vibration string equation, *IOP Conference Series: Materials Science and Engineering* 2021, *1129*(1), 012030.
[37] Bu P.W., Tang F.Y., Wu C.Y., Yang Y.J., Finite difference/finite element method for two-dimensional space and time fractional Bloch-Torrey equations, *Journal of Computational Physics* 2015, *293*, 264–279.
[38] Elsayed M.A., Orlov N.V., Numerical scheme for solving time–space vibration string equation of fractional derivative, *Mathematics* 2020, *8*(7), 1069.
[39] Aleroev T.S., Elsayed A.M., Analytical and approximate solution for solving the vibration string equation with a fractional derivative, *Mathematics* 2020, *8*(7), 1154.
[40] Aleroev T.S., Solving the boundary value problems for differential equations with fractional derivatives by the method of separation of variables, *Mathematics* 2020, *8*(11), 1877.
[41] Li X., Xu C., A space-time spectral method for the time fractional diffusion equation, *SIAM Journal on Numerical Analysis* 2009, *47*, 2108–2131.
[42] Ren J., Sun Z.Z., Zhao X., Compact difference scheme for the fractional subdiffusion equation with Neumann boundary conditions, *Journal of Computational Physics* 2013, *232*(1), 456–467.
[43] Guo M.S., Mei Q.L., Zhang Q.Z., Li C., Li J.M., Wang Y., A linearized finite difference/spectral-Galerkin scheme for three-dimensional distributed-order time-space fractional nonlinear reaction-diffusion-wave equation: Numerical simulations of Gordontype solitons, *Computer Physics Communications* 2020, *252*, 107144.
[44] Huang J., Tang Y., VÃ¡zquez L., Yang J., Two finite difference schemes for time fractional diffusion-wave equation, *Numerical Algorithms* 2013, *64*, 707–720.

[45] Zhao Z., Li C., Fractional difference/finite element approximations for the time-space fractional telegraph equation, *Applied Mathematics and Computation* 2012, *219*(6), 2975–2988.

[46] Sun Z.Z., Wu X.N., A fully discrete difference scheme for a diffusion-wave system, *Applied Numerical Mathematics* 2006, *56*, 193–209.

[47] Harendra S., Numerical simulation for fractional delay differential equations, *International Journal of Dynamic and Control* 2020, *20*(23–24), 103722.

[48] He D.D., Pan J.K., An unconditionally stable linearized difference scheme for the fractional Ginzburg-Landau equation, *Numerical Algorithms* 2018, *79*, 899–925.

[49] Chen A., Li P.C., Numerical solution of fractional diffusion-wave equation, *Numerical Functional Analysis and Optimization* 2016, *37*(1), 19–39.

[50] Harendra S., Jacobi collocation method for the fractional advection-dispersion equation arising in porous media, *Numerical Methods for Partial Differential Equations* 2020, *36*, 1–18.

[51] Wang Z., et al., Finite difference schemes for two-dimensional time-space fractional differential equations, *International Journal of Computer Mathematics* 2016, *93*(3), 578–595.

[52] Zhibo W., Seakweng V., A high order ADI scheme for the two-dimensional time fractional diffusion-wave equation, *International Journal of Computer Mathematics* 2015, *92*(5), 970–979.

[53] Harendra S., Srivastava M.H., Devendra K., A reliable algorithm for the approximate solution of the nonlinear Lane-Emden type equations arising in astrophysics, *Numerical Methods for Partial Differential Equations* 2018, *34*(5), 1524–1555.

[54] Huang, J.F., Arshad S., Jiao Y.D., Tang Y.F., Convolution quadrature methods for time-space fractional nonlinear diffusion wave equations, *East Asian Journal on Applied Mathematics* 2019, *9*, 538–557.

[55] Li C.P., Zeng F.Z., *Numerical methods for fractional calculus*, Chapman and Hall/ CRC, New York, NY, 2015.

[56] Lubich C., Discretized fractional calculus. *SIAM Journal on Mathematical Analysis* 1986, *17*, 704–719.

[57] Chen H., Lu S.J., Chen W.P., A unified numerical scheme for the multi-term time fractional diffusion and diffusion wave equations with variable coefficients, *Journal of Computational and Applied Mathematics* 2018, *330*, 380–397.

[58] Tian W.Y., Zhou, H., Deng W.H., A class of second order difference approximations for solving space fractional diffusion equations, *Mathematics of Computation* 2015, *84*, 1703–1727.

[59] Aleroev T.S., Aleroeva H., Problems of Sturm–Liouville type for differential equations with fractional derivatives. In A. Kochubei, Y. Luchko (Eds) *Fractional differential equations*, De Gruyter, Berlin, Boston, 2019, 21–46.

2

Analysis of a Nonlinear System Arising in a Helium-Burning Network with Mittag–Leffler Law

P. Veeresha
CHRIST (Deemed to be University), Bengaluru, India

Lanre Akinyemi
Lafayette College, Easton, USA

CONTENTS

2.1 Introduction

Differential equations are used to investigate real-world phenomena in order to study their basic properties, capture their behaviors, and predict future consequences. Most natural phenomena are nonlinear and complex in nature, and hence a reliable investigative tool is appropriate. However, scientists and mathematicians have pointed out the limitations of classical calculus to exemplify real-world problems, which can be examined with the help of generalized calculus concepts with fractional order known as fractional calculus (FC). Although this notion has attracted attention from many scholars recently, in fact it originated in Newton's time. Some of the most fascinating leaps in science and engineering began with FC. Due to its favorable characteristics, including the memory effect, non-locality, history, and

DOI: 10.1201/9781003263517-2

heredity, it has become the most influential tool in the examination and designation of complex and nonlinear mechanisms over the last few decades. The rapid evolution of computer software and mathematical algorithms has enabled many researchers to study the theory and fundamentals of FC with its associated notions and applications [1–16].

In most stars, energy is derived from the process of transformation of hydrogen into helium. The rapid temperature increase leads to hydrogen fusion. The stellar core contracts gravitationally due to the temperature which is high enough to start the process of helium burning. There are two types of helium burning, explosive and hydrostatic. Kinetic equations are used to describe the nucleo-synthesis of the elements in the stars. The following equation describes this phenomenon, with the change in the density number $\mathcal{N}_i$ of the species i over time t [17]:

$$\frac{d\mathcal{N}_i}{dt} = -\sum_j \mathcal{N}_i\mathcal{N}_j \langle \sigma v \rangle_{ij} + \sum_{k,l\neq i} \mathcal{N}_k\mathcal{N}_l \langle \sigma v \rangle_{kl},$$

where $\langle \sigma v \rangle pq$ denotes the interaction reaction involving species p and q. For a number density $\mathcal{N}_i$ and Avogadro's number $\mathcal{N}_A$ of the species i expressed in abundance $\mathcal{X}_i$ as $\mathcal{N}_i = \frac{\rho \mathcal{N}_A \mathcal{X}_i}{\mathcal{A}_i}$ where the mass density of gas ρ and $\mathcal{A}_i$ is the unit mass [18].

A triple-alpha collision occurring at high temperature (100 million degrees) in stellar helium core produces C^{12}. Later, C^{12} with He^4 produces O^{16} and continues the same processes [19]. These chemical processes are represented as follows:

$$3He^4 \to C^{12} + \gamma + 7.281Mev$$

$$C^{12} + He^4 \to O^{16} + \gamma + 7.15Mev$$

$$O^{16} + He^4 \to Ne^{20} + \gamma + 4.75Mev$$

The author in [20] used these chemical reactions to produce a mathematical model with a system of four equations which described the helium-burning phase as follows [21]:

$$\begin{aligned} \frac{dx}{dt} &= -3\beta x^3 - \mu xz - \delta xy, \\ \frac{dy}{dt} &= \beta x^3 - \delta xy, \\ \frac{dz}{dt} &= \delta xy - \mu xz, \\ \frac{dr}{dt} &= \mu xz. \end{aligned} \tag{2.1}$$

Here, x represents helium, y symbolizes carbon, z designates oxygen, and per unit mass of stellar material, r denotes a neon number of atoms. The rate of each reaction is defined by β, δ and μ.

Many scholars have produced definitions of differential and integral operators with fractional order, for example, Riemann, Caputo, Atangana, and Baleanu. However, each definition has its own limitations. Recently, many complex and vital models of these operators have been systematically and effectively studied to confirm their efficiency and effectiveness, illustrated by the appropriated operator. Most of the FC systems studied are explored with the Caputo operator. Many authors have modified these notions to accommodate exponential kernel and non-singularity. The Caputo–Fabrizio [22] and Atangana–Baleanu (AB) [23] are the most familiar and widely used fractional operators. More specifically, the AB operator using a generalized Mittag–Leffler function has been developed. The behavior and consequences of complex nonlinear phenomena, particularly in connection with super diffusive, random walk, discrete and statistical studies, the chaotic nature of dynamic systems, and many others, can be captured by the concept of fractional calculus [24–43].

Here, we consider the system (2.1) with a novel operator as follows:

$$\begin{aligned}
&{}^{ABC}_{a}D_t^{\alpha}x(t) = -3\beta x^3 - \delta xy - \mu xz,\\
&{}^{ABC}_{a}D_t^{\alpha}y(t) = \beta x^3 - \delta xy,\\
&{}^{ABC}_{a}D_t^{\alpha}z(t) = \delta xy - \mu xz,\\
&{}^{ABC}_{a}D_t^{\alpha}r(t) = \mu xz,
\end{aligned} \tag{2.2}$$

where α is AB fractional order.

Recently, numerous unconventional algorithms and schemes have been established to study differential systems having integer and arbitrary order. The homotopy analysis method (HAM) developed by Liao Shijun [44, 45] has been effectively used to investigate solutions to nonlinear models arising in engineering science without linearization, assuming any physical parameters and perturbation. However, as this system is heavy on time and computer memory, researchers have suggested combining HAM with previous transform schemes.

Here, we use q-HATM to assess the solution and analyze system behaviors of the four equations demonstrating the chemical reaction involved in the helium-burning network. The method used was made familiar by Singh et al. [46] using Laplace transform (LT) with the q-HAM concept. Due to the consistency and efficacy of this method, many researchers have been using it extensively to find solutions for numerous types of nonlinear models in various areas [47–54]. The method reduces computational work and time taken, while maintaining greater accuracy than other established techniques.

The most interesting current problem in astrophysics is nucleo-synthesis in the extremely hot, dense region at the center of a star. Helium and hydrogen burning and the CNO cycle are the essential phases in nuclear reactors. This phenomenon has been described and modelled using differential equations, and later analyzed with many analytical and numerical techniques [55–57]. The present study uses a novel technique with new fractional derivatives to apprehend the nature of the system.

2.2 Preliminaries

First, we recall the basic notions of FC and LT [23, 58–62].

Definition 1

In the Caputo sense for $f \in H^1(a,b)$, the AB fractional derivative is presented with normalization function $\mathcal{B}[\alpha](\mathcal{B}[0]=\mathcal{B}[1]=1)$, as

$$ {}_{a}^{ABC}D_{t}^{\alpha}\left(f(t)\right)=\frac{\mathcal{B}[\alpha]}{1-\alpha}\int_{a}^{t} f'(\vartheta)E_{\alpha}\left[-\alpha\frac{(t-\vartheta)^{\alpha}}{1-\alpha}\right]d\vartheta. \tag{2.3} $$

Definition 2

The AB fractional derivative in Riemann–Liouville (RL) for $f \in H^1(a,b)$ is presented as

$$ {}_{a}^{ABR}D_{t}^{\alpha}\left(f(t)\right)=\frac{\mathcal{B}[\alpha]}{1-\alpha}\frac{d}{dt}\int_{a}^{t} f(\vartheta)E_{\alpha}\left[\alpha\frac{(t-\vartheta)^{\alpha}}{\alpha-1}\right]d\vartheta. \tag{2.4} $$

Definition 3

The fractional AB integral is presented as

$$ {}_{a}^{AB}I_{t}^{\alpha}\left(f(t)\right)=\frac{1-\alpha}{\mathcal{B}[\alpha]}f(t)+\frac{\alpha}{\mathcal{B}[\alpha]\Gamma(\alpha)}\int_{a}^{t} f(\vartheta)(t-\vartheta)^{\alpha-1}d\vartheta. \tag{2.5} $$

Definition 4

The LT of $f(t)$ with fractional AB derivative is

$$ L\left[{}_{0}^{ABC}D_{t}^{\alpha}\left(f(t)\right)\right](s)=\frac{\mathcal{B}[\alpha]}{1-\alpha}\frac{s^{\alpha}L\left[f(t)\right]-s^{\alpha-1}f(0)}{s^{\alpha}+\left(\alpha/(1-\alpha)\right)}. \tag{2.6} $$

Theorem 1

For the RL and AB derivatives, the Lipschitz conditions hold the following results [23]

$$\left\| {}_{a}^{ABR}D_{t}^{\alpha} f_{1}(t) - {}_{a}^{ABR}D_{t}^{\alpha} f_{2}(t) \right\| < K_{2} \left\| f_{1}(x) - f_{2}(x) \right\|, \tag{2.7}$$

and

$$\left\| {}_{a}^{ABC}D_{t}^{\alpha} f_{1}(t) - {}_{a}^{ABC}D_{t}^{\alpha} f_{2}(t) \right\| < K_{1} \left\| f_{1}(x) - f_{2}(x) \right\|. \tag{2.8}$$

Theorem 2

The fractional differential equation ${}_{a}^{ABC}D_{t}^{\alpha} f_{1}(t) = s(t)$ has a unique solution as follows [23]

$$f(t) = \frac{1-\alpha}{\mathcal{B}[\alpha]} s(t) + \frac{\alpha}{\mathcal{B}[\alpha]\Gamma(\alpha)} \int_{0}^{t} s(\varsigma)(t-\varsigma)^{\alpha-1} d\varsigma. \tag{2.9}$$

2.3 q-HATM Solution Procedure

Here, we use the equation to demonstrate the rule of the method with initial condition [63, 64]

$${}_{a}^{ABC}D_{t}^{\alpha} v(x,t) + \mathcal{N} v(x,t) + \mathcal{R} v(x,t) = f(x,t), \quad 0 < \alpha \leq 1, \tag{2.10}$$

and

$$v(x,0) = \mathcal{G}(x). \tag{2.11}$$

By using LT on Eq. (2.10), then

$$\mathcal{L}[v(x,t)] - \frac{\mathcal{G}(x)}{s} + \frac{1}{\mathcal{B}[\alpha]}\left(1-\alpha+\frac{\alpha}{s^{\alpha}}\right)\left\{\mathcal{L}[\mathcal{R}v(x,t)] + \mathcal{L}[\mathcal{N}v(x,t)] - \mathcal{L}[f(x,t)]\right\} = 0. \tag{2.12}$$

For $\varphi(x,t;q)$, $\mathcal{N}$ is defined as

$$\mathcal{N}[\varphi(x,t;q)] = \mathcal{L}[\varphi(x,t;q)] - \frac{\mathcal{G}(x)}{s} + \frac{1}{\mathcal{B}[\alpha]}\left(1-\alpha+\frac{\alpha}{s^{\alpha}}\right) \left\{\mathcal{L}[\mathcal{R}\varphi(x,t;q)] + L[\mathcal{N}\varphi(x,t;q)] - L[f(x,t)]\right\} \tag{2.13}$$

where $q \in \left[0, \frac{1}{n}\right]$. Now, the homotopy is given by [45, 46]

$$(1-nq)\mathcal{L}\left[\varphi(x,t;q)-v_0(x,t)\right]=\hbar q\mathcal{N}\left[\varphi(x,t;q)\right] \tag{2.14}$$

where $q \in \left[0, \frac{1}{n}\right]$. Then, the homotopy is given as

$$\varphi(x,t;0)=v_0(x,t), \varphi\left(x,t;\frac{1}{n}\right)=v(x,t) \tag{2.15}$$

Using Taylor's theorem, we have

$$\varphi(x,t;q)=v_0(x,t)+\sum_{m=1}^{\infty} v_m(x,t)q^m, \tag{2.16}$$

where

$$v_m(x,t)=\frac{1}{m!}\frac{\partial^m \varphi(x,t;q)}{\partial q^m}\bigg|_{q=0}. \tag{2.17}$$

For the appropriate value of $v_0(x,t)$, Eq. (2.13) unites at $q=\frac{1}{n}, n$ and $\hbar$. Therefore

$$v(x,t)=v_0(x,t)+\sum_{m=1}^{\infty} v_m(x,t)\left(\frac{1}{n}\right)^m. \tag{2.18}$$

Then, we obtain

$$\mathcal{L}\left[v_m(x,t)-\mathsf{k}_m v_{m-1}(x,t)\right]=\hbar\mathfrak{R}_m(\vec{v}_{m-1}) \tag{2.19}$$

where

$$\vec{v}_m=\left\{v_0(x,t), v_1(x,t), \ldots, v_m(x,t)\right\} \tag{2.20}$$

Eq. (2.19) simplifies, after making use of inverse LT, to

$$v_m(x,t)=\mathsf{k}_m v_{m-1}(x,t)+\hbar\mathcal{L}^{-1}\left[\mathfrak{R}_m(\vec{v}_{m-1})\right] \tag{2.21}$$

where

$$\mathfrak{R}_m(\vec{v}_{m-1})=L\left[v_{m-1}(x,t)\right]-\left(1-\frac{\mathsf{k}_m}{n}\right)\left(\frac{\mathcal{G}(x)}{s}+\frac{1}{\mathcal{B}[\alpha]}\left(1-\alpha+\frac{\alpha}{s^\alpha}\right)L\left[f(x,t)\right]\right)$$
$$+\frac{1}{\mathcal{B}[\alpha]}\left(1-\alpha+\frac{\alpha}{s^\alpha}\right)L\left[\mathcal{H}_{m-1}+Rv_{m-1}\right] \tag{2.22}$$

and

$$\mathsf{k}_m = \begin{cases} 0, m \leq 1, \\ n, m > 1. \end{cases} \tag{2.23}$$

In Eq. (2.22), $\mathcal{H}_m$ is a homotopy polynomial which is

$$\mathcal{H}_m = \frac{1}{m!}\left[\frac{\partial^m \varphi(x,t;q)}{\partial q^m}\right]_{q=0} \text{ and } \varphi(x,t;q) = \varphi_0 + q\varphi_1 + q^2\varphi_2 + \ldots. \tag{2.24}$$

By using Eqs. (2.21) and (2.22), we have

$$v_m(x,t) = (\mathsf{k}_m + \hbar)v_{m-1}(x,t) - \left(1 - \frac{\mathsf{k}_m}{n}\right)\mathcal{L}^{-1}\left(\frac{\mathcal{G}(x)}{s} + \frac{1}{\mathcal{B}[\alpha]}\left(1 - \alpha + \frac{\alpha}{s^\alpha}\right)L\left[f(x,t)\right]\right)$$
$$+ \hbar\mathcal{L}^{-1}\left\{\frac{1}{\mathcal{B}[\alpha]}\left(1 - \alpha + \frac{\alpha}{s^\alpha}\right)L\left[Rv_{m-1} + \mathcal{H}_{m-1}\right]\right\}. \tag{2.25}$$

Using q-HATM, the series solution is

$$v(x,t) = v_0(x,t) + \sum_{m=1}^{\infty} v_m(x,t). \tag{2.26}$$

2.4 Solution for Projected System

The reliability of the projected scheme for ODEs, and PDEs with integer and fractional order, has been proved by researchers. For instance, authors in [49] presented a system of equations exemplifying India's experience of the Covid-19 pandemic, fractional Liénard's equation with convergence is analyzed in [52], a numerical simulation for fractional order-coupled Burgers' equations is presented in [46], and the nonlinear chaotic model is analyzed in [26] within the framework of the Mittag–Leffler law. The novelty of the present method is that it includes parameters that enhance accuracy of the result obtained with a limited number of series terms. It also offers a simple algorithm to solve nonlinear PDEs without converting the nonlinear to linear and PDE to ODE. The homotopy parameter can help to show the reliability and efficiency of the algorithm with curves.

The results from this scheme can be compared with the system cited in Eq. (2.2):

$$\begin{aligned}
{}_{a}^{ABC}D_t^{\alpha} x(t) &= -3\beta x^3 - \delta xy - \mu xz, \\
{}_{a}^{ABC}D_t^{\alpha} y(t) &= \beta x^3 - \delta xy, \\
{}_{a}^{ABC}D_t^{\alpha} z(t) &= \delta xy - \mu xz, \\
{}_{a}^{ABC}D_t^{\alpha} r(t) &= \mu xz,
\end{aligned} \tag{2.27}$$

with initial conditions

$$x(0) = x_0(t), y(0) = y_0(t), z(0) = z_0(t), r(0) = r_0(t) \tag{2.28}$$

Taking LT on Eq. (2.27) and then we have, using Eq. (2.28)

$$\begin{aligned}
L[x(t)] &= \frac{1}{s}(x_0(t)) + \frac{1}{B[\alpha]}\left(1 - \alpha + \frac{\alpha}{S^{\alpha}}\right) L\{3\beta x^3 + \delta xy + \mu xz\}, \\
L[y(t)] &= \frac{1}{s}(y_0(t)) - \frac{1}{B[\alpha]}\left(1 - \alpha + \frac{\alpha}{S^{\alpha}}\right) L\{\beta x^3 - \delta xy\}, \\
L[z(t)] &= \frac{1}{s}(z_0(t)) - \frac{1}{B[\alpha]}\left(1 - \alpha + \frac{\alpha}{S^{\alpha}}\right) L\{-\mu xz + \delta xy\}, \\
L[r(t)] &= \frac{1}{s}(r_0(t)) - \frac{1}{B[\alpha]}\left(1 - \alpha + \frac{\alpha}{S^{\alpha}}\right) L\{\mu xz\}.
\end{aligned} \tag{2.29}$$

The non-linear operator N is defined as

$$\begin{aligned}
&N^1[\varphi_1, \varphi_2, \varphi_3, \varphi_4] \\
&\quad = L[\varphi_1(t;q)] - \frac{1}{s}(x_0(t)) + \frac{1}{B[\alpha]}\left(1 - \alpha + \frac{\alpha}{s^{\alpha}}\right) L\{3\beta\varphi_1^3 + \delta\varphi_1\varphi_2 + \mu\varphi_1\varphi_3\}, \\
&N^2[\varphi_1, \varphi_2, \varphi_3, \varphi_4] \\
&\quad = L[\varphi_2(t;q)] - \frac{1}{s}(y_0(t)) + \frac{1}{B[\alpha]}\left(1 - \alpha + \frac{\alpha}{s^{\alpha}}\right) L\{3\beta\varphi_1^3 + \delta\varphi_1\varphi_2\}, \\
&N^3[\varphi_1, \varphi_2, \varphi_3, \varphi_4] \\
&\quad = L[\varphi_3(t;q)] - \frac{1}{s}(z_0(t)) + \frac{1}{B[\alpha]}\left(1 - \alpha + \frac{\alpha}{s^{\alpha}}\right) L\{\delta\varphi_1\varphi_2 + \mu\varphi_1\varphi_3\}, \\
&N^4[\varphi_1, \varphi_2, \varphi_3, \varphi_4] \\
&\quad = L[\varphi_4(t;q)] - \frac{1}{s}(r_0(t)) + \frac{1}{B[\alpha]}\left(1 - \alpha + \frac{\alpha}{s^{\alpha}}\right) L\{\mu\varphi_1\varphi_3\}.
\end{aligned} \tag{2.30}$$

The *m-th* order deformation equation at $\mathcal{H}(t)=1$ is given by the projected scheme by

$$
\begin{aligned}
L\left[x_m(t)-\mathrm{k}_m x_{m-1}(t)\right] &= \hbar \mathfrak{R}_{1,m}\left[\vec{x}_{m-1},\vec{y}_{m-1},\vec{z}_{m-1},\vec{r}_{m-1}\right],\\
L\left[y_m(t)-\mathrm{k}_m y_{m-1}(t)\right] &= \hbar \mathfrak{R}_{2,m}\left[\vec{x}_{m-1},\vec{y}_{m-1},\vec{z}_{m-1},\vec{r}_{m-1}\right],\\
L\left[z_m(t)-\mathrm{k}_m z_{m-1}(t)\right] &= \hbar \mathfrak{R}_{3,m}\left[\vec{x}_{m-1},\vec{y}_{m-1},\vec{z}_{m-1},\vec{r}_{m-1}\right],\\
L\left[r_m(t)-\mathrm{k}_m r_{m-1}(t)\right] &= \hbar \mathfrak{R}_{4,m}\left[\vec{x}_{m-1},\vec{y}_{m-1},\vec{z}_{m-1},\vec{r}_{m-1}\right],
\end{aligned}
\tag{2.31}
$$

where

$$
\begin{aligned}
&\mathfrak{R}_{1,m}\left[\vec{x}_{m-1},\vec{y}_{m-1},\vec{z}_{m-1},\vec{r}_{m-1}\right]=L\left[x_{m-1}(t)\right]-\left(1-\frac{\mathrm{k}_m}{n}\right)\left\{\frac{1}{s}\left(x_0(t)\right)\right\}\\
&\quad+\frac{1}{\mathcal{B}[\alpha]}\left(1-\alpha+\frac{\alpha}{s^\alpha}\right)L\left\{3\beta\sum_{i=0}^{m-1}\sum_{j=0}^{i}x_j x_{i-j}x_{m-1-i}+\delta\sum_{j=0}^{i}x_i y_{m-1-i}+\mu\sum_{j=0}^{i}x_i z_{m-1-i}\right\}\\
&\mathfrak{R}_{2,m}\left[\vec{x}_{m-1},\vec{y}_{m-1},\vec{z}_{m-1},\vec{r}_{m-1}\right]=L\left[y_{m-1}(t)\right]-\left(1-\frac{\mathrm{k}_m}{n}\right)\left\{\frac{1}{s}\left(y_0(t)\right)\right\}\\
&\quad-\frac{1}{\mathcal{B}[\alpha]}\left(1-\alpha+\frac{\alpha}{s^\alpha}\right)L\left\{\beta\sum_{i=0}^{m-1}\sum_{j=0}^{i}x_j x_{i-j}x_{m-1-i}-\delta\sum_{j=0}^{i}x_i y_{m-1-i}\right\},\\
&\mathfrak{R}_{3,m}\left[\vec{x}_{m-1},\vec{y}_{m-1},\vec{z}_{m-1},\vec{r}_{m-1}\right]=L\left[z_{m-1}(t)\right]-\left(1-\frac{\mathrm{k}_m}{n}\right)\left\{\frac{1}{s}\left(z_0(t)\right)\right\}\\
&\quad-\frac{1}{\mathcal{B}[\alpha]}\left(1-\alpha+\frac{\alpha}{s^\alpha}\right)L\left\{\delta\sum_{j=0}^{i}x_i y_{m-1-i}-\mu\sum_{j=0}^{i}x_i z_{m-1-i}\right\},\\
&\mathfrak{R}_{4,m}\left[\vec{x}_{m-1},\vec{y}_{m-1},\vec{z}_{m-1},\vec{r}_{m-1}\right]=L\left[r_{m-1}(t)\right]-\left(1-\frac{\mathrm{k}_m}{n}\right)\left\{\frac{1}{s}\left(r_0(t)\right)\right\}\\
&\quad-\frac{1}{\mathcal{B}[\alpha]}\left(1-\alpha+\frac{\alpha}{s^\alpha}\right)L\left\{\mu\sum_{j=0}^{i}x_i z_{m-1-i}\right\}.
\end{aligned}
\tag{2.32}
$$

By employing inverse *LT* on Eq. (2.31), we get

$$
\begin{aligned}
x_m(t)&=\mathrm{k}_m x_{m-1}(t)+\hbar L^{-1}\left\{\mathfrak{R}_{1,m}\left[\vec{x}_{m-1},\vec{y}_{m-1},\vec{z}_{m-1},\vec{r}_{m-1}\right]\right\},\\
y_m(t)&=\mathrm{k}_m y_{m-1}(t)+\hbar L^{-1}\left\{\mathfrak{R}_{2,m}\left[\vec{x}_{m-1},\vec{y}_{m-1},\vec{z}_{m-1},\vec{r}_{m-1}\right]\right\},\\
z_m(t)&=\mathrm{k}_m z_{m-1}(t)+\hbar L^{-1}\left\{\mathfrak{R}_{3,m}\left[\vec{x}_{m-1},\vec{y}_{m-1},\vec{z}_{m-1},\vec{r}_{m-1}\right]\right\},\\
r_m(t)&=\mathrm{k}_m r_{m-1}(t)+\hbar L^{-1}\left\{\mathfrak{R}_{4,m}\left[\vec{x}_{m-1},\vec{y}_{m-1},\vec{z}_{m-1},\vec{r}_{m-1}\right]\right\}.
\end{aligned}
\tag{2.33}
$$

By using $x_0(t) = 1$, $y_0(t) = z_0(t) = r_0(t) = 0$ and then solving the foregoing equations, we can obtain the terms of

$$\begin{aligned} x(t) &= x_0(t) + \sum_{m-1}^{\infty} x_m(t)\left(\frac{1}{n}\right)^m, \\ y(t) &= y_0(t) + \sum_{m-1}^{\infty} y_m(t)\left(\frac{1}{n}\right)^m, \\ z(t) &= z_0(t) + \sum_{m-1}^{\infty} z_m(t)\left(\frac{1}{n}\right)^m, \\ r(t) &= r_0(t) + \sum_{m-1}^{\infty} r_m(t)\left(\frac{1}{n}\right)^m. \end{aligned} \tag{2.34}$$

2.5 Existence of Solutions

This section examines the existence and uniqueness of the solutions. We consider Eq. (2.27) as

$$\begin{cases} \mathcal{G}_1(t,x) = {}^{ABC}_{\ \ 0}D_t^{\alpha}[x], \\ \mathcal{G}_1(t,y) = {}^{ABC}_{\ \ 0}D_t^{\alpha}[y], \\ \mathcal{G}_1(t,z) = {}^{ABC}_{\ \ 0}D_t^{\alpha}[z], \\ \mathcal{G}_1(t,r) = {}^{ABC}_{\ \ 0}D_t^{\alpha}[r]. \end{cases} \tag{2.35}$$

From Eq. (2.35) and Theorem 2 we have

$$\begin{cases} x(t) - x(0) = \dfrac{1}{\mathcal{B}(\alpha)}\left[(1-\alpha)\mathcal{G}_1(t,x) + \dfrac{\alpha}{\Gamma(\alpha)}\displaystyle\int_0^t \mathcal{G}_1(\zeta,x)(t-\zeta)^{\alpha-1}d\zeta\right], \\ y(t) - y(0) = \dfrac{1}{\mathcal{B}(\alpha)}\left[(1-\alpha)\mathcal{G}_2(t,y) + \dfrac{\alpha}{\Gamma(\alpha)}\displaystyle\int_0^t \mathcal{G}_2(\zeta,y)(t-\zeta)^{\alpha-1}d\zeta\right], \\ z(t) - z(0) = \dfrac{1}{\mathcal{B}(\alpha)}\left[(1-\alpha)\mathcal{G}_3(t,z) + \dfrac{\alpha}{\Gamma(\alpha)}\displaystyle\int_0^t \mathcal{G}_3(\zeta,z)(t-\zeta)^{\alpha-1}d\zeta\right], \\ r(t) - r(0) = \dfrac{1}{\mathcal{B}(\alpha)}\left[(1-\alpha)\mathcal{G}_4(t,r) + \dfrac{\alpha}{\Gamma(\alpha)}\displaystyle\int_0^t \mathcal{G}_4(\zeta,r)(t-\zeta)^{\alpha-1}d\zeta\right]. \end{cases} \tag{2.36}$$

Theorem 3

The kernel $\mathcal{G}_1$ admits the Lipschitz condition and contraction if $0 \le \left(2\delta^2 + \frac{1}{2}\lambda_2\delta(a+b) + \tau_2(2+\lambda_2\delta) + \xi_1 \right) < 1$ holds.

Proof. To prove the theorem, consider x and x_1 two functions, then

$$\begin{aligned}
\left\| \mathcal{G}_1(t,x) - \mathcal{G}_1(t,x_1) \right\| &= \left\| 3\beta\left[x^3(t) - x^3(t_1)\right] + \delta y(t)\left[x(t) - x(t_1)\right] - \mu z(t)\left[x(t) - x(t_1)\right] \right\| \\
&= \left\| \left(3\beta\left(x^2(t) + x^2(t_1) + x(t)x(t_1)\right) + \delta y(t) - \mu z(t)\right)\left[x(t) - x(t_1)\right] \right\| \\
&\le \left\| 3\beta\left(x^2(t) + x^2(t_1) + x(t)x(t_1)\right) + \delta y(t) - \mu z(t) \right\| \left\| x(t) - x(t_1) \right\| \\
&\le \left(3\beta\left(a^2 + b^2 + ab\right) + \delta\lambda_2 - \mu\lambda_3\right) \left\| x(t) - x(t_1) \right\| \\
&\le \eta_1 \left\| x(t) - x(t_1) \right\|,
\end{aligned} \tag{2.37}$$

where $\|y(t)\| \le \lambda_2$ and $\|z(t)\| \le \lambda_3$ are bounded functions and also $\|x(t)\| \le a$ and $\|x(t_1)\| \le b$. Putting $\eta_1 = 3\beta(a^2 + b^2 + ab) + \delta\lambda_2 - \mu\lambda_3$ in Eq. (2.37), one can get

$$\left\| \mathcal{G}_1(t,x) - \mathcal{G}_1(t,x_1) \right\| \le \eta_1 \left\| x(t) - x(t_1) \right\|. \tag{2.38}$$

Eq. (2.38) confirms the Lipschitz condition is attained for $\mathcal{G}_1$. If $0 \le (3\beta(a^2 + b^2 + ab) + \delta\lambda_2 - \mu\lambda_3) < 1$, then it leads to contraction. Also, we can prove

$$\begin{aligned}
\left\| \mathcal{G}_2(t,y) - \mathcal{G}_2(t,y_1) \right\| &\le \eta_2 \left\| y(t) - y(t_1) \right\|, \\
\left\| \mathcal{G}_3(t,z) - \mathcal{G}_3(t,z_1) \right\| &\le \eta_3 \left\| z(t) - z(t_1) \right\|, \\
\left\| \mathcal{G}_4(t,r) - \mathcal{G}_4(t,r_1) \right\| &\le \eta_4 \left\| r(t) - r(t_1) \right\|.
\end{aligned} \tag{2.39}$$

With the help of the foregoing equations, Eq. (2.36) gives

$$\begin{cases}
x_n(t) = \dfrac{(1-\alpha)}{\mathcal{B}(\alpha)}\mathcal{G}_1(t, x_{n-1}) + \dfrac{\alpha}{\mathcal{B}(\alpha)\Gamma(\alpha)}\displaystyle\int_0^t \mathcal{G}_1(\zeta, x_{n-1})(t-\zeta)^{\alpha-1}\, d\zeta, \\
y_n(t) = \dfrac{(1-\alpha)}{\mathcal{B}(\alpha)}\mathcal{G}_2(t, y_{n-1}) + \dfrac{\alpha}{\mathcal{B}(\alpha)\Gamma(\alpha)}\displaystyle\int_0^t \mathcal{G}_2(\zeta, y_{n-1})(t-\zeta)^{\alpha-1}\, d\zeta, \\
z_n(t) = \dfrac{(1-\alpha)}{\mathcal{B}(\alpha)}\mathcal{G}_3(t, z_{n-1}) + \dfrac{\alpha}{\mathcal{B}(\alpha)\Gamma(\alpha)}\displaystyle\int_0^t \mathcal{G}_3(\zeta, z_{n-1})(t-\zeta)^{\alpha-1}\, d\zeta, \\
r_n(t) = \dfrac{(1-\alpha)}{\mathcal{B}(\alpha)}\mathcal{G}_4(t, r_{n-1}) + \dfrac{\alpha}{\mathcal{B}(\alpha)\Gamma(\alpha)}\displaystyle\int_0^t \mathcal{G}_4(\zeta, r_{n-1})(t-\zeta)^{\alpha-1}\, d\zeta.
\end{cases} \tag{2.40}$$

The initial conditions are

$$x(0)=x_0(t),y(0)=y_0(t),z(0)=z_0(t),\text{ and } r(0)=r_0(t). \tag{2.41}$$

Between the successive difference terms, we have

$$\begin{cases}\phi_{1n}(t)=x_n(t)-x_{n-1}(t)\\ \quad =\dfrac{(1-\alpha)}{\mathcal{B}(\alpha)}\left(\mathcal{G}_1(t,x_{n-1})-\mathcal{G}_1(t,x_{n-2})\right)+\dfrac{\alpha}{\mathcal{B}(\alpha)\Gamma(\alpha)}\displaystyle\int_0^t \mathcal{G}_1(\zeta,x_{n-1})(t-\zeta)^{\alpha-1}d\zeta,\\ \phi_{2n}(t)=y_n(t)-y_{n-1}(t)\\ \quad =\dfrac{(1-\alpha)}{\mathcal{B}(\alpha)}\left(\mathcal{G}_2(t,y_{n-1})-\mathcal{G}_2(t,y_{n-2})\right)+\dfrac{\alpha}{\mathcal{B}(\alpha)\Gamma(\alpha)}\displaystyle\int_0^t \mathcal{G}_2(\zeta,y_{n-1})(t-\zeta)^{\alpha-1}d\zeta,\\ \phi_{3n}(t)=z_n(t)-z_{n-1}(t)\\ \quad =\dfrac{(1-\alpha)}{\mathcal{B}(\alpha)}\left(\mathcal{G}_3(t,z_{n-1})-\mathcal{G}_3(t,z_{n-2})\right)+\dfrac{\alpha}{\mathcal{B}(\alpha)\Gamma(\alpha)}\displaystyle\int_0^t \mathcal{G}_3(\zeta,z_{n-1})(t-\zeta)^{\alpha-1}d\zeta,\\ \phi_{4n}(t)=r_n(t)-r_{n-1}(t)\\ \quad =\dfrac{(1-\alpha)}{\mathcal{B}(\alpha)}\left(\mathcal{G}_4(t,r_{n-1})-\mathcal{G}_4(t,r_{n-2})\right)+\dfrac{\alpha}{\mathcal{B}(\alpha)\Gamma(\alpha)}\displaystyle\int_0^t \mathcal{G}_4(\zeta,r_{n-1})(t-\zeta)^{\alpha-1}d\zeta.\end{cases} \tag{2.42}$$

Notice that

$$\begin{cases}x_n(t)=\displaystyle\sum_{i=1}^n \phi_{1i}(t),\\ y_n(t)=\displaystyle\sum_{i=1}^n \phi_{2i}(t),\\ z_n(t)=\displaystyle\sum_{i=1}^n \phi_{3i}(t),\\ r_n(t)=\displaystyle\sum_{i=1}^n \phi_{4i}(t).\end{cases} \tag{2.43}$$

With the assistance of Eq. (2.38) and plugging the norm on Eq. (2.42), we get

$$\left\|\phi_{1n}(t)\right\|\le\frac{(1-\alpha)}{\mathcal{B}(\alpha)}\eta_1\left\|\phi_{1(n-1)}(t)\right\|+\frac{\alpha}{\mathcal{B}(\alpha)\Gamma(\alpha)}\eta_1\int_0^t\left\|\phi_{1(n-1)}(\zeta)\right\|d\zeta. \tag{2.44}$$

Similarly, we have

$$\|\phi_{2n}(x,t)\| \leq \frac{(1-\alpha)}{\mathcal{B}(\alpha)}\eta_2 \|\phi_{2(n-1)}(t)\| + \frac{\alpha}{\mathcal{B}(\alpha)\Gamma(\alpha)}\eta_2 \int_0^t \|\phi_{2(n-1)}(\zeta)\| d\zeta,$$
$$\|\phi_{3n}(x,t)\| \leq \frac{(1-\alpha)}{\mathcal{B}(\alpha)}\eta_3 \|\phi_{3(n-1)}(t)\| + \frac{\alpha}{\mathcal{B}(\alpha)\Gamma(\alpha)}\eta_3 \int_0^t \|\phi_{3(n-1)}(\zeta)\| d\zeta, \quad (2.45)$$
$$\|\phi_{4n}(x,t)\| \leq \frac{(1-\alpha)}{\mathcal{B}(\alpha)}\eta_4 \|\phi_{4(n-1)}(t)\| + \frac{\alpha}{\mathcal{B}(\alpha)\Gamma(\alpha)}\eta_4 \int_0^t \|\phi_{4(n-1)}(\zeta)\| d\zeta.$$

Next, we demonstrate subsequent results with the aid of the above results.

Theorem 4

If we have specific t_0, then the solution for Eq. (2.27) will exist and be unique. Further,

$$\frac{(1-\alpha)}{\mathcal{B}(\alpha)}\eta_i + \frac{\alpha}{\mathcal{B}(\alpha)\Gamma(\alpha)}\eta_i < 1.$$

for $i = 1, 2, 3, 4$.

Proof. Consider bounded functions x, y, z, and r admitting the Lipschitz condition. Then, using Eq. (2.43) and Eq. (2.45), one can get

$$\|\phi_{1i}(t)\| \leq \frac{\|x_n(0)\|}{\mathcal{B}(\alpha)}\left[\frac{\alpha}{\Gamma(\alpha)}\eta_1 + (1-\alpha)\eta_1\right]^n,$$
$$\|\phi_{2i}(t)\| \leq \frac{\|y_n(0)\|}{\mathcal{B}(\alpha)}\left[\frac{\alpha}{\Gamma(\alpha)}\eta_2 + (1-\alpha)\eta_2\right]^n,$$
$$\|\phi_{3i}(t)\| \leq \frac{\|z_n(0)\|}{\mathcal{B}(\alpha)}\left[\frac{\alpha}{\Gamma(\alpha)}\eta_3 + (1-\alpha)\eta_3\right]^n, \quad (2.46)$$
$$\|\phi_{4i}(t)\| \leq \frac{\|r_n(0)\|}{\mathcal{B}(\alpha)}\left[\frac{\alpha}{\Gamma(\alpha)}\eta_4 + (1-\alpha)\eta_4\right]^n.$$

The existence and continuity of the results achieved are thus verified. To verify that Eq. (2.46) is a solution for Eq. (2.27), consider

$$x = x(0) + x_n(t) - \mathcal{K}_{1n}(t),$$
$$y = y(0) + y_n(t) - \mathcal{K}_{2n}(t),$$
$$z = z(0) + z_n(t) - \mathcal{K}_{3n}(t), \quad (2.47)$$
$$r = r(0) + r_n(t) - \mathcal{K}_{4n}(t).$$

Let us consider

$$\|\mathcal{K}_{1n}(t)\| = \left\| \frac{(1-\alpha)}{\mathcal{B}(\alpha)}\left(\mathcal{G}_1(t,x)-\mathcal{G}_1(t,x_{n-1})\right) + \frac{\alpha}{\mathcal{B}(\alpha)\Gamma(\alpha)}\int_0^t (t-\zeta)^{\mu-1}\left(\mathcal{G}_1(\zeta,x)-\mathcal{G}_1(\zeta,x_{n-1})\right)d\zeta \right\|$$
$$\leq \frac{(1-\alpha)}{\mathcal{B}(\alpha)}\left\|\left(\mathcal{G}_1(t,x)-\mathcal{G}_1(t,x_{n-1})\right)\right\| + \frac{\alpha}{\mathcal{B}(\alpha)\Gamma(\alpha)}\int_0^t \left\|\left(\mathcal{G}_1(\zeta,x)-\mathcal{G}_1(\zeta,x_{n-1})\right)\right\| d\zeta \quad (2.48)$$
$$\leq \frac{(1-\alpha)}{\mathcal{B}(\alpha)}\eta_1\|x-x_{n-1}\| + \frac{\alpha}{\mathcal{B}(\alpha)\Gamma(\alpha)}\eta_1\|x-x_{n-1}\|t.$$

Similarly, at t_0 we have

$$\|\mathcal{K}_{1n}(t)\| \leq \left(\frac{1-\alpha}{\mathcal{B}(\alpha)} + \frac{\alpha\, t_0}{\mathcal{B}(\alpha)\Gamma(\alpha)}\right)^{n+1} \eta_1^{n+1}M. \quad (2.49)$$

As n tends to ∞, then $\|\mathcal{K}_{1n}(t)\|$ approaches to 0 with respect to Eq. (2.49). Similarly, we can prove for $\|\mathcal{K}_{2n}(t)\|$, $\|\mathcal{K}_{3n}(x,t)\|$ and $\|\mathcal{K}_{4n}(x,t)\|$.

Next, we prove the uniqueness of the projected system result. Suppose, considering the set of other solutions $x^*(t)$, $y^*(t)$, $z^*(t)$ and $r^*(t)$, then

$$x(t)-x^*(t) = \frac{1}{\mathcal{B}(\alpha)}\left[(1-\alpha)\left(\mathcal{G}_1(t,x)-\mathcal{G}_1(t,x^*)\right) + \frac{\alpha}{\Gamma(\alpha)}\int_0^t \left(\mathcal{G}_1(\zeta,x)-\mathcal{G}_1(\zeta,x^*)\right)d\zeta\right]. \quad (2.50)$$

Now, Eq. (2.50) simplifies on plugging the norm

$$\|x(t)-x^*(t)\| = \left\| \frac{\alpha}{\mathcal{B}(\alpha)\Gamma(\alpha)}\int_0^t \left(\mathcal{G}_1(\zeta,x)-\mathcal{G}_1(\zeta,x^*)\right)d\zeta + \frac{(1-\alpha)}{\mathcal{B}(\alpha)}\left(\mathcal{G}_1(t,x)-\mathcal{G}_1(t,x^*)\right)\right\|$$
$$\leq \frac{\alpha}{\mathcal{B}(\alpha)\Gamma(\alpha)}\eta_1 t\|x(t)-x^*(t)\| + \frac{(1-\alpha)}{\mathcal{B}(\alpha)}\eta_1\|x(t)-x^*(t)\|. \quad (2.51)$$

On simplification

$$\|x(t)-x^*(t)\|\left(1-\frac{\alpha}{\mathcal{B}(\alpha)\Gamma(\alpha)}\eta_1 t - \frac{(1-\alpha)}{\mathcal{B}(\alpha)}\eta_1\right) \leq 0. \quad (2.52)$$

Clearly $x(t) - x^*(t) = 0$, if

$$\left(1-\frac{\alpha}{\mathcal{B}(\alpha)\Gamma(\alpha)}\eta_1 t-\frac{(1-\alpha)}{\mathcal{B}(\alpha)}\eta_1\right)\geq 0. \tag{2.53}$$

Therefore, Eq. (2.53) proves the theorem.

Theorem 5

Suppose xn, yn, zn, rn, x, y, z, and r presented in the Banach space $\left(\mathfrak{B}[0,T],\|\cdot\|\right)$. If $0 < \lambda i < 1$, for $i = 1, 2, 3, 4$, then Eq. (2.26) converges to Eq. (2.10).

Proof: Let $\{\mathfrak{D}_n\}$ be the sequence and partial sum of Eq. (2.26). Now, we need to illustrate that $\{\mathcal{S}_n\}$ is Cauchy sequence in $\left(\mathfrak{B}[0,T],\|\cdot\|\right)$. Let

$$\begin{aligned}\|\mathfrak{D}_{n+1}(t)-\mathfrak{D}_n(t)\| &= \|x_{n+1}(t)\| \\ &\leq \lambda_1\|x_n(t)\| \\ &\leq \lambda_1^2\|x_{n-1}(t)\|\leq\ldots\leq\lambda_1^{n+1}\|x_0(t)\|.\end{aligned}$$

Now, for every $n, m \in N$ $(m \leq n)$, we have

$$\begin{aligned}\|\mathfrak{D}_n-\mathfrak{D}_m\| &= \|(\mathfrak{D}_n-\mathfrak{D}_{n-1})+(\mathfrak{D}_{n-1}-\mathfrak{D}_{n-2})+\ldots+(\mathfrak{D}_{m+1}-\mathfrak{D}_m)\| \\ &\leq\|\mathfrak{D}_n-\mathfrak{D}_{n-1}\|+\|\mathfrak{D}_{n-1}-\mathfrak{D}_{n-2}\|+\ldots+\|\mathfrak{D}_{m+1}-\mathfrak{D}_m\| \\ &\leq\left(\lambda_1^{n}+\lambda_1^{n-1}+\ldots+\lambda_1^{m+1}\right)\|x_0\| \\ &\leq\lambda_1^{m+1}\left(\lambda_1^{n-m-1}+\lambda_1^{n-m-2}+\ldots+\lambda_1+1\right)\|x_0\| \\ &\leq\lambda_1^{m+1}\left(\frac{1-\lambda_1^{n-m}}{1-\lambda_1}\right)\|x_0\|.\end{aligned} \tag{2.54}$$

But $0 < \lambda_1 < 1$, therefore $\lim_{n,m\to\infty}\|\mathfrak{D}_n-\mathfrak{D}_m\|=0$. Hence, $\{\mathfrak{D}_n\}$ is the Cauchy sequence.

2.6 Results and Discussion

This study has explored solutions for fractional-order differential systems illustrating chemical reactions in the helium-burning network, with the help of q-HATM in the framework of the Mittag–Leffler law. In this section, we establish the behavior of the results achieved. In Figure 2.1, the response of q-HATM results in distinct order. The considered method is associated with the homotopy parameter, which can help us to regulate the convergence providence of the result obtained. In association with this, we draw an $\hbar$-curve for $n = 1$ and 2 in Figure 2.2; similar behavior can be expected for the remaining

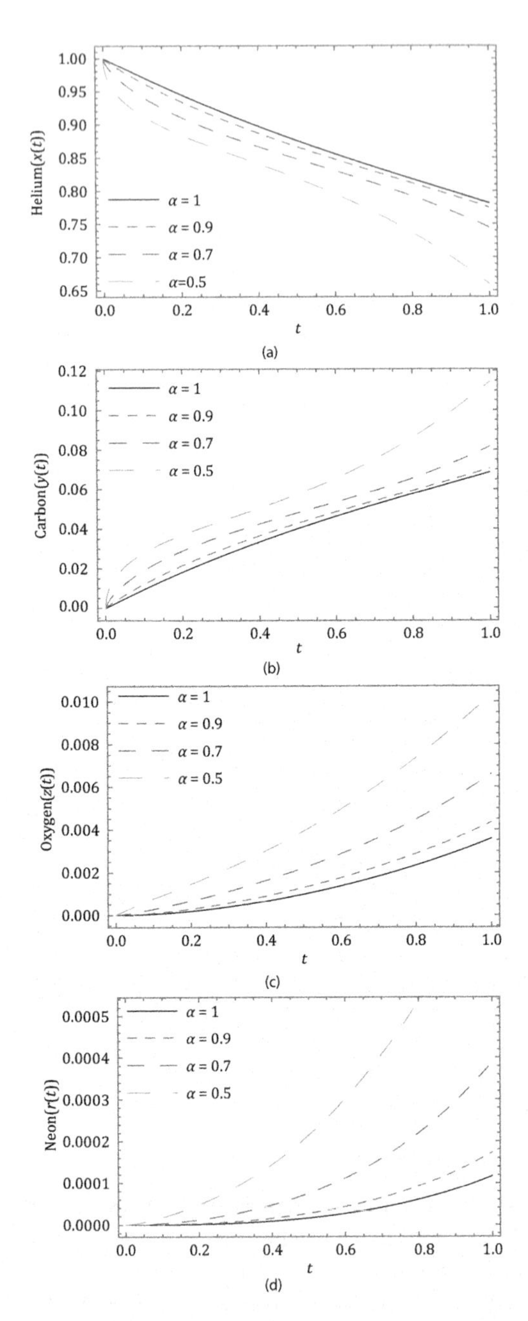

FIGURE 2.1
Nature of solution obtained for (a) helium ($x(t)$), (b) carbon ($y(t)$), (c) oxygen ($z(t)$) and (d) neon ($r(t)$) with different α at $\hbar = -1$ and $n = 1$.

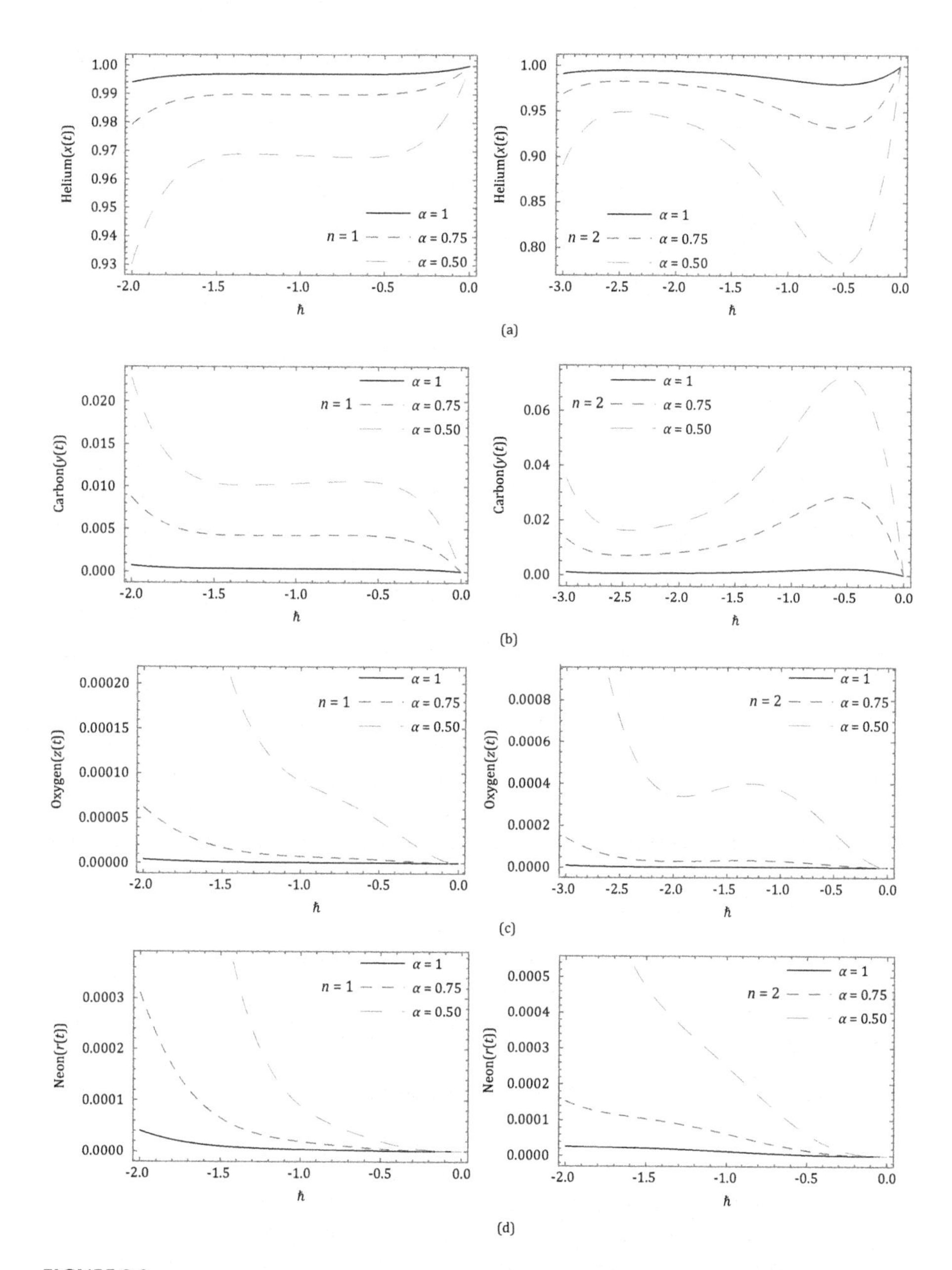

FIGURE 2.2
ħ-curves for (a) helium ($x(t)$), (b) carbon ($y(t)$), (c) oxygen ($z(t)$) and (d) neon ($r(t)$) with distinct α at $t = 0.01$ for $n = 1$ and 2.

cases. The solid and straight-line segment illustrates the convergence of the result. The $\hbar$-curve enables us to control and adjust the convergence domain of the solution obtained. For apt $\hbar$, the solution rapidly congregates to the exact result. In complex and nonlinear models, the slight discrepancy leads to close examination of the results and prediction of the response of the relevant phenomena in a systematic and better way. Further, from all the captured figures, we can observe that the projected algorithm is very effective in investigating the coupled fractional system.

2.7 Conclusion

In this study, the q-HATM is effectively borrowed to achieve a solution for a fractional-order helium-burning network system. The existence as well as the uniqueness of the solution is established using fixed-point hypothesis. The borrowed fractional operator is demarcated with generalized Mittag–Leffler function which leads to non-local and non-singular kernel. The projected solution procedure assesses the solution for differential equations where conversion, discretization, or perturbation cannot be used. We considered five different examples to confirm the exactness as well as the reliability of the method. The plots confirm the effect of fractional order in real-world problems, rendering the results obtained with the algorithm the more motivating and fascinating. These types of models, which contribute to scholarship on nonlinear phenomena, represent a fresh and innovative approach to the study of real-world difficulties.

References

[1] J. Liouville, Mémoire surquelques questions de géometrie et de mécanique, et sur un nouveau genre de calcul pour résoudre ces questions, *J. Ecole. Polytech.*, 13 (1832) 1–69.
[2] G. F. B. Riemann, *Versuch einer allgemeinen Auffassung der Integration und Differentiation, Gesammelte Mathematische Werke*, Leipzig, (1896).
[3] M. Caputo, *Elasticita e Dissipazione*, Zanichelli, Bologna, (1969).
[4] K. S. Miller, B. Ross, *An Introduction to Fractional Calculus and Fractional Differential Equations*, A Wiley, New York, (1993).
[5] I. Podlubny, *Fractional Differential Equations*, Academic Press, New York, (1999).
[6] A. A. Kilbas, H. M. Srivastava, J. J. Trujillo, *Theory and Applications of Fractional Differential Equations*, Elsevier, Amsterdam, (2006).

[7] D. Baleanu, Z. B. Guvenc, J. A. Tenreiro Machado, *New Trends in Nanotechnology and Fractional Calculus Applications*, Springer Dordrecht Heidelberg, London New York, (2010).
[8] C. Baishya, S. J. Achar, P. Veeresha, D. G. Prakasha, Dynamics of a fractional epidemiological model with disease infection in both the populations, *Chaos*, 31 (043130) (2021). DOI: 10.1063/5.0028905.
[9] S. W. Yao, E. Ilhan, P. Veeresha, H. M. Baskonus, A powerful iterative approach for quintic complex Ginzburg-Landau equation within the frame of fractional operator, *Fractals*, 25 (5) (2021). DOI: 10.1142/S0218348X21400235.
[10] D. Baleanu, G. C. Wu, S. D. Zeng, Chaos analysis and asymptotic stability of generalized Caputo fractional differential equations, *Chaos Soliton. Fractal.*, 102 (2017), 99–105.
[11] P. Veeresha, E. Ilhan, D. G. Prakasha, H. M. Baskonus, W. Gao, Regarding on the fractional mathematical model of Tumour invasion and metastasis, *Comp. Model. Eng. Sci.*, 127 (3) (2021), 1013–1036.
[12] H. M. Baskonus, T. A. Sulaiman, H. Bulut, On the new wave behavior to the Klein-Gordon-Zakharov equations in plasma physics, *Indian J. Phys.*, 93 (3) (2019), 393–399.
[13] D. G. Prakasha, P. Veeresha, H. M. Baskonus, Analysis of the dynamics of hepatitis E virus using the Atangana-Baleanu fractional derivative, *Eur. Phys. J. Plus*, 134 (241) (2019). DOI: 10.1140/epjp/i2019-12590-5.
[14] L. Akinyemi, S. N. Huseen, A powerful approach to study the new modified coupled Korteweg–de Vries system, *Math. Comput. Simul.*, 177 (2020), 556–567.
[15] P. Veeresha, D. G. Prakasha, Solution for fractional Kuramoto–Sivashinsky equation using novel computational technique. *Int. J. Appl. Comput. Math.*, 7 (33) (2021). DOI: 10.1007/s40819-021-00956-0.
[16] L. Akinyemi, A fractional analysis of Noyes–Field model for the nonlinear Belousov–Zhabotinsky reaction, *Comput. Appl. Math.*, 39 (2020), 1–34.
[17] V. Kourganoff, *Introduction to the Physics of Stellar Interiors*, D. Reidel Publishing Company, Dordrecht, (1973).
[18] H. J. Haubold, A. M. Mathai, On thermonuclear reaction rates, *Astrophysics Space Sci.*, 258 (1998), 185–199.
[19] M. I. Nouh, M. A. Sharaf, A. S. Saad, Symbolic analytical solutions for the abundances differential equations of the Helium burning phase, *Astron. Nachr.*, 324 (5) (2003), 432–436.
[20] D. D. Clayton, *Principles of Stellar Evolution and Nucleosynthesis*, University of Chicago Press, Chicago, (1983).
[21] M. I. Nouh, Computational method for a fractional model of the Helium burning network, *New Astron.*, 66 (2019), 40–44.
[22] M. Caputo, M. Fabrizio, A new definition of fractional derivative without singular kernel, *Prog. Fract. Differ. Appl.*, 1 (2) (2015), 73–85.
[23] A. Atangana, D. Baleanu, New fractional derivatives with non-local and non-singular kernel theory and application to heat transfer model, *Therm. Sci.*, 20 (2016), 763–769.
[24] H. Singh, A. M. Wazwaz, Computational method for reaction diffusion-model arising in a spherical catalyst, *Int. J. Appl. Comput. Math.*, 7 (65) (2021). DOI: 10.1007/s40819-021-00993-9.

[25] H. Singh, H. M. Srivastava, D. Kumar, A reliable algorithm for the approximate solution of the nonlinear Lane-Emden type equations arising in astrophysics, *Numer. Methods Partial Differ. Equ.*, 34 (5) (2018), 1524–1555.

[26] P. Veeresha, D. G. Prakasha, A.-H. Abdel-Aty, H. Singh, E. E. Mahmoud, S. Kumar, An efficient approach for fractional nonlinear chaotic model with Mittag-Leffler law, *J. King Saud Univ. Sci.*, 33 (2) (2021). DOI: 10.1016/j.jksus.2021.101347.

[27] H. Singh, H. M. Srivastava, Numerical investigation of the fractional-order Liénard and duffing equations arising in oscillating circuit theory, *Front. Phys.*, 8 (120). DOI: 10.3389/fphy.2020.00120.

[28] C. Baishya, Dynamics of a fractional stage structured predator-prey model with prey refuge, *Indian J. Ecol.*, 47 (4) (2020), 1118–1124.

[29] L. Akinyemi, P. Veeresha, M. Senol, Numerical solutions for coupled nonlinear Schrodinger-Korteweg-de Vries and Maccari's systems of equations, *Mod. Phys. Lett. B*, (2021). DOI: 10.1142/s0217984921503395.

[30] H. Singh, Analysis for fractional dynamics of Ebola virus model, *Chaos Soliton. Fractal.*, 138 (109992) (2020). DOI: 10.1016/j.chaos.2020.109992.

[31] H. Singh, J. Singh, S. D. Purohit, D. Kumar, *Advanced Numerical Methods for Differential Equations: Applications in Science and Engineering*, CRC Press Taylor and Francis, Boca Raton, 2021.

[32] H. Singh, D. Kumar, *D. Baleanu, Methods of Mathematical Modelling: Fractional Differential Equations*, CRC Press Taylor and Francis, Boca Raton, 2019.

[33] G. Yel, C. Cattani, H. M. Baskonus, W. Gao, On the complex simulations with dark-bright to the Hirota-Maccari system, *J. Comput. Nonlinear Dynam.*, 16 (6) (2021). DOI: 10.1115/1.4050677.

[34] W. Gao, H. M. Baskonus, L. Shi, New investigation of bats-hosts-reservoir-people coronavirus model and application to 2019-nCoV system, *Adv. Differ. Equ.*, 391 (2020). DOI: 10.1186/s13662-020-02831-6.

[35] H. Singh, Jacobi collocation method for the fractional advection-dispersion equation arising in porous media, *Numer. Methods Partial Differ. Equ.*, (2020). DOI: 10.1002/num.22674.

[36] D. G. Prakasha, N. S. Malagi, P. Veeresha, New approach for fractional Schrödinger–Boussinesq equations with Mittag-Leffler kernel, *Math. Meth. Appl. Sci.*, 43 (2020), 9654–9670.

[37] H. Singh, Numerical simulation for fractional delay differential equations, *Int. J. Dynam. Control*, 9 (2021), 463–474.

[38] A. Kumar, E. Ilhan, A. Ciancio, G. Yel, H. M. Baskonus, Extractions of some new travelling wave solutions to the conformable Date-Jimbo-Kashiwara-Miwa equation, *AIMS Math.*, 6 (5), 4238–4264.

[39] M. Senol, L. Akinyemi, A. Ata, O. S. Iyiola, Approximate and generalized solutions of conformable type Coudrey–Dodd–Gibbon–Sawada–Kotera equation, *Int. J. Mod. Phys. B*, 35(2) (2021), 2150021.

[40] P. Veeresha, D. G. Prakasha, S. Kumar, A fractional model for propagation of classical optical solitons by using non-singular derivative, *Math. Meth. Appl. Sci.*, (2020). DOI: 10.1002/mma.6335.

[41] H. Singh, Analysis of drug treatment of the fractional HIV infection model of CD4+ T-cells, *Chaos Soliton. Fractal.* 146(110868) (2021). DOI: 10.1016/j.chaos.2021.110868.

[42] L. Akinyemi, M. Senol, S. N. Huseen, Modified homotopy methods for generalized fractional perturbed Zakharov–Kuznetsov equation in dusty plasma, *Adv. Diff. Equ.*, 45 (2021), 1–27.
[43] D. G. Prakasha, et al., An efficient computational technique for time-fractional Kaup-Kupershmidt equation, *Numer. Methods Partial Differ. Equ.*, 37 (2) (2021), 1299–1316.
[44] S. J. Liao, Homotopy analysis method and its applications in mathematics, *J. Basic Sci. Eng.*, 5(2) (1997), 111–125.
[45] S. J. Liao, Homotopy analysis method: a new analytic method for nonlinear problems, *Appl. Math. Mech.*, 19 (1998), 957–962.
[46] J. Singh, D. Kumar, R. Swroop, Numerical solution of time- and space-fractional coupled Burgers' equations via homotopy algorithm, *Alexandria Eng. J.*, 55(2) (2016) 1753–1763.
[47] H. M. Srivastava, D. Kumar, J. Singh, An efficient analytical technique for fractional model of vibration equation, *Appl. Math. Model.*, 45 (2017), 192–204.
[48] P. Veeresha, D. G. Prakasha, D. Baleanu, An efficient technique for fractional coupled system arisen in magneto-thermoelasticity with rotation using Mittag-Leffler kernel, *J. Comput. Nonlinear Dynam.*, 16 (1) (2021). DOI: 10.1115/1.4048577.
[49] M. S. Kiran, et al., A mathematical analysis of ongoing outbreak COVID-19 in India through nonsingular derivative, *Numer. Meth. Partial Diffe. Equa.*, (2020). DOI: 10.1002/num.22579
[50] P. Veeresha, E. Ilhan, H. M. Baskonus, Fractional approach for analysis of the model describing wind-influenced projectile motion, *Phys. Scr.*, 96 (2021). DOI: 10.1088/1402-4896/abf868.
[51] P. Veeresha, D. G. Prakasha, Novel approach for modified forms of Camassa–Holm and Degasperis–Procesi equations using fractional operator, *Commun. Theor. Phys.*, 72 (10) (2020). DOI: 10.1088/1572-9494/aba24b.
[52] D. Kumar, R. P. Agarwal, J. Singh, A modified numerical scheme and convergence analysis for fractional model of Lienard's equation, *J. Comput. Appl. Math.*, 399 (2018), 405–413.
[53] P. Veeresha, D. G. Prakasha, A reliable analytical technique for fractional Caudrey-Dodd-Gibbon equation with Mittag-Leffler kernel, *Nonlinear Eng.*, 9 (1) (2020), 319–328.
[54] P. Veeresha, D. G. Prakasha, H. M. Baskonus, New numerical surfaces to the mathematical model of cancer chemotherapy effect in Caputo fractional derivatives, *Chaos* 29 (013119) (2019). DOI: 10.1063/1.5074099.
[55] H. L. Duorah, R.S. Kushwaha, Helium-burning reaction products and the rate of energy generation, *Astrophysical J.*, 137 (1963), 566–571.
[56] R. K. Saxena, A.M. Mathai, H.J. Haubold, On fractional kinetic equations, *Astrophys. Space Sci.*, 282 (2002), 281–287.
[57] W. R. Hix, F.K. Thielemann, Computational methods for nucleosynthesis and nuclear energy generation, *J. Comput. Appl. Math.*, 109 (1) (1999), 321–351.
[58] J. Singh, D. Kumar, Z. Hammouch, A. Atangana, A fractional epidemiological model for computer viruses pertaining to a new fractional derivative, *Appl. Math. Comput.*, 316 (2018), 504–515.

[59] P. Veeresha, D. G. Prakasha, Z. Hammouch, *An efficient approach for the model of thrombin receptor activation mechanism with Mittag-Leffler function, The International Congress of the Moroccan Society of Applied Mathematics* (pp. 44–60). Springer, Cham.

[60] A. Atangana, B. T. Alkahtani, Analysis of the Keller-Segel model with a fractional derivative without singular kernel, *Entropy*, 17 (2015), 4439–4453.

[61] A. Atangana, B.T. Alkahtani, Analysis of non- homogenous heat model with new trend of derivative with fractional order, *Chaos Soliton. Fractal.*, 89 (2016), 566–571.

[62] A. Atangana, D. Baleanu, New fractional derivatives with non-local and non-singular kernel theory and application to heat transfer model, *Therm. Sci.*, 20 (2016), 763–769.

[63] D. Kumar, R. P. Agarwal, J. Singh, A modified numerical scheme and convergence analysis for fractional model of Lienard's equation, *J. Comput. Appl. Math.* 339 (2018), 405–413.

[64] P. Veeresha, D. G. Prakasha, J. Singh, D. Kumar, D. Baleanu, Fractional Klein-Gordon-Schrödinger equations with Mittag-Leffler memory, *Chinese J. Phy.*, 68 (2020), 65–78.

3

Computational Study of Constant and Variable Coefficients Time-Fractional PDEs via Reproducing Kernel Hilbert Space Method

Ali Akgül
Siirt University, Siirt, Turkey

Nourhane Attia
University M'hamed Bougara of Boumerdes, Boumerdes, Algeria

CONTENTS

3.1 Introduction

Many phenomena in several fields of science and engineering can be defined successfully using fractional-order PDEs [1]. The fractional derivative can be interpreted in many ways such as Caputo's definition [2], Riemann-Liouville's definition [3], Grunwald-Letnikov's definition [4], Klimek [5],

DOI: 10.1201/9781003263517-3

Jumarie [6], Atangana-Baleanu [7], Conformable [8], etc. The solutions of FPDEs play a vital role in understanding natural phenomena [9], although it is difficult to find exact solutions for them, especially nonlinear equations. Developing analytical and numerical methods to solve complex nonlinear problems is therefore a major research task [10–16].

The aim of this research is to provide an efficient numerical technique based on the RK theory for solving classes of time FPDEs such as those found in engineering and applied science.

In this chapter, the RKHSM is used to construct numerical solutions for the following set of time FPDEs:

- The time-fractional heat-like equation (TFHLE)

$$ {}_{0}^{C}\mathrm{D}_{\eta}^{\alpha}\upsilon(\zeta,\eta)=\frac{\zeta^{2}}{2}\partial_{\zeta^{2}}^{2}\upsilon(\zeta,\eta). \tag{3.1}$$

- The time-fractional Navier-Stokes equation (TFNSE)

$$ {}_{0}^{C}\mathrm{D}_{\eta}^{\alpha}\upsilon(\zeta,\eta)=\mathrm{P}+\mathrm{v}\left[\partial_{\zeta^{2}}^{2}\upsilon(\zeta,\eta)+\frac{1}{\zeta}\partial_{\zeta}\upsilon(\zeta,\eta)\right]. \tag{3.2}$$

Both equations are subject to the following initial and boundary conditions (ICs, BCs):

- The IC:

$$\upsilon(\zeta,0)=\mathrm{f}(\zeta),\quad \zeta\in[0,1]. \tag{3.3}$$

- The Dirichlet boundary conditions:

$$\upsilon(0,\eta)=\varphi_{1}(\eta),\quad \text{and}\quad \upsilon(1,\eta)=\varphi_{2}(\eta),\quad \eta\in(0,\mathrm{d}]. \tag{3.4}$$

In the above equation, the following will be assumed:

- $\zeta\in[0,1]$ and $\eta\in[0,\mathrm{d}],\mathrm{d}\in\mathbb{R}_{+}^{*}$.
- υ is an unknown function to be calculated.
- $\partial_{\zeta^{2}}^{2}=\dfrac{\partial^{2}}{\partial\zeta^{2}}$ and $\partial_{\zeta}=\dfrac{\partial}{\partial\zeta}$.

- ${}_{0}^{C}\mathrm{D}_{\eta}^{\alpha}$ represents the Caputo time-fractional derivative of order $\alpha \in (0,1]$ and it is defined by:

$$\left({}_{0}^{C}\mathrm{D}_{\eta}^{\alpha}\upsilon\right)(\zeta,\eta)=\frac{1}{\Gamma(1-\alpha)}\int_{0}^{\eta}(\eta-t)^{-\gamma}\,\partial_{\eta}\upsilon(\zeta,\eta)\mathrm{d}t,\quad 0<\alpha<1,\eta>0. \tag{3.5}$$

- $f(\zeta)$, $\varphi_1(\eta)$ and $\varphi_2(\eta)$ are known functions
- v and P are constant viscosity and pressure respectively.

The TFHLE, which has an important role in describing many phenomena in engineering, has been investigated using the variational iterational method (VIM) [17]. The vital role of the TFNSE is its ability to define several computational fluid dynamics problems. The Adomian decomposition method was applied to solve the TFNSE by Momani and Odibat [18]. A mesh-free technique for TFNSE is presented by Haq and Hussain [19].

The RKHSM was first proposed in 1908 [20]. It is an effective numerical method for complex nonlinear problems without discretization. This method has been applied to solve several types of equations, including time-fractional diffusion equation [21], generalized Schamel equation [22], fractional gas dynamics equation [23], fractional Bagley-Torvik equation [24], Riccati differential equations [25], second-order PDEs [26], fractional Bratu-type equations [27], nonlinear fractional Volterra integro-differential equations [28], nonlocal fractional boundary value problems [29], forced Duffing equations [30], and others [31–33].

The second section of this paper gives some basic definitions and theorems concerning RK theory. The description of the RKHSM and its application to the proposed problem are presented in the third section. The fourth section provides error estimations. The effectiveness of the RKHSM and the accuracy of the solutions are validated through three applications in the fifth section. The final section provides conclusions.

3.2 Excerpts from RK Theory

In this section, we first recall some essentials from RK theory concerning reproducing kernel Hilbert spaces and their RK functions.

Definition 3.2.1

A RK of the space $\mathcal{H}$ is a function $\mathcal{K}:\mathcal{E}\times\mathcal{E}\to\mathbb{C}$ satisfying [34]

$$\mathcal{K}(\cdot,\eta)\in\mathcal{H},\quad \forall\eta\in\mathcal{E},$$

and the reproducing property (RP) which says that

$$\langle \mathfrak{f}, \mathcal{K}(\cdot,\eta)\rangle = \mathfrak{f}(\eta), \quad \forall \mathfrak{f} \in \mathcal{H} \text{ and } \forall \eta \in \mathcal{E},$$

$\mathcal{H}$ is a Hilbert space over $\mathcal{E}$ and $\mathcal{E} \neq \varnothing$.

Definition 3.2.2

Suppose $\mathcal{H}$ is a Hilbert space. If there exists a RK "$\mathcal{K}$" of $\mathcal{H}$, then $\mathcal{H}$ is called a reproducing kernel Hilbert space (RKHS).

Next, we define some function spaces which are useful for our purpose.

3.2.1 The RKHS of the Form $\mathcal{W}_2^m[a,b]$

Definition 3.2.3

We denote by $\mathcal{W}_2^m[a,b]$ the function space [34]

$$\mathcal{W}_2^m[a,b] = \{\mathfrak{f}(\zeta)|,\ \mathfrak{f}^{(i)}|,\ i = 0, 1, \ldots, (m-1) \text{ are abs.cont.funcs.on}[a,b]|, \mathfrak{f}^{(m)} \in L^2[a,b]\}, \tag{3.6}$$

which is supplied with the inner product:

$$\langle \mathfrak{f}, \mathfrak{g}\rangle_{\mathcal{W}_2^m} = \sum_{i=0}^{m-1} \mathfrak{f}^{(i)}(a)\mathfrak{g}^{(i)}(a) + \int_a^b \mathfrak{f}^{(m)}(\zeta)\mathfrak{g}^{(m)}(\zeta)d\zeta,$$

and norm:

$$\|\mathfrak{f}\|_{\mathcal{W}_2^m} = \sqrt{\langle \mathfrak{f}, \mathfrak{f}\rangle_{\mathcal{W}_2^m}},$$

for all $\mathfrak{f}, \mathfrak{g} \in \mathcal{W}_2^m[a,b]$. In (3.6), "Absolutely continuous function(s)" is abbreviated as "abs. cont. func(s)."

We note next some RKHSs of the form $\mathcal{W}_2^m[a,b]$ with the expression of RK function of each space.

1. The function space $\mathcal{W}_2^3[0,1]$:

$$\mathcal{W}_2^3[0,1] = \{\mathfrak{f}(\zeta)|\mathfrak{f}, \mathfrak{f}' \text{ and } \mathfrak{f}'' \text{ are abs.cont.funcs.on } [0,1], \mathfrak{f}^{(3)} \in L^2[0,1], \text{ and } \mathfrak{f}(0) = \mathfrak{f}(1) = 0\},$$

with its RK function (RKF) that it is defined by the following theorem [35].

Theorem 3.1

We obtain the RK function $\mathrm{R}_x(\zeta)$ *of the space* $\mathcal{W}_2^3[0,1]$ *as:*

$$\mathcal{R}_x(\zeta)=\begin{cases}\frac{3\zeta x}{13}-\frac{\zeta x^5}{156}+\frac{5\zeta x^4}{156}-\frac{5\zeta x^3}{78}-\frac{5\zeta x^2}{26}+\frac{21\zeta^2x^2}{104}-\frac{\zeta^2x^5}{624}+\frac{5\zeta^2x^4}{624}-\frac{5\zeta^2x^3}{312}-\frac{5\zeta^2x}{26}\\+\frac{7\zeta^3x^2}{104}-\frac{\zeta^3x^5}{1872}+\frac{5\zeta^3x^4}{1872}-\frac{5\zeta^3x^3}{936}-\frac{5\zeta^3x}{78}-\frac{\zeta^4x}{104}+\frac{\zeta^4x^5}{3744}-\frac{5\zeta^4x^4}{3744}+\frac{5\zeta^4x^3}{1872}+\frac{5\zeta^4x^2}{624}\\-\frac{\zeta^5x^5}{18720}+\frac{\zeta^5x^4}{3744}-\frac{\zeta^5x^3}{1872}-\frac{\zeta^5x^2}{624}-\frac{\zeta^5x}{156}+\frac{\zeta^5}{120},\zeta\le x,\\\frac{x^5}{120}-\frac{\zeta x^4}{104}+\frac{3\zeta x}{13}-\frac{\zeta x^5}{156}-\frac{5\zeta x^3}{78}-\frac{5\zeta x^2}{26}+\frac{7\zeta^2x^3}{104}+\frac{21\zeta^2x^2}{104}-\frac{\zeta^2x^5}{624}+\frac{5\zeta^2x^4}{624}-\frac{5\zeta^2x}{26}\\-\frac{\zeta^3x^5}{1872}+\frac{5\zeta^3x^4}{1872}-\frac{5\zeta^3x^3}{936}-\frac{5\zeta^3x^2}{312}-\frac{5\zeta^3x}{78}+\frac{\zeta^4x^5}{3744}-\frac{5\zeta^4x^4}{3744}+\frac{5\zeta^4x^3}{1872}+\frac{5\zeta^4x^2}{624}+\frac{5\zeta^4x}{156}\\-\frac{\zeta^5x^5}{18720}+\frac{\zeta^5x^4}{3744}-\frac{\zeta^5x^3}{1872}-\frac{\zeta^5x^2}{624}-\frac{\zeta^5x}{156},\zeta>x.\end{cases}\tag{3.7}$$

Proof. We must prove

$$\langle \mathfrak{f},\mathcal{R}_x\rangle_{\mathcal{W}_2^3}=\mathfrak{f}(x).$$

We have

$$\langle \mathfrak{f},\mathcal{R}_x\rangle_{\mathcal{W}_2^3}=\mathfrak{f}(0)\mathcal{R}_x(0)+\mathfrak{f}'(0)\mathcal{R}_x'(0)+\mathfrak{f}''(0)\mathcal{R}_x''(0)+\int_0^1\mathfrak{f}^{(3)}(\zeta)\mathcal{R}_x^{(3)}(\zeta)d\zeta.$$

Applying integration by parts, we obtain:

$$\begin{aligned}\langle \mathfrak{f},\mathcal{R}_x\rangle_{\mathcal{W}_2^3}&=\mathfrak{f}(0)\mathcal{R}_x(0)+\mathfrak{f}'(0)\mathcal{R}_x'(0)+\mathfrak{f}''(0)\mathcal{R}_x''(0)+\mathfrak{f}''(1)\mathcal{R}_x^{(3)}(1)-\mathfrak{f}''(0)\mathcal{R}_x^{(3)}(0)\\&\quad-\mathfrak{f}'(1)\mathcal{R}_x^{(4)}(1)+\mathfrak{f}'(0)\mathcal{R}_x^{(4)}(0)+\int_0^1\mathfrak{f}'(\zeta)\mathcal{R}_x^{(5)}(\zeta)d\zeta.\end{aligned}$$

Since $\mathfrak{f}(\zeta)\in\mathcal{W}_2^3[0,1]$, we have

$$\mathfrak{f}(0)=\mathfrak{f}(1)=0.\tag{3.8}$$

Then:

$$\langle \mathfrak{f}, \mathcal{R}_x \rangle_{\mathcal{W}_2^3} = \mathfrak{f}'(0)\mathcal{R}_x'(0) + \mathfrak{f}''(0)\mathcal{R}_x''(0) + \mathfrak{f}''(1)\mathcal{R}_x^{(3)}(1) - \mathfrak{f}''(0)\mathcal{R}_x^{(3)}(0)$$
$$-\mathfrak{f}'(1)\mathcal{R}_x^{(4)}(1) + \mathfrak{f}'(0)\mathcal{R}_x^{(4)}(0) + \int_0^1 \mathfrak{f}'(\zeta)\mathcal{R}_x^{(5)}(\zeta)\,d\zeta.$$

We need to compute $\mathcal{R}_x^{i}(0)$, $i = 1,2,\dots,4$, $\mathcal{R}_x^{(3)}(1)$ and $\mathcal{R}_x^{(4)}(1)$:

$$\mathcal{R}_x'(0) = \frac{3x}{13} - \frac{x^5}{156} + \frac{5x^4}{156} - \frac{5x^3}{78} - \frac{5x^2}{26},$$

$$\mathcal{R}_x''(0) = -\frac{5x}{13} - \frac{x^5}{312} + \frac{5x^4}{312} - \frac{5x^3}{156} + \frac{21x^2}{52},$$

$$\mathcal{R}_x^{(3)}(0) = -\frac{5x}{13} - \frac{x^5}{312} + \frac{5x^4}{312} - \frac{5x^3}{156} + \frac{21x^2}{52},$$

$$\mathcal{R}_x^{(4)}(0) = \frac{5x^2}{26} + \frac{x^5}{156} - \frac{5x^4}{156} + \frac{5x^3}{78} - 3/13x,$$

$$\mathcal{R}_x^{(3)}(1) = \mathcal{R}_x^{(4)}(1) = 0.$$

By using the above equations, we obtain:

$$\langle \mathfrak{f}, \mathcal{R}_x \rangle_{\mathcal{W}_2^3} = \int_0^1 \mathfrak{f}'(\zeta)\mathcal{R}_x^{(5)}(\zeta)\,d\zeta. \tag{3.9}$$

We have

$$\mathcal{R}_x^{(5)}(\zeta) = \begin{cases} \mathcal{A}(x)+1, & ,\zeta \le x, \\ \mathcal{A}(x), & ,\zeta > x. \end{cases}$$

where $\mathcal{A}(x) = -\frac{x^5}{156} + \frac{5x^4}{156} - \frac{5x^3}{78} - \frac{5x^2}{26} - \frac{10x}{13}.$

Thus

$$\begin{aligned}\langle \mathfrak{f}, \mathcal{R}_x \rangle_{\mathcal{W}_2^3} &= \int_0^x \mathfrak{f}'(\zeta)\mathcal{R}_x^{(5)}(\zeta)\mathrm{d}\zeta + \int_x^1 \mathfrak{f}'(\zeta)\mathcal{R}_x^{(5)}(\zeta)\mathrm{d}\zeta \\ &= \int_0^x \mathfrak{f}'(\zeta)(\mathcal{A}(x)+1)\mathrm{d}\zeta + \int_x^1 \mathfrak{f}'(\zeta)\mathcal{A}(x)\mathrm{d}\zeta \\ &= \mathfrak{f}(x).\end{aligned}$$

and, since $\mathfrak{f}(x) \in \mathcal{W}_2^3[0,1]$, we deduce

$$\langle \mathfrak{f}, \mathcal{R}_x \rangle_{\mathcal{W}_2^3} = \mathfrak{f}(x).$$

2. The function space $\mathcal{W}_2^2[0,1]$:

$$\mathcal{W}_2^2[0,1] = \{\mathfrak{f}(\eta) |, \ \mathfrak{f} \text{ and } \mathfrak{f}' \text{ are abs.cont.funcs.on } [0,1]|, \\ \mathfrak{f}'' \in \mathrm{L}^2[0,1]|, \text{ and } \mathfrak{f}(0)=0\},$$

with its RK function (RKF) that it is defined by the following theorem [35].

Theorem 3.2

We obtain the RK function $\mathcal{S}_t(\eta)$ of the space $\mathcal{W}_2^2[0,1]$ as:

$$\mathcal{S}_t(\eta) = \begin{cases} t\eta + \frac{1}{2}t\eta^2 - \frac{1}{6}\eta^3, & \eta \le t, \\ -\frac{1}{6}t^3 + \frac{1}{2}\eta t^2 + \eta t, & \eta > t. \end{cases} \tag{3.10}$$

Proof. We must prove

$$\langle \mathfrak{f}, \mathcal{S}_t \rangle_{\mathcal{W}_2^2} = \mathfrak{f}(t).$$

We have

$$\langle \mathfrak{f}, \mathcal{S}_t \rangle_{\mathcal{W}_2^2} = \mathfrak{f}(0)\mathcal{S}_t(0) + \mathfrak{f}'(0)\mathcal{S}_t'(0 + \int_0^1 \mathfrak{f}''(\eta)\mathcal{S}_t''(\eta)\mathrm{d}\eta.$$

Applying integration by parts, we obtain:

$$\begin{aligned}\langle \mathfrak{f}, \mathcal{S}_t \rangle_{\mathcal{W}_2^2} &= \mathfrak{f}(0)\mathcal{S}_t(0) + \mathfrak{f}'(0)\mathcal{S}_t'(0) + \mathfrak{f}''(1)\mathcal{S}_t^{(3)}(1) - \mathfrak{f}''(0)\mathcal{S}_t^{(3)}(0) \\ &\quad - \int_0^1 \mathfrak{f}'(\eta)\mathcal{S}_t^{(3)}(\eta)\mathrm{d}\eta.\end{aligned}$$

Since $\mathfrak{f}(\eta) \in \mathcal{W}_2^2[0,1]$, we have

$$\mathfrak{f}(0) = 0. \tag{3.11}$$

Then:

$$\langle \mathfrak{f}, \mathcal{S}_t \rangle_{\mathcal{W}_2^2} = \mathfrak{f}'(0)\mathcal{S}_t'(0) + \mathfrak{f}'(1)\mathcal{S}_t''(1) - \mathfrak{f}'(0)\mathcal{S}_t''(0) - \int_0^1 \mathfrak{f}'(\eta)\mathcal{S}_t^{(3)}(\eta)\,d\eta.$$

We need to compute $\mathcal{S}_t'(0)$, $\mathcal{S}_t''(0)$ and $\mathcal{S}_t''(1)$:

$$\begin{aligned} \mathcal{S}_t'(0) &= t, \\ \mathcal{S}_t''(0) &= t, \\ \mathcal{S}_t''(1) &= 0. \end{aligned}$$

By using the above equations, we obtain:

$$\langle \mathfrak{f}, \mathcal{S}_t \rangle_{\mathcal{W}_2^2} = -\int_0^1 \mathfrak{f}'(\eta)\mathcal{S}_t^{(3)}(\eta)\,d\eta. \tag{3.12}$$

We have

$$\mathcal{S}_t^{(3)}(\eta) = \begin{cases} -1, & ,\eta \le t, \\ 0, & ,\eta > t. \end{cases}$$

Thus

$$\begin{aligned} \langle \mathfrak{f}, \mathcal{S}_t \rangle_{\mathcal{W}_2^2} &= -\int_0^t \mathfrak{f}'(\eta)\mathcal{S}_t^{(3)}(\eta)\,d\eta - \int_t^1 \mathfrak{f}'(\eta)\mathcal{S}_t^{(3)}(\eta)\,d\eta \\ &= \int_0^t \mathfrak{f}'(\eta)\,d\eta \\ &= \mathfrak{f}(t). \end{aligned}$$

and, since $\mathfrak{f}(t) \in \mathcal{W}_2^2[0,1]$, we deduce

$$\langle \mathfrak{f}, \mathcal{S}_t \rangle_{\mathcal{W}_2^2} = \mathfrak{f}(t).$$

3. The function space $\mathcal{W}_2^1[0,1]$:

$$\mathcal{W}_2^1[0,1] = \{\mathfrak{f}(\zeta) \mid \mathfrak{f} \text{ is abs.cont.func.on } [0,1] \text{ and } \mathfrak{f}' \in L^2[0,1]\},$$

with its RK function (RKF) that it is defined by the following theorem [35]

Theorem 3.3

We obtain the RK function $\mathcal{Z}_x(\zeta)$ of the space $\mathcal{W}_2^1[0,1]$ as:

$$\mathcal{Z}_x(\zeta)=\begin{cases} 1+\zeta, & \zeta \le x, \\ x+1, & \zeta > x. \end{cases} \tag{3.13}$$

Proof. We must prove

$$\langle \mathfrak{f}, \mathcal{Z}_x \rangle_{\mathcal{W}_2^1} = \mathfrak{f}(x).$$

We have

$$\langle \mathfrak{f}, \mathcal{Z}_x \rangle_{\mathcal{W}_2^1} = \mathfrak{f}(0)\mathcal{Z}_x(0)+\int_0^1 \mathfrak{f}'(\zeta)\mathcal{Z}_x'(\zeta)\mathrm{d}\zeta.$$

We need to compute $\mathcal{Z}_x(0)$:

$$\mathcal{Z}_x(0)=1,$$

By using the above equation, we obtain:

$$\langle \mathfrak{f}, \mathcal{Z}_x \rangle_{\mathcal{W}_2^1} = \mathfrak{f}(0)+\int_0^1 \mathfrak{f}'(\zeta)\mathcal{Z}_x'(\zeta)\mathrm{d}\zeta.$$

We have

$$\mathcal{Z}_x'(\zeta)=\begin{cases} 1 & ,\zeta \le x, \\ 0 & ,\zeta > x. \end{cases}$$

Thus

$$\begin{aligned} \langle \mathfrak{f}, \mathcal{Z}_x \rangle_{\mathcal{W}_2^1} &= \mathfrak{f}(0)+\int_0^x \mathfrak{f}'(\zeta)\mathcal{Z}_x'(\zeta)\mathrm{d}\zeta+\int_x^1 \mathfrak{f}'(\zeta)\mathcal{Z}_x'(\zeta)\mathrm{d}\zeta \\ &= \mathfrak{f}(0)+\int_0^\mu \mathfrak{f}'(\zeta)\mathrm{d}\zeta \\ &= \mathfrak{f}(0)+\mathfrak{f}(x)-\mathfrak{f}(0), \end{aligned}$$

so, we deduce

$$\langle \mathfrak{f}, \mathcal{Z}_x \rangle_{\mathcal{W}_2^1} = \mathfrak{f}(x).$$

3.2.2 The RKHS of the Form $\mathcal{B}_2^{(m,n)}(\mathcal{D})$

$$\text{Throughout}\, \mathcal{D} = [a,b]\times[c,d] \subset \mathbb{R}^2.$$

Definition 3.2.4

We denote by $\mathcal{B}_2^{(m,n)}(\mathcal{D})$ the binary function space [34]

$$\mathcal{B}_2^{(m,n)}(\mathcal{D}) = \left\{ \upsilon(\zeta,\eta) \mid, \frac{\partial^{m+n-2}}{\partial\zeta^{m-1}\partial\eta^{n-1}}\upsilon(\zeta,\eta) \text{ is compl.cont. func.in } \mathcal{D}, \frac{\partial^{m+n}}{\partial\zeta^{m}\partial\eta^{n}}\upsilon(\zeta,\eta) \in L^2(\mathcal{D}) \right\}, \tag{3.14}$$

which is supplied with the inner product:

$$\begin{aligned}\langle \upsilon,\varkappa \rangle_{\mathcal{B}_2^{(m,n)}} = &\sum_{i=0}^{m-1}\int_c^d \left[\frac{\partial^n}{\partial\eta^n}\frac{\partial^i}{\partial\zeta^i}\upsilon(a,\eta)\frac{\partial^n}{\partial\eta^n}\frac{\partial^i}{\partial\zeta^i}\varkappa(a,\eta)\right] d\eta \\ &+\sum_{i=0}^{n-1}\left\langle \frac{\partial^i}{\partial\eta^i}\upsilon(\zeta,c), \frac{\partial^i}{\partial\eta^i}\varkappa(\zeta,c)\right\rangle_{\mathcal{W}_2^m} \\ &+\int_c^d\int_a^b \frac{\partial^m}{\partial\zeta^m}\frac{\partial^n}{\partial\eta^n}\upsilon(\zeta,\eta)\frac{\partial^m}{\partial\zeta^m}\frac{\partial^n}{\partial\eta^n}\varkappa(\zeta,\eta)\,d\zeta\, d\eta,\end{aligned} \tag{3.15}$$

and norm:

$$\|\upsilon\|_{\mathcal{B}_2^{(m,n)}} = \sqrt{\langle \upsilon,\upsilon \rangle_{\mathcal{B}_2^{(m,n)}}}, \tag{3.16}$$

for all $\upsilon,\varkappa \in \mathcal{B}_2^{(m,n)}(\mathcal{D})$. In (3.14), *"completely continuous function" is abbreviated as "compl. cont. func."*

We note next some RKHSs of the form $\mathcal{B}_2^{(m,n)}(\mathcal{D})$ with the expression of RK function of each space.

1. The function space $\mathcal{B}_2^{(3,2)}(\mathcal{I})$:

$$\begin{aligned}\mathcal{B}_2^{(3,2)}(\mathcal{I}) = \{&\upsilon(\zeta,\eta) \mid \frac{\partial^3}{\partial\zeta^2\partial\eta}\upsilon(\zeta,\eta) \text{ is compl.cont.func.in} \\ &\mathcal{I} = [0,1]\times[0,1], \frac{\partial^5}{\partial\zeta^3\partial\eta^2}\upsilon(\zeta,\eta) \in L^2(\mathcal{I}), \\ &\text{and } \upsilon(\zeta,0) = \upsilon(0,\eta) = \upsilon(1,\eta) = 0\},\end{aligned}$$

with its RKF which is defined as [35]

$$\mathcal{F}_{(x,t)}(\zeta,\eta)=\mathcal{R}_x(\zeta)\ \mathcal{S}_t(\eta). \tag{3.17}$$

2. The function space $\mathcal{B}_2^{(1,1)}(\mathcal{I})$:

$$\mathcal{B}_2^{(1,1)}(\mathcal{I})=\left\{\upsilon(\zeta,\eta)|\ \upsilon(\zeta,\eta)\text{ is compl.cont.func.in }\mathcal{I}\text{ and }\frac{\partial^2}{\partial\zeta\partial\eta}\upsilon(\zeta,\eta)\in \mathrm{L}^2(\mathcal{I})\right\},$$

with its RKF that it is defined as [35]

$$\mathcal{G}_{(x,t)}(\zeta,\eta)=\mathcal{Z}_x(\zeta)\ \mathcal{Z}_t(\eta). \tag{3.18}$$

3.3 Methodology for RKHSM

In this section, we deal with the proposed fractional PDEs subject to non-homogeneous initial and boundary conditions, which are abbreviated "ICs" and "BCs" respectively. We begin by considering the general problem completely represented by Eqs (3.1)–(3.2):

$${}_0^C\mathrm{D}_\eta^\alpha\upsilon(\zeta,\eta)+\mathcal{A}_1(\zeta,\eta)\partial_{\zeta^2}^2\upsilon(\zeta,\eta)+\mathcal{A}_2(\zeta,\eta)\partial_\zeta\upsilon(\zeta,\eta)=h(\zeta,\eta), \tag{3.19}$$

with the BCs:

$$\upsilon(0,\eta)=\varphi_1(\eta),\quad\text{and}\quad \upsilon(1,\eta)=\varphi_2(\eta), \tag{3.20}$$

and ICs:

$$\upsilon(\zeta,0)=\mathrm{f}(\zeta), \tag{3.21}$$

where $(\zeta,\eta)\in[0,1]\times[0,\mathrm{d}]$, $\mathrm{d}\in\mathbb{R}_+^*$. $\mathcal{A}_i(\zeta,\eta)$, $\mathrm{i}=1,\ 2$ are variable coefficients. $h(\zeta,\eta)$ is an inhomogeneous term. Note that when $\mathcal{A}_1(\zeta,\eta)=-\zeta^2/2$ and $\mathcal{A}_2(\zeta,\eta)=h(\zeta,\eta)=0$, then Eq. (3.19) gives Problem (3.1). While, when $\mathcal{A}_1(\zeta,\eta)=-\upsilon$ and $\mathcal{A}_2(\zeta,\eta)=-\upsilon/\zeta$, and $h(\zeta,\eta)$ = P then Eq. (3.19) gives Problem (3.2).

3.3.1 Reformulation of the Problem

A good and simple way to apply the RKHS method to Eq. (3.19) is to homogenize the BCs and ICs (3.20)–(3.21). The correct method for doing so is to consider the transformation:

$$\varkappa(\zeta,\eta)=\upsilon(\zeta,\eta)+\mathrm{U}(\zeta,\eta), \tag{3.22}$$

where

$$\mathrm{U}(\zeta,\eta)=-\varphi_1(\eta)(1-\zeta)-\varphi_2(\eta)\zeta-\mathrm{f}_1(\zeta)+\varphi_1(0)(1-\zeta)+\varphi_2(0)\zeta$$

Then, (3.19)–(3.21) become

$${}_0^C\mathrm{D}_\eta^\alpha\varkappa(\zeta,\eta)+\mathcal{A}_1(\zeta,\eta)\partial_{\zeta^2}^2\varkappa(\zeta,\eta)+\mathcal{A}_2(\zeta,\eta)\partial_\zeta\varkappa(\zeta,\eta)=\hbar(\zeta,\eta), \tag{3.23}$$

with BCs

$$\varkappa(0,\eta)=\varkappa(1,\eta)=0, \tag{3.24}$$

and ICs

$$\varkappa(\zeta,0)=0, \tag{3.25}$$

where

$$\hbar(\zeta,\eta)=h(\zeta,\eta)+{}_0^C\mathrm{D}_\eta^\alpha\mathrm{U}(\zeta,\eta)+\mathcal{A}_1(\zeta,\eta)\partial_{\zeta^2}^2\mathrm{U}(\zeta,\eta)+\mathcal{A}_2(\zeta,\eta)\partial_\zeta\mathrm{U}(\zeta,\eta).$$

3.3.2 Construction of a Numerical Solution

We next show how to build the RKHS method solution of our problem (3.23)–(3.25). As a second step after the homogenization of the conditions (3.20)–(3.21), let us consider the following linear operator $\Im:\mathcal{B}_2^{(3,2)}(\mathcal{I})\to\mathcal{B}_2^{(1,1)}(\mathcal{I})$ defined by

$$\Im\varkappa(\zeta,\eta)={}_0^C\mathrm{D}_\eta^\alpha\varkappa(\zeta,\eta)+\mathcal{A}_1(\zeta,\eta)\partial_{\zeta^2}^2\varkappa(\zeta,\eta)+\mathcal{A}_2(\zeta,\eta)\partial_\zeta\varkappa(\zeta,\eta). \tag{3.26}$$

We next verify the linearity and boundedness of the chosen operator:

Lemma 3.1

$\Im:\mathcal{B}_2^{(3,2)}(\mathcal{I})\to\mathcal{B}_2^{(1,1)}(\mathcal{I})$ *is a bounded linear operator.*

Proof. The linearity of $\mathfrak{J}$ is easy to check. We begin directly by verifying its boundedness. To demonstrate this, it suffices to verify:

$$\|\mathfrak{J}\varkappa\|_{\mathcal{B}_2^{(1,1)}} \le C\|\varkappa\|_{\mathcal{B}_2^{(3,2)}}, \text{ with } C>0.$$

From the expressions (3.15) and (3.16) we have

$$\begin{aligned}
\|\mathfrak{J}\varkappa(\zeta,\eta)\|^2_{\mathcal{B}_2^{(1,1)}} &= \langle \mathfrak{J}\varkappa(\zeta,\eta), \mathfrak{J}\varkappa(\zeta,\eta)\rangle_{\mathcal{B}_2^{(1,1)}} \\
&= \int_0^d \left[\frac{\partial}{\partial\eta}\mathfrak{J}\varkappa(0,\eta)\right]^2 d\eta + \langle \mathfrak{J}\varkappa(\zeta,0), \mathfrak{J}\varkappa(\zeta,0)\rangle_{W_2^1} \\
&\quad + \int_0^d\int_0^1 \left[\frac{\partial}{\partial\zeta}\frac{\partial}{\partial\eta}\mathfrak{J}\varkappa(\zeta,\eta)\right]^2 d\zeta d\eta \\
&= [\mathfrak{J}\varkappa(0,0)]^2 + \int_0^1\left[\frac{\partial}{\partial\zeta}\mathfrak{J}\varkappa(\zeta,0)\right]^2 d\zeta + \int_0^d\left[\frac{\partial}{\partial\eta}\mathfrak{J}\varkappa(0,\eta)\right]^2 d\eta \\
&\quad + \int_0^d\int_0^1 \left[\frac{\partial}{\partial\zeta}\frac{\partial}{\partial\eta}\mathfrak{J}\varkappa(\zeta,\eta)\right]^2 d\zeta d\eta.
\end{aligned}$$

In view of RP,

$$\varkappa(\zeta,\eta) = \langle \varkappa(\circledast,\bullet), \mathcal{F}_{(\zeta,\eta)}(\circledast,\bullet)\rangle_{\mathcal{B}_2^{(3,2)}}.$$

Similarly,

$$\frac{\partial^i}{\partial\zeta^i}\frac{\partial^j}{\partial\eta^j}\mathfrak{J}\varkappa(\zeta,\eta) = \left\langle \varkappa(\circledast,\bullet), \frac{\partial^i}{\partial\zeta^i}\frac{\partial^j}{\partial\eta^j}\mathfrak{J}\mathcal{F}_{(\zeta,\eta)}(\circledast,\bullet)\right\rangle_{\mathcal{B}_2^{(3,2)}}, \quad i,j=0,1.$$

Then,

$$\begin{aligned}
\left|\frac{\partial^i}{\partial\zeta^i}\frac{\partial^j}{\partial\eta^j}\mathfrak{J}\varkappa(\zeta,\eta)\right| &= \left|\left\langle \varkappa(\circledast,\bullet), \frac{\partial^i}{\partial\zeta^i}\frac{\partial^j}{\partial\eta^j}\mathfrak{J}\mathcal{F}_{(\zeta,\eta)}(\circledast,\bullet)\right\rangle_{\mathcal{B}_2^{(3,2)}}\right| \\
&\left| \le \varkappa_{\mathcal{B}_2^{(3,2)}} \frac{\partial^i}{\partial\zeta^i}\frac{\partial^j}{\partial\eta^j}\mathfrak{J}\mathcal{F}_{(\zeta,\eta)}(\circledast,\bullet)_{\mathcal{B}_2^{(3,2)}}\right|.
\end{aligned}$$

We used Schwarz inequality to get the above result. Therefore the continuity of $F_{(\zeta,\eta)}(\circledast, \bullet)$ implies

$$\left| \frac{\partial^{i}}{\partial \zeta^{i}} \frac{\partial^{j}}{\partial \eta^{j}} \Im \varkappa(\zeta,\eta) \right| \le C_{i,j} \|\varkappa\|_{\mathcal{B}_2^{(3,2)}}, \quad i,j = 0,1.$$

Hence

$$\|\Im \varkappa(\zeta,\eta)\|^2_{\mathcal{B}_2^{(1,1)}} \le C_{0,0}^2 \|\varkappa\|^2_{\mathcal{B}_2^{(3,2)}} + \int_0^1 C_{1,0}^2 \|\varkappa\|^2_{\mathcal{B}_2^{(3,2)}} \, d\zeta + \int_0^d C_{0,1}^2 \|\varkappa\|^2_{\mathcal{B}_2^{(3,2)}} \, d\eta$$
$$+ \int_0^d \int_0^1 C_{1,1}^2 \|\varkappa\|^2_{\mathcal{B}_2^{(3,2)}} \, d\zeta \, d\eta \le C \|\varkappa\|^2_{\mathcal{B}_2^{(3,2)}}.$$

where $C = C_{0,0}^2 + C_{1,0}^2 + C_{0,1}^2 d + C_{1,1}^2 d.$

So, (3.26) allows us to rewrite (3.23)–(3.25) as

$$\begin{cases} \Im \varkappa(\zeta,\eta) = \hbar(\zeta,\eta), & 0 \le \zeta \le 1,\ 0 \le \eta \le d, \\ \varkappa(0,\eta) = \upsilon(1,\eta) = 0, & \\ \varkappa(\zeta,0) = 0, & \end{cases} \tag{3.27}$$

where

$$\hbar(\zeta,\eta) = h(\zeta,\eta) + {}_0^C D_\eta^\alpha U(\zeta,\eta) + \mathcal{A}_1(\zeta,\eta)\partial^2_{\zeta^2} U(\zeta,\eta) + \mathcal{A}_2(\zeta,\eta)\partial_\zeta U(\zeta,\eta).$$

At this point, the main issue is to construct the orthonormal system in $\mathcal{B}_2^{(3,2)}(\mathcal{I})$. To do so, define the functions

$$\rho_i(\zeta,\eta) = \mathcal{G}_{(\zeta_i,\eta_i)}(\zeta,\eta) \quad \text{and} \quad \phi_i(\zeta,\eta) = \Im^* \rho_i(\zeta,\eta),$$

where

- $\mathcal{G}_{(\zeta_i,\eta_i)}(\zeta,\eta)$ is the RKF associated with $\mathcal{B}_2^{(1,1)}(\mathcal{I})$.
- $\Im^*$ is the formal adjoint of $\Im$.
- The countable set $\{(\zeta_i,\eta_i)\}_{i=1}^{\infty}$ is dense in $\mathcal{I}$.

Remark

It is absolutely essential for us to establish the expression of the function system $\{\phi_i\}_{i=1}^{\infty}$. So

$$
\begin{aligned}
\phi_i(\zeta,\eta) &= \mathfrak{I}^* \rho_i(\zeta,\eta) \\
&= \left\langle \mathfrak{I}^* \rho_i(x,t), \mathcal{F}_{(\zeta,\eta)}(x,t) \right\rangle_{\mathcal{B}_2^{(3,2)}} \\
&= \left\langle \rho_i(x,t), \mathfrak{I}_{(x,t)} \mathcal{F}_{(\zeta,\eta)}(x,t) \right\rangle_{\mathcal{B}_2^{(1,1)}} \\
&= \left\langle \mathcal{G}_{(x_i,t_i)}(x,t), \mathfrak{I}_{(x,t)} \mathcal{F}_{(\zeta,\eta)}(x,t) \right\rangle_{\mathcal{B}_2^{(1,1)}} \\
&= \mathfrak{I}_{(x,t)} \mathcal{F}_{(\zeta,\eta)}(x,t) \Big|_{(x,t)=(\zeta_i,\eta_i)} \\
&= \mathfrak{I}_{(x,t)} \mathcal{F}_{(x,t)}(\zeta,\eta) \Big|_{(x,t)=(\zeta_i,\eta_i)} \\
&= \left\{ {}_0^C\mathrm{D}_t^\alpha \mathcal{F}_{(x,t)}(\zeta,\eta) + \mathcal{A}_1(x,t)\partial_{x^2}^2 \mathcal{F}_{(x,t)}(\zeta,\eta) + \mathcal{A}_2(x,t)\partial_x \mathcal{F}_{(x,t)}(\zeta,\eta) \right\} \Big|_{(x,t)=(\zeta_i,\eta_i)}.
\end{aligned}
$$

Therefore, as a result of using Gram-Schmidt's process, the orthonormal system $\{\bar{\Phi}_i\}_{i=1}^{\infty}$ in $\mathcal{B}_2^{(3,2)}(\mathcal{I})$ should be given by

$$
\bar{\Phi}_i(\zeta,\eta) = \sum_{k=1}^{i} \sigma_{ik} \phi_k(\zeta,\eta), \quad \sigma_{ii} > 0,\ i = 1,2,\ldots.
$$

where the orthogonalization coefficients σ_{ik} are given by:

$$
\sigma_{ij} = \begin{cases} \dfrac{1}{\|\phi_1\|}, & \text{for } \ i = j = 1, \\ \dfrac{1}{\ell_i}, & \text{for } \ i = j \neq 1, \\ -\dfrac{1}{\ell_i} \sum_{k=j}^{i-1} C_{ik}\sigma_{kj}, & \text{for } \ i > j, \end{cases}
$$

with $\ell_i = \sqrt{\|\phi_i\|^2 - \sum_{k=1}^{i-1} C_{ik}^2}$ and $C_{ik} = \langle \phi_i, \bar{\Phi}_k \rangle_{\mathcal{B}_2^{(3,2)}}$.

Theorem 3.1

Assume $\{(\zeta_i,\eta_i)\}_{i=1}^{\infty}$ *is dense on* $\mathcal{I}$*, then* $\{\phi_i\}_{i=1}^{\infty}$ *is the complete system of the space* $\mathcal{B}_2^{(3,2)}(\mathcal{I})$.

Proof. We see that $\phi_i(\zeta,\eta) \in \mathcal{B}_2^{(3,2)}(\mathcal{I})$. And, for a fixed $\varkappa(\zeta,\eta) \in \mathcal{B}_2^{(3,2)}(\mathcal{I})$, then

$$\langle \varkappa(\zeta,\eta), \phi_i(\zeta,\eta) \rangle_{\mathcal{B}_2^{(3,2)}} = 0, \; i = 1, 2, \ldots.$$

Since

$$\langle \varkappa(\zeta,\eta), \phi_i(\zeta,\eta) \rangle_{\mathcal{B}_2^{(3,2)}} = \langle \varkappa(\zeta,\eta), \mathfrak{I}^* \rho_i(\zeta,\eta) \rangle_{\mathcal{B}_2^{(3,2)}} = \langle \mathfrak{I}\varkappa(\zeta,\eta), \rho_i(\zeta,\eta) \rangle_{\mathcal{B}_2^{(1,1)}}$$
$$= \mathfrak{I}\varkappa(\zeta_i,\eta_i) = 0,$$

and since $\{(\zeta_i,\eta_i)\}_{i=1}^{\infty}$ is dense on $\mathcal{I}$,

$$\mathfrak{I}\varkappa(\zeta,\eta) = 0. \tag{3.28}$$

So, if we apply the inverse operator $\mathfrak{I}^{-1}$, we get

$$\varkappa(\zeta,\eta) = 0.$$

Theorem 3.2

Assume $\{(\zeta_i,\eta_i)\}_{i=1}^{\infty}$ *is a dense set on* $\mathcal{I}$ *and there exists a unique solution* $\varkappa(\zeta,\eta)$ *of (3.27) on* $\mathcal{B}_2^{(3,2)}(\mathcal{I})$. *Then*

1. *The solution of (3.27) is*

$$\varkappa(\zeta,\eta) = \sum_{i=1}^{\infty}\sum_{k=1}^{i} \sigma_{ik}\hbar(\zeta_k,\eta_k)\bar{\Phi}_i(\zeta,\eta). \tag{3.29}$$

2. *And, the solution of the problem (3.19)–(3.21) is*

$$\upsilon(\zeta,\eta) = \sum_{i=1}^{\infty}\sum_{k=1}^{i} \sigma_{ik}\hbar(\zeta_k,\eta_k)\bar{\Phi}_i(\zeta,\eta) - U(\zeta,\eta). \tag{3.30}$$

Proof.

1. In view of the fact that $\{\bar{\Phi}_i(\zeta,\eta)\}_{i=1}^{\infty}$ is a complete orthonormal basis in $\mathcal{B}_2^{(3,2)}(\mathcal{I})$, it follows

$$\begin{aligned}
\varkappa(\zeta,\eta) &= \sum_{i=1}^{\infty}\left\langle \varkappa(\zeta,\eta),\bar{\Phi}_i(\zeta,\eta)\right\rangle_{\mathcal{B}_2^{(3,2)}}\bar{\Phi}_i(\zeta,\eta)\\
&= \sum_{i=1}^{\infty}\left\langle \varkappa(\zeta,\eta),\sum_{k=1}^{i}\sigma_{ik}\phi_k(\zeta,\eta)\right\rangle_{\mathcal{B}_2^{(3,2)}}\bar{\Phi}_i(\zeta,\eta)\\
&= \sum_{i=1}^{\infty}\sum_{k=1}^{i}\sigma_{ik}\left\langle \varkappa(\zeta,\eta),\phi_k(\zeta,\eta)\right\rangle_{\mathcal{B}_2^{(3,2)}}\bar{\Phi}_i(\zeta,\eta)\\
&= \sum_{i=1}^{\infty}\sum_{k=1}^{i}\sigma_{ik}\left\langle \varkappa(\zeta,\eta),\mathfrak{I}^*\rho_k(\zeta,\eta)\right\rangle_{\mathcal{B}_2^{(3,2)}}\bar{\Phi}_i(\zeta,\eta)\\
&= \sum_{i=1}^{\infty}\sum_{k=1}^{i}\sigma_{ik}\left\langle \mathfrak{I}\varkappa(\zeta,\eta),\rho_k(\zeta,\eta)\right\rangle_{\mathcal{B}_2^{(1,1)}}\bar{\Phi}_i(\zeta,\eta)\\
&= \sum_{i=1}^{\infty}\sum_{k=1}^{i}\sigma_{ik}\left\langle \mathfrak{I}\varkappa(\zeta,\eta),\mathcal{G}_{(\zeta_k,\eta_k)}(\zeta,\eta)\right\rangle_{\mathcal{B}_2^{(1,1)}}\bar{\Phi}_i(\zeta,\eta)\\
&= \sum_{i=1}^{\infty}\sum_{k=1}^{i}\sigma_{ik}\hbar(\zeta_k,\eta_k)\bar{\Phi}_i(\zeta,\eta),
\end{aligned}$$

with $\hbar(\zeta_k,\eta_k) = \mathfrak{I}\varkappa(\zeta_k,\eta_k)$.

2. The second assertion is easy to get. Substituting the formula (3.29) into the considered transformation (3.22), we obtain (3.30).

Remarks

1. As a result, the solution of (3.29) is approximated by

$$\varkappa_p(\zeta,\eta) = \sum_{i=1}^{p}\sum_{k=1}^{i}\sigma_{ik}\hbar(\zeta_k,\eta_k)\bar{\Phi}_i(\zeta,\eta).$$

2. Due to the fact that $\mathcal{B}_2^{(3,2)}(\mathcal{I})$ is a Hilbert space,

$$\sum_{i=1}^{\infty}\sum_{k=1}^{i}\sigma_{ik}\hbar(\zeta_k,\eta_k)\bar{\Phi}_i(\zeta,\eta) < \infty.$$

And this says that the approximate solution $\varkappa_n(\zeta,\eta)$ is convergent in the norm.

3.4 Efficient Implementation

The $\mathfrak{p}$-term approximate solution of problem (3.29) obtained via the RKHS method is given by

$$\varkappa_{\mathfrak{p}}(\zeta,\eta)=\sum_{i=1}^{\mathfrak{p}}\xi_i\bar{\Phi}_i(\zeta,\eta), \tag{3.31}$$

where

$$\xi_i=\sum_{k=1}^{i}\sigma_{ik}\hbar(\zeta_k,\eta_k). \tag{3.32}$$

Here, by letting $(\zeta_1,\eta_1)=(0,0)$, the values of $\varkappa(\zeta_1,\eta_1)$ will be known from the ICs and BCs. And, $\varkappa_0(\zeta_1,\eta_1)=\varkappa(\zeta_1,\eta_1)$.

We next demonstrate that the $\mathfrak{p}$-term approximate solution $\varkappa_{\mathfrak{p}}(\zeta,\eta)$ is uniformly convergent to its exact solution $\varkappa(\zeta,\eta)$.

Theorem 3.3

Assume $\varkappa_{\mathfrak{p}_{\mathcal{B}_2^{(3,2)}}}$ *is bounded in (3.27), let* $\{(\zeta_i,\eta_i)\}_{i=1}^{\infty}$ *be a dense set on* $\mathcal{I}$*, and* (3.31) *has at most one solution* $\varkappa(\zeta,\eta)$*. Then*

1. $\varkappa_{\mathfrak{p}}(\zeta,\eta)$ *converges to* $\varkappa(\zeta,\eta)$
2. $\varkappa_{\mathfrak{p}}(\zeta,\eta)=\sum_{i=1}^{\mathfrak{p}}\xi_i\bar{\Phi}_i(\zeta,\eta)$, where ξ_i is given by (3.32).

Proof.

1. We derive from (3.31) that

$$\varkappa_{\mathfrak{p}+1}(\zeta,\eta)=\varkappa_{\mathfrak{p}}(\zeta,\eta)+\xi_{\mathfrak{p}+1}\bar{\Phi}_{\mathfrak{p}+1}(\zeta,\eta),$$

the orthogonality of $\{\bar{\Phi}_i(\zeta,\eta)\}_{i=1}^{\infty}$ implies

$$\begin{aligned}\|\varkappa_{\mathfrak{p}+1}\|^2_{\mathcal{B}_2^{(3,2)}}&=\|\varkappa_{\mathfrak{p}}\|^2_{\mathcal{B}_2^{(3,2)}}+\xi_{\mathfrak{p}+1}^2\\&=\|\varkappa_{\mathfrak{p}-1}\|^2_{\mathcal{B}_2^{(3,2)}}+\xi_{\mathfrak{p}}^2+\xi_{\mathfrak{p}+1}^2\\&\;\;\vdots\\&=\|\varkappa_0\|^2_{\mathcal{B}_2^{(3,2)}}+\sum_{i=1}^{\mathfrak{p}+1}\xi_i^2,\end{aligned} \tag{3.33}$$

and so

$$\left\|\varkappa_{p}\right\|_{\mathcal{B}_2^{(3,2)}} \leq \left\|\varkappa_{p+1}\right\|_{\mathcal{B}_2^{(3,2)}}.$$

The convergence of $\left\|\varkappa_{p}\right\|_{\mathcal{B}_2^{(3,2)}}$ follows easily from the fact that $\left\|\varkappa_{p}\right\|_{\mathcal{B}_2^{(3,2)}}$ is bounded. Thus, there exists a positive constant Θ such that

$$\sum_{i=1}^{\infty}\xi_i^2 = \Theta.$$

Consequently

$$\left\{\Theta_i^2\right\}_{i=1}^{\infty} \in \ell^2.$$

Since $\left(\varkappa_q(\zeta,\eta)-\varkappa_{q-1}(\zeta,\eta)\right)\perp\cdots\perp\left(\varkappa_{p+1}(\zeta,\eta)-\varkappa_p(\zeta,\eta)\right)$, for $q>p$ we have

$$\begin{aligned}\left\|\varkappa_q-\varkappa_p\right\|^2_{\mathcal{B}_2^{(3,2)}} &= \left\|\varkappa_q-\varkappa_{q-1}+\varkappa_{q-1}-\cdots+\varkappa_{p+1}-\varkappa_p\right\|^2_{\mathcal{B}_2^{(3,2)}}\\ &= \left\|\varkappa_q-\varkappa_{q-1}\right\|^2_{\mathcal{B}_2^{(3,2)}}+\left\|\varkappa_{q-1}-\varkappa_{q-2}\right\|^2_{\mathcal{B}_2^{(3,2)}}+\cdots+\left\|\varkappa_{p+1}-\varkappa_p\right\|^2_{\mathcal{B}_2^{(3,2)}}.\end{aligned} \tag{3.34}$$

Furthermore,

$$\left\|\varkappa_q-\varkappa_{q-1}\right\|^2_{\mathcal{B}_2^{(3,2)}} = \xi_q^2.$$

Consequently, we have

$$\left\|\varkappa_q-\varkappa_p\right\|^2_{\mathcal{B}_2^{(3,2)}} = \sum_{\ell=p+1}^{q}\xi_\ell^2 \to 0, \quad \text{as} \quad n\to\infty.$$

On account of the completeness of $\mathcal{B}_2^{(3,2)}(\mathcal{I})$, we deduce that $\varkappa_p \to \tilde{\varkappa}$ as $p\to\infty$.

2. To demonstrate that $\tilde{\varkappa}(\zeta,\eta)$ is the representation solution of (3.27), we must first take the limits in (3.31), to get

$$\tilde{\varkappa}(\zeta,\eta) = \sum_{i=1}^{\infty}\xi_i\bar{\Phi}_i(\zeta,\eta). \tag{3.35}$$

Applying the linear operator $\mathfrak{I}$ to (3.35), we find

$$\mathfrak{I}\tilde{\varkappa}(\zeta,\eta)=\sum_{i=1}^{\infty}\xi_i\mathfrak{I}\bar{\Phi}_i(\zeta,\eta),$$

it follows that

$$\begin{aligned}\mathfrak{I}\tilde{\varkappa}(\zeta_\ell,\eta_\ell)&=\sum_{i=1}^{\infty}\xi_i\left\langle\mathfrak{I}\bar{\Phi}_i(\zeta,\eta),\rho_\ell(\zeta,\eta)\right\rangle_{B_2^{(1,1)}}\\&=\sum_{i=1}^{\infty}\xi_i\left\langle\bar{\Phi}_i(\zeta,\eta),\mathfrak{I}^*\rho_\ell(\zeta,\eta)\right\rangle_{B_2^{(3,2)}}\\&=\sum_{i=1}^{\infty}\xi_i\left\langle\bar{\Phi}_i(\zeta,\eta),\phi_\ell(\zeta,\eta)\right\rangle_{B_2^{(3,2)}}.\end{aligned}$$

Multiplying the latter equality by $\sigma_{j\ell}$ and taking the summation $\sum_{\ell=1}^{j}$, we get

$$\begin{aligned}\sum_{\ell=1}^{j}\sigma_{j\ell}\mathfrak{I}\tilde{\varkappa}(\zeta_\ell,\eta_\ell)&=\sum_{i=1}^{\infty}\xi_i\left\langle\bar{\Phi}_i(\zeta,\eta),\sum_{\ell=1}^{j}\sigma_{j\ell}\phi_\ell(\zeta,\eta)\right\rangle_{B_2^{(3,2)}}\\&=\sum_{i=1}^{\infty}\xi_i\left\langle\bar{\Phi}_i(\zeta,\eta),\bar{\Phi}_j(\zeta,\eta)\right\rangle_{B_2^{(3,2)}}\\&=\xi_j.\end{aligned}$$

Thanks to (3.32), we find

$$\mathfrak{I}\tilde{\varkappa}(\zeta_\ell,\eta_\ell)=\hbar(\zeta_\ell,\eta_\ell).$$

For all $(x,t)\in\mathcal{I}$, there exists a subsequence $\left\{(\zeta_{pj},\eta_{pj})\right\}_{j=1}^{\infty}$ such that

$$(\zeta_{pj},\eta_{pj})\to(x,t),\quad \text{as}\quad j\to\infty.$$

This latter follows from the density of $\left\{(\zeta_i,\eta_i)\right\}_{i=1}^{\infty}$ on $\mathcal{I}$.
On the other hand, it is well known to us that

$$\mathfrak{I}\tilde{\varkappa}(\zeta_{pj},\eta_{pj})=\hbar(\zeta_{pj},\eta_{pj}).$$

Let $j \to \infty$, we utilize the the continuity of $\hbar$ to deduce

$$\Im \tilde{\varkappa}(x,t) = \hbar(x,t).$$

We obtain therefore the desired result for $\varkappa(\zeta,\eta)$.

3.5 Numerical Simulation

In this section, we confirm the theoretical predictions by applying them to two engineering problem applications. The time-fractional heat-like equation is discussed in Problem 3.1 and the time-fractional Navier-Stokes equation is considered in Problem 3.2. The Maple 18 software package is used for all computations.

Problem 3.1

Considering the following time-fractional heat-like equation, abbreviated "TFHLE":

$$\begin{cases} {}_0^C D_\eta^\alpha \upsilon(\zeta,\eta) = \frac{1}{2}\zeta^2 \partial_{\zeta^2}^2 \upsilon(\zeta,\eta), & \alpha \in (0,1],\ \zeta \in [0,1],\ \eta \in [0,d],\ d \in \mathbb{R}_+^*, \\ \upsilon(\zeta,0) = \zeta^2, \\ \upsilon(0,\eta) = 0, \quad and \quad \upsilon(1,\eta) = M_{\alpha,1}(\eta^\alpha). \end{cases} \tag{3.36}$$

The exact solution is

$$\upsilon(\zeta,\eta) = \zeta^2 M_{\alpha,1}(\eta^\alpha),$$

where $M\alpha_{,1}(z)$ is the Mittag-Leffler function which is given by:

$$M_{\alpha,1}(z) = \sum_{k=0}^{\infty} \frac{z^k}{\Gamma(k\alpha+1)}.$$

As discussed in Section 3.3, before applying the RKHS method, we must, first of all, homogenize the IC and BCs of problem (3.36). To do this, we use the following transformation:

$$\varkappa(\zeta,\eta)=\upsilon(\zeta,\eta)+\mathrm{U}(\zeta,\eta),$$

where

$$\mathrm{U}(\zeta,\eta)=-\zeta\left(-1+\zeta+\mathrm{M}_{\alpha,1}\left(\eta^{\alpha}\right)\right).$$

Thus, (3.36) becomes

$$\begin{cases} {}_{0}^{C}\mathrm{D}_{\eta}^{\alpha}\varkappa(\zeta,\eta)-\frac{1}{2}\zeta^{2}\partial_{\zeta^{2}}^{2}\varkappa(\zeta,\eta)=\hbar(\zeta,\eta), & \alpha\in(0,1],\ \zeta\in[0,1],\ \eta\in[0,\mathrm{d}],\ \mathrm{d}\in\mathbb{R}_{+}^{*}, \\ \varkappa(\zeta,0)=0, \\ \varkappa(0,\eta)=0, \quad \textit{and} \quad \varkappa(1,\eta)=0. \end{cases} \tag{3.37}$$

where,

$$\hbar(\zeta,\eta)={}_{0}^{C}\mathrm{D}_{\eta}^{\alpha}\mathrm{U}(\zeta,\eta)-\frac{1}{2}\zeta^{2}\partial_{\zeta^{2}}^{2}\mathrm{U}(\zeta,\eta). \tag{3.38}$$

After the change of variable, we introduce the bounded linear operator:

$$\begin{aligned} \mathfrak{I}:\quad \mathcal{B}_{2}^{(3,2)}(\mathcal{I}) &\rightarrow \mathcal{B}_{2}^{(1,1)}(\mathcal{I}) \\ \varkappa(\zeta,\eta) &\rightarrow \mathfrak{I}\varkappa(\zeta,\eta)=\frac{C}{0}\mathrm{D}_{\eta}^{\alpha}\varkappa(\zeta,\eta)-\frac{1}{2}\zeta^{2}\partial_{\zeta^{2}}^{2}\varkappa(\zeta,\eta) \end{aligned}$$

So, we can rewrite (3.37) as

$$\begin{cases} \mathfrak{I}\varkappa(\zeta,\eta)=\hbar(\zeta,\eta), & \alpha\in(0,1],\ \zeta\in[0,1],\ \eta\in[0,\mathrm{d}],\ \mathrm{d}\in\mathbb{R}_{+}^{*}, \\ \varkappa(\zeta,0)=0, \\ \varkappa(0,\eta)=0, \quad \textit{and} \quad \varkappa(1,\eta)=0. \end{cases} \tag{3.39}$$

where $\hbar(\zeta,\eta)$ is given above by (3.42).

Problem 3.2

Considering the following time-fractional Navier-Stokes equation, abbreviated "TFNSE":

$$\begin{cases} {}_{0}^{C}D_{\eta}^{\alpha}\upsilon(\zeta,\eta)=P+v\left[\partial_{\zeta^2}^{2}\upsilon(\zeta,\eta)+\dfrac{1}{\zeta}\partial_{\zeta}\upsilon(\zeta,\eta)\right], \quad \alpha\in(0,1],\ \zeta\in[0,1],\ \eta\in[0,d],\ d\in\mathbb{R}_{+}^{*}, \\ \upsilon(\zeta,0)=1-\zeta^2, \\ \upsilon(0,\eta)=1+(P-4)\dfrac{t^{\alpha}}{\Gamma(1+\alpha)}, \quad \mathit{and} \quad \upsilon(1,\eta)=(P-4)\dfrac{t^{\alpha}}{\Gamma(1+\alpha)}. \end{cases} \tag{3.40}$$

The exact solution is

$$\upsilon(\zeta,\eta)=1-\zeta^2+(P-4)\frac{t^{\alpha}}{\Gamma(1+\alpha)},$$

As discussed in Section 3.3, before applying the RKHS method, we must, first of all, homogenize the IC and BCs of problem (3.40). To do this, we use the following transformation:

$$\varkappa(\zeta,\eta)=\upsilon(\zeta,\eta)+U(\zeta,\eta),$$

where

$$U(\zeta,\eta)=\zeta^2-1-\frac{(p-4)\left(1+(-1+x)x^2\right)\eta^{\alpha}}{\Gamma(1+\alpha)}.$$

Thus, (3.40) becomes

$$\begin{cases} {}_{0}^{C}D_{\eta}^{\alpha}\varkappa(\zeta,\eta)-v\left[\partial_{\zeta^2}^{2}\varkappa(\zeta,\eta)+\dfrac{1}{\zeta}\partial_{\zeta}\varkappa(\zeta,\eta)\right]=\hbar(\zeta,\eta), \quad \alpha\in(0,1],\ \zeta\in[0,1],\ \eta\in[0,d],\ d\in\mathbb{R}_{+}^{*}, \\ \varkappa(\zeta,0)=0, \\ \varkappa(0,\eta)=0, \quad \mathit{and} \quad \varkappa(1,\eta)=0. \end{cases} \tag{3.41}$$

where

$$\hbar(\zeta,\eta)= {}_0^C D_\eta^\alpha U(\zeta,\eta)+P-v\left[\partial_{\zeta^2}^2 U(\zeta,\eta)+\frac{1}{\zeta}\partial_\zeta U(\zeta,\eta)\right]. \quad (3.42)$$

After the change of variable, we introduce the bounded linear operator:

$$\begin{aligned}&\mathfrak{I}:\mathcal{B}_2^{(3,2)}(\mathcal{I})\to\mathcal{B}_2^{(1,1)}(\mathcal{I})\\ &\varkappa(\zeta,\eta)\to\mathfrak{I}\varkappa(\zeta,\eta)={}_0^C D_\eta^\alpha \varkappa(\zeta,\eta)-v\left[\partial_{\zeta^2}^2\varkappa(\zeta,\eta)+\frac{1}{\zeta}\partial_\zeta\varkappa(\zeta,\eta)\right]\end{aligned}$$

So, we can rewrite (3.41) as

$$\begin{cases}\mathfrak{I}\varkappa(\zeta,\eta)=\hbar(\zeta,\eta), & \alpha\in(0,1],\ \zeta\in[0,1],\ \eta\in[0,d],\ d\in\mathbb{R}_+^*,\\ \varkappa(\zeta,0)=0, & \\ \varkappa(0,\eta)=0, \text{ and } \varkappa(1,\eta)=0. & \end{cases} \quad (3.43)$$

where $\hbar(\zeta,\eta)$ is given above by (3.42).

3.6 Computational Results and Discussion

We apply the RKHS method in the same way as described in Sections 3.3 and 3.4. With p collocation points in which $\zeta_i = i/\ell, i=1,2\ldots,\ell$ and $\eta_j = j/q, i=1,2\ldots,q$ and $p=\ell\times q$, the approximate solution of both problems is obtained and compared with its exact solution. These types of equations appear in a variety of applied sciences and engineering problems.

For Problem 3.1, we take $p=100$ collocation points and ζ, $\eta \in [0,1]$. Figure 3.1 demonstrates the surface graphs of the numerical solution of this problem. From this figure, we can see and understand the geometric behavior of the obtained solutions when $\alpha = 1, 0.9, 0.8, 0.7$. To compare between the exact and RKHS method solutions of (3.36), we computed the absolute errors for various values of α and showed them in Table 3.1. To highlight more comparisons between results, we added some existing results to this table. From the results gained, we can see a good agreement between the RKHS

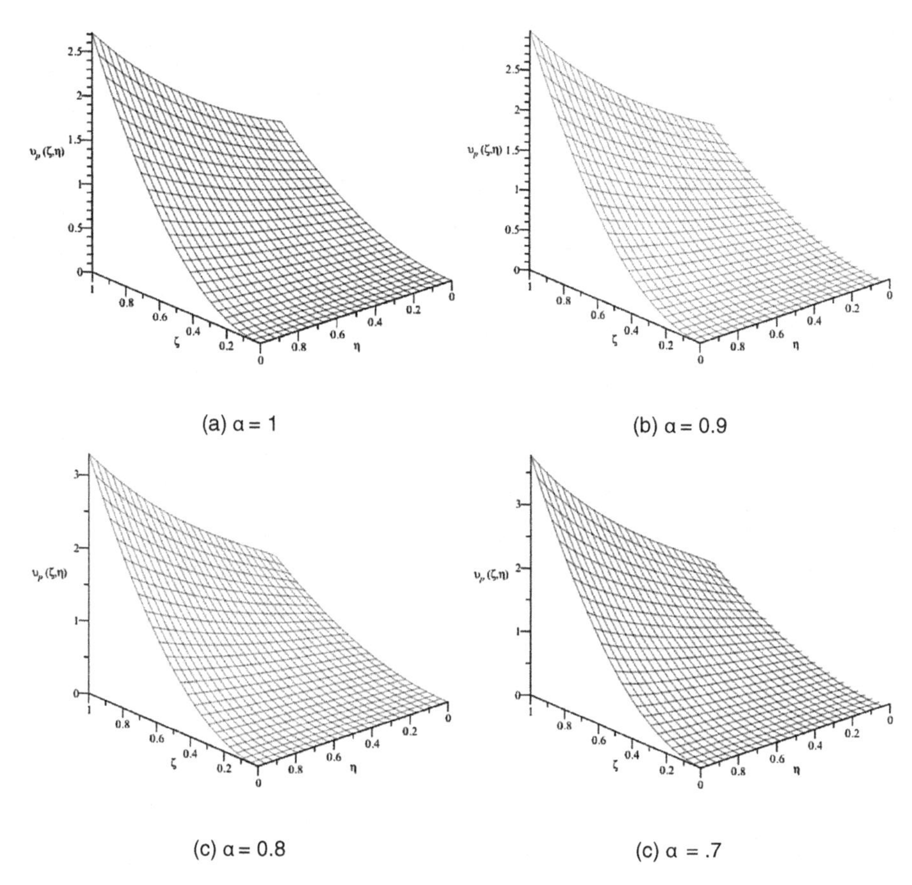

(a) $\alpha = 1$ (b) $\alpha = 0.9$

(c) $\alpha = 0.8$ (c) $\alpha = .7$

FIGURE 3.1
Graphical results of approximate solution of Problem 3.1 for $\gamma = 1, 0.9, 0.8, 0.7$ and $\mathfrak{p} = 100$ respectively at various values of $(\zeta,\eta) \in [0,1] \times [0,1]$ using RKHS method.

method's solutions and exact solutions. And we observe that the RKHS method is more accurate than the cited one.

For Problem 3.2, we take $\mathfrak{p} = 100$ collocation points and $\zeta, \eta \in [0,1]$. Figure 3.2 demonstrates the surface graphs of the numerical solution of this problem. From this figure, we can see and understand the geometric behavior of the obtained solutions when $\alpha = 1, 0.9, 0.8, 0.7$. To compare between the exact and RKHS method solutions of (3.40), we computed the absolute errors for various values of α and showed them in Table 3.2. To highlight more comparisons between results, we added some existing results to this table. From the results gained, we can see good agreement between the RKHS methods solutions and exact solutions. And we observe that the RKHS method is more accurate than the cited one.

TABLE 3.1
Absolute Errors of Approximating the Solution in Problem 3.1 at Different $(\xi,\tau) \in [0,1] \times [0,1]$ and with Various Values of α

		$\gamma = 0.9$			$\gamma = 0.75$		
η	ζ	**RKHSM**	**[36]**	**[19]**	**RKHSM**	**[36]**	**[19]**
0.25	0.3	7.8×10^{-4}	7.0×10^{-4}	1.6×10^{-3}	2.3×10^{-3}	3.7×10^{-3}	2.7×10^{-3}
	0.6	6.7×10^{-4}	1.1×10^{-2}	4.3×10^{-3}	1.9×10^{-3}	4.4×10^{-2}	6.1×10^{-3}
	0.9	1.9×10^{-4}	5.1×10^{-2}	2.5×10^{-3}	5.2×10^{-4}	1.1×10^{0}	3.3×10^{-3}
0.5	0.3	5.3×10^{-4}	3.1×10^{-3}	3.3×10^{-3}	1.5×10^{-3}	1.4×10^{-2}	4.5×10^{-3}
	0.6	3.7×10^{-4}	3.0×10^{-2}	7.0×10^{-3}	1.1×10^{-3}	9.8×10^{-2}	9.0×10^{-3}
	0.9	9.3×10^{-5}	9.4×10^{-2}	3.6×10^{-3}	3.0×10^{-4}	2.9×10^{-1}	4.4×10^{-3}
0.75	0.3	4.8×10^{-4}	7.7×10^{-3}	5.1×10^{-3}	1.1×10^{-3}	2.8×10^{-2}	6.5×10^{-3}
	0.6	3.0×10^{-4}	5.2×10^{-2}	9.8×10^{-3}	7.3×10^{-4}	1.6×10^{-1}	1.2×10^{-2}
	0.9	8.0×10^{-5}	1.4×10^{-1}	4.8×10^{-3}	1.9×10^{-3}	4.3×10^{-1}	5.8×10^{-3}
1.0	0.3	3.0×10^{-3}	1.5×10^{-2}	7.2×10^{-3}	6.3×10^{-4}	4.9×10^{-2}	8.9×10^{-3}
	0.6	2.9×10^{-3}	7.7×10^{-2}	1.3×10^{-2}	2.6×10^{-4}	2.8×10^{-1}	1.6×10^{-2}
	0.9	8.9×10^{-4}	2.7×10^{-1}	6.3×10^{-3}	5.3×10^{-5}	8.1×10^{-1}	7.4×10^{-3}

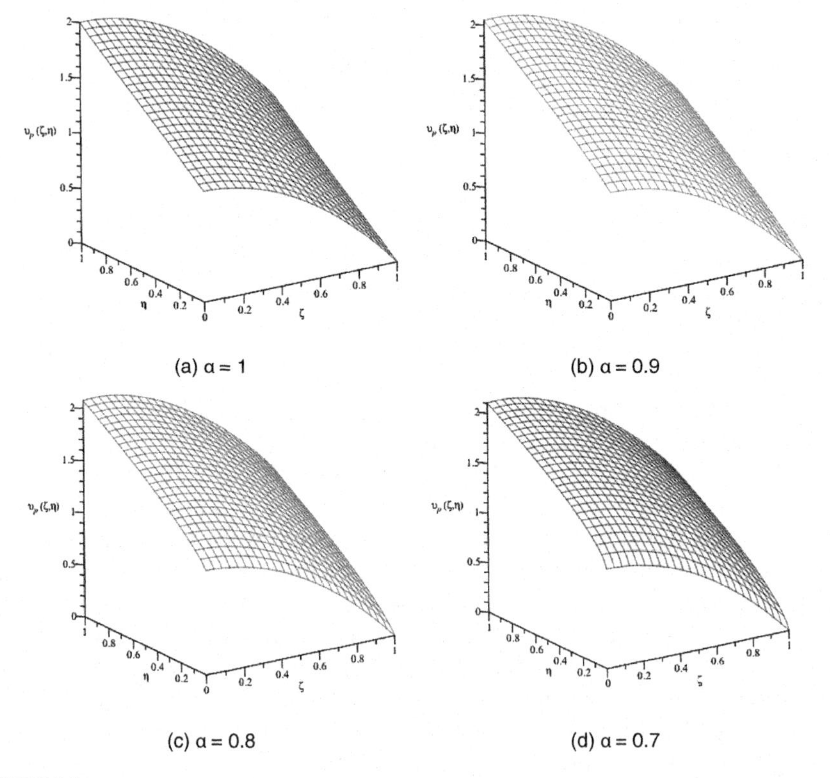

FIGURE 3.2
Graphical results of approximate solution of Problem 3.1 for $\gamma = 1, 0.9, 0.8, 0.7$ and $\mathfrak{p} = 100$ respectively at various values of $(\zeta,\eta) \in [0,1] \times [0,1]$ using RKHS method.

TABLE 3.2

Numerical Outcomes for TFNSE Presented in Problem 3.2 at Different $(\xi,\tau) \in [0,1] \times [0,1]$ for Various Values of α

		$\gamma = 1$		$\gamma = 0.9$		$\gamma = 0.75$	
η	ζ	**RKHSM**	**ES**	**RKHSM**	**ES**	**RKHSM**	**ES**
0.25	0.3	1.160508432	1.16	1.209641631	1.208590966	1.264838070	1.264179435
	0.6	0.890382701	0.89	0.939341878	0.938590966	0.994530383	0.994179435
	0.9	0.440102104	0.44	0.488781932	0.488590966	0.544240470	0.544179435
0.5	0.3	1.410967574	1.41	1.468571011	1.467190444	1.527099073	1.526662213
	0.6	1.140699620	1.14	1.198096097	1.197190444	1.256728506	1.256662213
	0.9	0.690181431	0.69	0.747414553	0.747190444	0.806645645	0.806662213
0.75	0.3	1.661432047	1.66	1.714438070	1.712575306	1.763293575	1.762943658
	0.6	1.391021893	1.39	1.443766238	1.442575306	1.492820646	1.492943658
	0.9	0.940262214	0.94	0.992865080	0.992575306	1.042860721	1.042943658
1.0	0.3	1.911899489	1.91	1.952107286	21.94975413	1.983781290	1.983671274
	0.6	1.641345712	1.64	1.681237361	1.679754134	1.713229353	1.713671274
	0.9	1.190343188	1.19	1.230111147	1.229754134	1.263484618	1.263671274

3.7 Conclusion

In this study, an efficient method of solving time FPDEs, which govern important phenomena in engineering and the physical sciences, has been applied successfully. The accuracy and applicability of the proposed method are validated by computing the numerical solutions at many grid points. The boundedness of the linear operator is demonstrated. The results obtained demonstrate the convenience and efficiency of the RKHS method for the problems presented. From the results, it can be concluded that the proposed method is a powerful tool to approximate many other types of problems that arise in a variety of engineering and physical science contexts. As a future research direction, we can use the RKHSM to obtain the numerical solutions of different kinds of fractional stochastic differential equations.

References

[1] H. Singh, D. Kumar, D. Baleanu (Eds), *Methods of Mathematical Modelling: Fractional Differential Equations* (1st ed.), CRC Press (2019). doi:10.1201/9780429274114

[2] C. Li, D. Qian, Y. Chen, On Riemann-Liouville and Caputo derivatives, *Discrete Dyn. Nat. Soc.* vol. 2011 2011 p. 562494.

[3] R. Hilfer, Y. Luchko, Z. Tomovski, Operational method for the solution of fractional differential equations with generalized Riemann-Liouville fractional derivatives, *Fract. Calc. Appl. Anal.* vol. 12 (2009) pp. 299–318.

[4] M. D. Ortigueira, L. Rodriguez-Germá, J. J. Trujillo, Complex Grünwald-Letnikov, Liouville, Riemann-Liouville, and Caputo derivatives for analytic functions, *Comm. Nonlinear Sci. Numer. Simulat.* vol. 16 (2011) pp. 4174–4182.

[5] B. S. H. Kashkari, M. I. Syam, Reproducing kernel method for solving nonlinear fractional Fredholm integrodifferential equation, *Complexity* vol. 2018 (2018) p. 2304858.

[6] G. Jumarie, Table of some basic fractional calculus formulae derived from a modified Riemann-Liouville derivative for non-differentiable functions, *Appl. Math. Lett.* vol. 22 (2009) pp. 378–385.

[7] A. Atangana and D. Baleanu, New fractional derivatives with nonlocal and non-singular kernel: theory and application to heat transfer model, *Therm. Sci.* vol. 20 (2016) pp. 763–769.

[8] R. Khalil, M. Al Horani, A. Yousef, M. Sababheh, A new definition of fractional derivative, *J. Comput. Appl. Math.* vol. 264 (2014) pp. 65–70.

[9] H. Singh, J. Singh, S. D. Purohit, D. Kumar (Eds), *Advanced Numerical Methods for Differential Equations: Applications in Science and Engineering* (1st ed.), CRC Press (2021). doi:10.1201/9781003097938

[10] H. Singh, Jacobi collocation method for the fractional advection-dispersion equation arising in porous media, *Numer. Methods Partial Differ. Equ.* (2020) pp. 1–18. doi:10.1002/num.22674.

[11] H. Singh, Analysis of drug treatment of the fractional HIV infection model of CD4+ T-cells, *Chaos Soliton. Fract.* vol. 146 (2021) pp. 110868.

[12] H. Singh, Numerical simulation for fractional delay differential equations, *Int. J. Dyn. Control.* vol. 9 (2021) pp. 463–474.

[13] H. Singh, H. M. Srivastava, D. Kumar, A reliable algorithm for the approximate solution of the nonlinear Lane-Emden type equations arising in astrophysics, *Numer. Methods Partial Differ. Equ.* vol. 34 (2018) pp. 1524–1555.

[14] H. Singh, H. M. Srivastava, Numerical investigation of the fractional-Order Liénard and Duffing equations arising in oscillating circuit theory, *Front. Phys.* vol. 8 (2020) p. 120.

[15] H. Singh, Analysis for fractional dynamics of Ebola virus model, *Chaos Soliton. Fract.* vol. 138 (2020) p. 109992.

[16] H. Singh, A.-M. Wazwaz, Computational method for reaction diffusion-model arising in a spherical catalyst, *Inter. J. Appl. Comput. Math.* vol. 7 (2021) p. 65.

[17] R. Y. Molliq, M.S.M. Noorani, I. Hashim, Variational iteration method for fractional heat- and wave-like equations, *Nonlinear Anal. Real World Appl.* vol. 10 (2009) pp. 1854–1869.

[18] S. Momani, Z. Odibat, Analytical solution of a time-fractional Navier-Stokes equation by Adomian decomposition method, *Appl. Math. Comput.* vol. 177 (2006) pp. 488–494.

[19] S. Haq, M. Hussain, The meshless Kansa method for time-fractional higher order partial differential equations with constant and variable coefficients, *RACSAM* vol. 113 (2019) pp. 1935–1954.

[20] S. Zaremba, Sur le calcul numérique des fonctions demandées dans le problème de Dirichlet et le problème hydrodynamique, *Bull. Int. Acad. Sci. Cracovie* vol. 68 (1908) pp. 125–195.

[21] F. Hemati, M. Ghasemi, G. R. Khoshsiar, Numerical solution of the multiterm time-fractional diffusion equation based on reproducing kernel theory, *Numer. Methods Partial Differ. Equ.* vol. 37 (2020) pp. 44–68.

[22] B. Ghanbari, A. Akgül, Abundant new analytical and approximate solutions to the generalized Schamel equation, *Phys. Scr.* vol. 95 (2020) pp. 075201.

[23] A. Akgül, A. Cordero, J. R. Torregrosa, Solutions of fractional gas dynamics equation by a new technique, *Math. Meth. Appl. Sci.* vol. 43 (2020) pp. 1349–1358.

[24] M. G. Sakar, O. Salıdr, A. Akgül, *A novel technique for fractional Bagley-Torvik equation, Proc. Natl. Acad. Sci., India, Sect. A Phys. Sci.* vol. 89 (2019) pp. 539–545.

[25] M. G. Sakar, Iterative reproducing kernel Hilbert spaces method for Riccati differential equation, *J. Comput. Appl. Math.* vol. 309 (2017) pp. 163–174.

[26] M. Modanli, M. A. Akgül, On solutions to the second-order partial differential equations by two accurate methods, *Numer. Methods Partial Differ. Equ.* vol. 34 (2018) pp. 1678–1692.

[27] E. Babolian, S. Javadi, E. Moradi, RKM for solving Bratu-type differential equations of fractional order, *Math. Meth. Appl. Sci.* vol. 39 (2016) pp. 1548–1557.

[28] W. Jiang, T. Tian, Numerical solution of nonlinear Volterra integro-differential equations of fractional order by the reproducing kernel method, *Appl. Math. Model.* vol. 39 (2015) pp. 4871–4876.

[29] F. Geng, M. Cui, A reproducing kernel method for solving nonlocal fractional boundary value problems, *Appl. Math. Lett.* vol. 25 (2012) pp. 818–823.

[30] F. Geng, M. Cui, New method based on the HPM and RKHSM for solving forced Duffing equations with integral boundary conditions, *J. Comput. Appl. Math.* vol. 233 (2009) pp. 165–172.

[31] N. Attia, A. Akgül, D. Seba, A. Nour, An efficient numerical technique for a biological population model of fractional order, *Chaos Soliton. Fract.* vol. 141 (2020) p. 110349.

[32] N. Attia, A. Akgül, D. Seba, A. Nour, Reproducing kernel Hilbert space method for the numerical solutions of fractional cancer tumor models, *Math. Meth. Appl. Sci.* (2020). doi:10.1002/mma.6940

[33] N. Attia, D. Seba, A. Akgül, A. Nour, Solving Duffing-Van der Pol oscillator equations of fractional order by an accurate technique, *J. Appl. Comput. Mech.* vol. 7 (2021) pp. 1480–1487.

[34] M. Cui, Y. Lin, *Nonlinear Numerical Analysis in the Reproducing Kernel Space*, Nova Science Publishers, New York, 2009.

[35] N. Attia, A. Akgül, D. Seba, A. Nour, On solutions of time-fractional advection-diffusion equation, *Numer. Methods Partial Differ. Equ.* (2020). doi:10.1002/num.22621

[36] A. Saadatmandi, M. Dehghan, M. R. Azizi, The Sinc-Legendre collocation method for a class of fractional convection-diffusion equations with variable coefficients, *Comm. Nonlinear Sci. Numer. Simulat.* vol. 17 (2012) pp. 4125–4136.

4

Spectral Collocation Method Based Upon Special Functions for Fractional Partial Differential Equations[*]

H. M. Srivastava
University of Victoria, Victoria, British Columbia, Canada
China Medical University, Taichung, Taiwan, Republic of China
Azerbaijan University, Baku, Azerbaijan
International Telematic University Uninettuno, Rome, Italy

Khaled M. Saad
Najran University, Najran, Kingdom of Saudi Arabia
Taiz University, Taiz, Yemen

M. M. Khader
Ibn Saud Islamic University (IMSIU), Riyadh, Kingdom of Saudi Arabia
Benha University, Benha, Egypt

Harendra Singh
Post Graduate College, Ghazipur, India

CONTENTS

[*] 2010 *Mathematics Subject Classification*: Primary 41A50, 65L12; Secondary 65N12, 65N35.

DOI: 10.1201/9781003263517-4

4.1 Introduction

Fractional calculus is an important branch of mathematics that focuses on the properties of derivatives and integrals of non-integer orders. In particular, this branch includes finding the solution of fractional differential equations as well as fractional integral equations. This science was first mentioned by Leibniz in 1695 in his famous letter to L'Hôpital, when he raised the question: What is the value of the derivative in the case of the half order? Liouville, Grünwald, Riemann, Euler, Lagrange, Heaviside, Fourier, and Abel, among others, have also played a major role in building the construction of fractional calculus as well as introducing new methods [1].

There are many applications in the fields of science and engineering that cannot be modeled with classical differential equations, which highlights the importance of fractional differential equations, through which many applications can be modeled [2, 3]. For most of the previous applications, in the case of fractional derivatives no analytical solutions exist, so approximate and numerical methods were needed for these applications. Remarkable efforts to find approximate solutions have recently been made by researchers studying the dynamic behavior of these applications. Examples include the modified cubic B-spline differential [4, 5], A hybrid numerical [6, 7], finite element method [8, 9], weighted average differential quadrature [10], q-homotopy analysis transform [11], fractional homotopy perturbation Laplace transform scheme [12], Sumudu transform approach and homotopy polynomials [13], homotopy analysis transform method [14], Laplace transform and Padé, and the spectral collocation method based on the shifted Chebyshev polynomials of type three utilized to solve fractional Korteweg–de Vries, Korteweg–de Vries–Burgers and fractional Burgers equations [15–18]. For more applications on fractional derivatives see [19–33].

Spectral methods have played an important and widespread role in solving fractional differential equations, as they give numerical results with high accuracy and efficiency [34]. A method for solving multi-term fractional-order differential equations known as the efficient Chebyshev spectral method was developed by [35]. The pseudo-spectral method is also used to introduce direct solution of a special family of fractional initial value problems [36]. Collocation schemes can be used to facilitate the handling of fractional partial differential equations of multi-term fractionals [15, 17, 26]. For more recent work related to the spectral methods, see [37–39].

This section contains two definitions that will be used in the rest of the chapter [2, 3].

Definition 1

The Riemann–Liouville fractional integral of order ω, denoted by J_0^{ω}, is given by

$$J_0^{\omega}\varphi(\eta) = \frac{1}{\Gamma(\omega)}\int_0^{\eta}(\eta-\tau)^{\omega-1}\varphi(\tau)d\tau. \\ (\omega > 0, \text{ and } \varphi(\eta) \in L_1(a,b)), \tag{4.1}$$

where $L_1(a, b)$ the space of all integrable functions on (a,b).

Definition 2

The Liouville–Caputo fractional derivative of order ω, denoted by ${}^{LC}D_0^{\omega}$, is given by

$$ {}_{0}^{LC}\mathcal{D}_{\xi}^{\omega}\varphi(\eta) = \frac{1}{\Gamma(n-\omega)}\int_0^{\eta}(\eta-\tau)^{n-v-1}\mathcal{D}^{n}\varphi(\tau)d\tau \\ (n-1 < \omega < n; n \in \mathbb{N} = \{1,2,3,\cdots\}). \tag{4.2}$$

4.2 Shifted Chebyshev Polynomials and LC-Fractional Derivatives

4.2.1 Shifted Chebyshev Polynomials

Chebyshev polynomials are known as the family of orthogonal polynomials [40]. As a result of their good properties, they are widely used in approximation of functions and can be determined using the following successive relations:

$$T_{n+1}(z) = 2zT_n(z) - T_{n-1}(z), \quad T_0(z) = 1,\ T_1(z) = z, \quad n = 1,2,\ldots.$$

When we introduce a new variable, $z = 2\eta - 1$, we can redefine these polynomials over the interval $[0,1]$. The resulting functions are called shifted Chebyshev polynomials and are given by

$$\bar{\zeta}_n(\eta) = T_n(2\eta - 1) = T_{2n}(\sqrt{\eta}).$$

The analytic form of $\bar{\zeta}_n(\eta)$ of degree n is given by

$$\bar{\zeta}_n(\eta) = n\sum_{k=0}^{n}(-1)^{n-k}\frac{2^{2k}(n+k-1)!}{(2k)!(n-k)!}\eta^k, \qquad n = 2,3,\ldots. \tag{4.3}$$

We can express the function $\varphi(\eta) \in L_2[0,1]$ as a linear combination of $\bar{\zeta}_n(\eta)$ as

$$\varphi(\eta) = \sum_{i=0}^{\infty} \chi_i \bar{\zeta}_i(\eta), \tag{4.4}$$

where the coefficients χi are given by

$$\chi_0 = \frac{1}{\pi}\int_0^1 \frac{\varphi(\eta)\bar{\zeta}_0(\eta)}{\sqrt{\eta-\eta^2}} d\eta, \tag{4.5}$$

$$\chi_i = \frac{2}{\pi}\int_0^1 \frac{\varphi(\eta)\bar{\zeta}_i(\eta)}{\sqrt{\eta-\eta^2}} d\eta, \quad i = 1,2,.... \tag{4.6}$$

To obtain the approximation form, we consider the first $(m + 1)$-terms of $\bar{\zeta}_m(\eta)$

$$\varphi_m(\eta) = \sum_{i=0}^{m} \chi_i \bar{\zeta}_i(\eta). \tag{4.7}$$

Theorem 2.1

(Chebyshev truncation theorem) [40]
To obtain the error of the approximation $\Omega m(t)$

$$E_T(m) \equiv |\Omega(t) - \Omega_m(t)| \leq \sum_{k=m+1}^{\infty} |a_k|, \qquad \forall \Omega(t), m, \forall t \in [-1,1]. \tag{4.8}$$

we approximate the function $\varphi(\eta)$ in the following form

$$\Omega_m(\chi) = \sum_{k=0}^{m} \chi_k T_k(\eta). \tag{4.9}$$

Theorem 2.2 introduces the basic approximation formula of $D\omega\varphi m(\eta)$ that will be used in the following calculations.

Theorem 2.2

[41, 42]

In view of (4.7), we define $D\omega(\varphi m(\eta))$ in the form

$$D^{\omega}(\varphi_m(\eta)) = \sum_{i=\lceil \omega \rceil}^{m} \sum_{k=\lceil \omega \rceil}^{i} a_i \Upsilon_{i,k}^{(\omega)} \eta^{k-\omega}, \tag{4.10}$$

where $\Upsilon_{i,k}^{(\omega)}$ is given by

$$\Upsilon_{i,k}^{(\omega)} = (-1)^{i-k} \frac{2^{2k} i(i+k-1)!\Gamma(k+1)}{(i-k)!(2k)!\Gamma(k+1-\omega)}. \tag{4.11}$$

Theorem 2.3

The amount of error between $D\nu\Omega(t)$ by $D\nu\Omega m(t)$ can be expressed as [42]

$$|E_{\mathbb{T}}(m)| = |D^{\omega}\varphi(\eta) - D^{\omega}\varphi_m(\eta)| \leq \left| \sum_{i=m+1}^{\infty} c_i \left(\sum_{k=\lceil \omega \rceil}^{i} \sum_{j=0}^{k-\lceil \omega \rceil} \Theta_{i,j,k} \right) \right|. \tag{4.12}$$

4.2.2 Implementation of the Proposed Method and Examples Based on the Shifted Chebyshev Polynomials

In this section, we provide two examples and explain the proposed method for one of them in detail.

Example 1
Multi-space fractional Korteweg–de Vries–Burgers equation (MSFKDVBE)
The MSFKDVBE is given by:

$$\varphi_\eta + \varphi D_\xi^\theta \varphi - \varepsilon D_\xi^\rho \varphi + \frac{1}{2} D_\xi^\vartheta \varphi = 0, 0 \leq \eta \leq T, 0 < \xi < 1, \varepsilon \in \Re^+,$$
$$(0 < \theta \leq 1, 1 < \rho \leq 2, 2 < \vartheta < 3), \tag{4.13}$$

The initial and boundary conditions are given by

$$\varphi(\xi, 0) = g(\xi), \quad 0 < \xi < 1, \tag{4.14}$$

$$\varphi(0, \eta) = C_1(\eta), \tag{4.15}$$

$$\varphi(1,\eta) = C_2(\eta), \tag{4.16}$$

$$\varphi_\xi(0,\eta) = C_3(\eta). \tag{4.17}$$

The Korteweg–de Vries equation is extended to the MSFKDVBE which include the viscous loss $\varepsilon\varphi\xi\xi$, further to the convective non-linearity $\varphi\varphi\xi$ and the dispersion $\varphi\xi\xi\xi$ ([43, 44]). The exact solution of the MSFKDVBE in the case of integer order, according to ([45, 46]), is given by

$$\varphi(\xi,\eta) = \frac{12}{25}\epsilon^2 - \frac{12}{25}\epsilon^2 \tanh(\kappa) + \frac{6}{25}\epsilon^2 \operatorname{sech}^2(\kappa), \tag{4.18}$$

where $\kappa = \frac{1}{2}\left(\frac{2}{5}\epsilon\xi - \frac{24}{125}\epsilon^3\eta\right)$.

The approximate solution of the MSFKDVBE is obtained in the following basic steps:

(1) By taking the first $(m+1)$-terms of the sum of the shifted Chebyshev polynomials $\bar{\zeta}_s(\xi)$, we can approximate the function $\varphi(\xi, \eta)$ as:

$$\varphi_m(\xi,\eta) = \sum_{i=0}^{m} a_i(\eta)\bar{\zeta}_i(\xi). \tag{4.19}$$

(2) To get the system of differential equations, we substitute (4.10) and (4.19) into Eq. (4.13) as:

$$\begin{aligned}
&\sum_{i=0}^{m} \frac{da_i(\eta)}{dt}\bar{\zeta}_i(\xi) + \left(\sum_{i=0}^{m} a_i(\eta)\bar{\zeta}_i(\xi)\right)\left(\sum_{i=\lceil\theta\rceil}^{m}\sum_{j=\lceil\theta\rceil}^{i} a_i(\eta)\Upsilon_{i,j}^{\theta}(\xi)\right) \\
&-\epsilon\left(\sum_{i=\lceil\rho\rceil}^{m}\sum_{j=\lceil\rho\rceil}^{i} a_i(\eta)\Upsilon_{i,j}^{\rho}(\xi)\right) \\
&+\frac{1}{2}\left(\sum_{i=\lceil\vartheta\rceil}^{m}\sum_{j=\lceil\vartheta\rceil}^{i} a_i(\eta)\Upsilon_{i,j}^{\vartheta}(\xi)\right) = 0.
\end{aligned} \tag{4.20}$$

(3) At $(m+1-\lceil\rho\rceil)$ points ξr, we collocate the Eq. (4.20) as:

$$
\begin{aligned}
&\sum_{i=0}^{m} \frac{da_i(\eta)}{dt}\bar{\zeta}_i(\xi_r) + \left(\sum_{i=0}^{m} a_i(\eta)\bar{\zeta}_i(\xi_r)\right)\left(\sum_{i=\lceil\theta\rceil}^{m}\sum_{j=\lceil\theta\rceil}^{i} a_i(\eta)\Upsilon_{i,j}^{\theta}(\xi_r)\right) \\
&\quad -\epsilon\left(\sum_{i=\lceil\rho\rceil}^{m}\sum_{j=\lceil\rho\rceil}^{i} a_i(\eta)\Upsilon_{i,j}^{\rho}(\xi_r)\right) \\
&\quad +\frac{1}{2}\left(\sum_{i=\lceil\vartheta\rceil}^{m}\sum_{j=\lceil\vartheta\rceil}^{i} a_i(\eta)\Upsilon_{i,j}^{\vartheta}(\xi_r)\right) = 0
\end{aligned}
\tag{4.21}
$$

(4) By substituting Eq. (4.19) in (4.15)–(4.17), we obtain the related initial and boundary conditions of this system

$$\sum_{i=0}^{m} \bar{\zeta}_i(0)a_i(\eta) = B_1(\eta), \tag{4.22}$$

$$\sum_{i=0}^{m} \bar{\zeta}_i(1)a_i(\eta) = B_2(\eta), \tag{4.23}$$

$$\sum_{i=0}^{m} \bar{\zeta}_i'(0)a_i(\eta) = B_3(\eta), \tag{4.24}$$

where $\bar{\zeta}_i(0) = (-1)^i, \bar{\zeta}_i(1) = 1, \bar{\zeta}_i'(0) = (-2(-1)^i i^2), i = 0,1,2,\cdots$. We solve the equation $\bar{\zeta}_{m+1-\rho}(\xi) = 0$ to find the roots and then set it as collocation points.

(5) Now, to evaluate the unknowns $ai(\eta)$, $i = 0, 1, \ldots, m$, we solve the system of the ordinary differential Eqs (4.21)–(4.24). To this end, we divide the interval $[0, T]$ into subintervals of equal length by the points $0 = \eta_0 \le \eta_1 \le \eta_2 \le \ldots \le \eta s = T$, $s = 0, 1, \ldots, N$, $\eta s = \tau s$, $\tau = T/N$ ($N \in \mathrm{N} := \{1,2,\cdots\}$) and set $a_i^s = a_i(\eta_s)$. Hence, the system (4.21–4.24) transforms to a set of nonlinear algebraic equations as:

$$
\begin{aligned}
&\sum_{i=0}^{m}\left(\frac{a_i^s - a_i^{s-1}}{\tau}\right)\bar{\zeta}_i(\xi_r) + \left(\sum_{i=0}^{m} a_i^s\bar{\zeta}_i(\xi_r)\right)\left(\sum_{i=\lceil\theta\rceil}^{m}\sum_{j=\lceil\theta\rceil}^{i} a_i^s\Upsilon_{i,j}^{\theta}(\xi_r)\right) \\
&\quad -\epsilon\left(\sum_{i=\lceil\rho\rceil}^{m}\sum_{j=\lceil\rho\rceil}^{i} a_i^s\Upsilon_{i,j}^{\rho}(\xi_r)\right) + \frac{1}{2}\left(\sum_{i=\lceil\vartheta\rceil}^{m}\sum_{j=\lceil\vartheta\rceil}^{i} a_i^s\Upsilon_{i,j}^{\vartheta}(\xi_r)\right) = 0
\end{aligned}
\tag{4.25}
$$

$$\sum_{i=0}^{m} \bar{\zeta}_i(0)a_i^s = B_1^s, \tag{4.26}$$

$$\sum_{i=0}^{m} \bar{\zeta}_i(1)a_i^s = B_2^s, \tag{4.27}$$

$$\sum_{i=0}^{m} \bar{\zeta}_i'(0)a_i^s = B_3^s. \tag{4.28}$$

(6) To convert the system (4.25)–(4.28) into matrix form, we apply the Newton iteration method with $m = 4$ to obtain

$$\Phi^{s+1} = \Phi^s - J^{-1}(\Phi^s)S(\Phi^s), \qquad \Phi^s = (a_0^s, a_1^s, a_2^s, a_3^s, a_4^s)^T, \tag{4.29}$$

where $S(\Phi s)$ is the vector which represents the nonlinear equations and $J^{-1}(\Phi s)$ is the inverse of the Jacobian matrix. To obtain the initial solution Φ^0, we set $s = 0$ and from the initial condition (4.15), we can get

(a) To obtain

$$\varphi(\xi,0) = f(\xi) \simeq \sum_{i=0}^{4} a_i(0)\bar{\zeta}_i(\xi). \tag{4.30}$$

substitute (4.19) into the initial condition (4.15).

(b) Collocate Eq. (4.30)

$$f(\xi_r) \simeq \sum_{i=0}^{4} a_i(0)\bar{\zeta}_i(\xi_r), r = 0,1,2,3,4, \tag{4.31}$$

where the ξr are the roots of $\bar{\zeta}_5(\xi)$. Now, by solving the linear system of Eq. (4.31), we obtain the components of the initial solution φ^0.

After obtaining the numerical approximate solutions to this system, we substitute it into (4.19) to get the numerical approximate solutions of Eq. (4.13).

Example 2

Generalized fractional Fisher equation (GFFE)

In this example, we consider the nonlinear generalized fractional Fisher equation in the form [47] in the Liouville–Caputo operator sense:

$$\begin{gathered}\varphi_\eta = D_\xi^\nu \varphi + \varphi(1-\varphi)(\varphi-\beta), 0 \le \eta \le T, 0 < \xi < 1, \\ (0 < \beta < 1, \quad 0 < \nu \le 2)\end{gathered} \tag{4.32}$$

The exact solution of (4.32) at the integer order is given by [47]:

$$\varphi(\xi,\eta) = \frac{1}{2}(1+\beta) + \frac{1}{2}(1-\beta)\tanh\left(\sqrt{\frac{1}{8}}(1-\beta)x + \frac{1}{4}(1-\beta^2)\eta\right). \tag{4.33}$$

We can get the boundary and initial conditions from the exact solution as:

$$\varphi(0,\eta) = g_1(\eta), \tag{4.34}$$

$$\varphi(1,\eta) = g_2(\eta), \tag{4.35}$$

$$\varphi(\xi,0) = \bar{u}(\xi). \tag{4.36}$$

Now, we approximate the function $\varphi(\xi,t)$ as:

$$\varphi_m(\xi,\eta) = \sum_{i=0}^{m} b_i(\eta)\bar{\zeta}_i(\xi). \tag{4.37}$$

and then we can solve Eq. (4.32) numerically as in Example 1, using the Chebyshev collocation method with FDM. By following the same steps through which the approximate solution was found, as in Example 1, we can obtain the following nonlinear algebraic equations:

$$\begin{aligned}\sum_{i=0}^{m}\left(\frac{b_i^s - b_i^{s-1}}{\tau}\right)\bar{\zeta}_i(\xi_r) - \left(\sum_{i=\lceil v\rceil}^{m}\sum_{j=\lceil v\rceil}^{i} b_i^s \Upsilon_{i,j}^{\kappa}(\xi_r)\right) \\ - \left(\sum_{i=0}^{m} b_i^s \bar{\zeta}_i(\xi_r)\right)\left(1 - \sum_{i=0}^{m} b_i^s \bar{\zeta}_i(\xi_r)\right) \\ \times\left(\sum_{i=0}^{m} b_i^s \bar{\zeta}_i(\xi_r) - \beta\right) = 0.\end{aligned} \tag{4.38}$$

Now, MRM can be used to find the approximate solutions to a set of non-linear algebraic equations (4.38). By substituting these solutions into (4.37) we obtain the numerical approximate solutions of (4.32).

4.3 Shifted Legendre Polynomials and LC-Fractional Derivatives

In this section, we introduce the spectral collocation method, but this time with the help of Legendre polynomials. We follow the same procedure as described for the Chebyshev polynomials.

4.3.1 Shifted Legendre Polynomials

Legendre polynomials can be defined over the interval $[-1,1]$ and determined through the following successive iterative [48]:

$$L_{n+1}(z)=\frac{2n+1}{n+1}zL_n(z)-\frac{n}{n+1}L_{n-1}(z), L_0(z)=1, L_1(z)=z, \quad (n=1,2,\cdots). \tag{4.39}$$

The analytic form of the Legendre polynomials $Ln(z)$ of degree n is given by

$$L_n(z)=\sum_{k=0}^{[\frac{n}{2}]}(-1)^k\frac{(2n-2k)!}{2^n k!(n-k)!(n-2k)!}z^{n-2k}, \tag{4.40}$$

In order to obtain the shifted Legendre polynomials on the interval $[0,1]$, we introduce the new variable $z = 2\xi - 1$. The Legendre polynomials thus shifted are defined as follows:

$$P_n(\xi)=L_n(2\xi-1)=L_{2n}\left(\sqrt{\xi}\right),$$

where the set

$$\left\{L_n(z) \quad \left(n\in\mathbb{N}_0=\{0,1,2,\cdots\}\right)\right\}$$

forms a family of orthogonal Legendre polynomials on the interval $[-1,1]$. The shifted Legendre polynomial $Pn(\xi)$ of degree n has the expansion given by [49].

$$P_n(\xi)=\sum_{k=0}^{n}\frac{(-1)^{n+k}\,(n+k)!}{(k!)^2(n-k)!}\xi^k \qquad (n\in\mathbb{N}_0), \tag{4.41}$$

so that, clearly, $P_0(\xi) = 1$, $P_1(\xi) = 2\xi - 1$, and so on. To derive approximate solutions, we can approximate the function $\Omega(\xi) \in L_2[0,1]$ as a linear combination of the following $(m + 1)$ terms of $Pn(\xi)$ given by (4.41):

$$\Omega(\xi)\simeq\Omega_m(\xi)=\sum_{i=0}^{m}a_iP_i(\xi), \tag{4.42}$$

where the coefficients ai are given by

$$a_i=(2i+1)\int_0^1\bar{\Theta}_i(\xi)\Omega(\xi)\mathrm{d}\xi \qquad (i\in\mathbb{N}_0=\{0,1,2,\cdots\}). \tag{4.43}$$

Now, we state the following useful theorem.

Theorem 3.1.

[50] *Let $\beta(t)$ be approximated by the shifted Legendre polynomials in* (42). *Suppose also that $\lambda > 0$. Then*

$$D^\lambda\beta(t)=\sum_{i=\lceil\lambda\rceil}^{m}\sum_{k=\lceil\lambda\rceil}^{i}a_iH_{i,k}^{(\lambda)}t^{k-\lambda}, \tag{4.44}$$

where

$$H_{i,k}^{(\lambda)}=\frac{(i+k)!\,\Gamma(k+1)}{(k!)^2\,(i-k)!\,\Gamma(k-\lambda+1)}.$$

4.3.2 Implementation of the Proposed Method and Examples Based on Shifted Chebyshev Polynomials

Example 3

Multi-space fractional Korteweg–de Vries equation (MSFKDVE)
The MSFKDVE is given by:

$$\varphi_t + \varphi D_\xi^\theta \varphi + \frac{1}{2} D_\xi^\rho \varphi = 0, \quad 0 \le t \le T, 0 < \xi < 1,, \quad (0 < \theta \le 1, 2 < \rho \le 3). \tag{4.45}$$

The initial and boundary conditions are given by

$$\varphi(\xi,0) = f(\xi), \tag{4.46}$$

$$\varphi(0,t) = B_1(t), \tag{4.47}$$

$$\varphi(1,t) = B_2(t), \tag{4.48}$$

$$\varphi_\xi(0,t) = B_3(t). \tag{4.49}$$

The KdV equation is a mathematical model of waves on shallow water surfaces. The soliton solution of (4.45) in the classical differential, according to [51, 52], is given by

$$\varphi(\xi,t) = 6\sigma^2 sech^2(\sigma\xi - 2\sigma^3 t), \tag{4.50}$$

where $6\sigma^2$ amplitude and $\sigma > 0$. Now, we approximate the solution of (4.45) by taking $\varphi(\xi,\eta)$ as

$$\varphi_m(\xi,\eta) = \sum_{i=0}^{m} b_i(\eta) P_i(\xi), \tag{4.51}$$

and then we can solve Eq. (4.45) numerically as in Example 1, using the Legendre collocation method with FDM.

Substitute from Eqs (4.44) and (4.51) in Eq. (4.45) to get:

$$\sum_{i=0}^{m}\frac{dc_i(\eta)}{dt}P_i(\xi)+\left(\sum_{i=0}^{m}c_i(\eta)P_i(\xi)\right)\left(\sum_{i=\lceil\theta\rceil}^{m}\sum_{j=\lceil\theta\rceil}^{i}c_i(\eta)H_{i,j}^{\theta}(\xi)\right)+\frac{1}{2}\left(\sum_{i=\lceil\rho\rceil}^{m}\sum_{j=\lceil\rho\rceil}^{i}c_i(\eta)H_{i,j}^{\rho}(\xi)\right)=0. \tag{4.52}$$

We treat this system as in Example 1, so we can obtain the nonlinear algebraic equations as

$$\sum_{i=0}^{m}\left(\frac{c_i^s-c_i^{s-1}}{\tau}\right)P_i(\xi_r)+\left(\sum_{i=0}^{m}c_i^sP_i(\xi_r)\right)\left(\sum_{i=\lceil\theta\rceil}^{m}\sum_{j=\lceil\theta\rceil}^{i}c_i^sH_{i,j}^{\theta}(\xi)\right)+\frac{1}{2}\left(\sum_{i=\lceil\rho\rceil}^{m}\sum_{j=\lceil\rho\rceil}^{i}c_i^sH_{i,j}^{\rho}(\xi)\right)=0. \tag{4.53}$$

In light of the system (4.53) and using NRM, we can get the approximate numerical solutions of (4.45).

Example 4

Multi-space fractional Kuramoto–Sivashinsky (MSFKS)

The generalized Kuramoto–Sivashinsky (GKS) equation is a model of nonlinear partial differential equation (NLPDE) frequently encountered in the study of continuous media which exhibits a chaotic behavior form [53]. We can rewrite GKS by replacing the classical derivative with fractional derivative in the sense of the Liouville–Caputo operator.

$$\varphi_\eta+\varphi D_\xi^\theta\varphi+\alpha D_\xi^\rho\varphi+\beta D_\xi^\vartheta\varphi=0 \qquad 0<\xi<1,\quad 0\le t\le T, \tag{4.54}$$
$$(0<\theta\le 1, 0<\rho\le 2, 0<\vartheta\le 4, \alpha>0, \beta>0).$$

The Kuramoto–Sivashinsky is one of the very important equations that arise in a variety of physical contexts, such as long waves on the interface between two viscous fluids, long waves on thin films [54], flame front instability [55], unstable drift waves in plasmas, and reaction diffusion systems [56]. It represents models of pattern formation on unstable flame fronts and thin hydrodynamic films [57]. The GKS has thus been studied extensively [58, 59]. In the case of $\theta = 1$, $\rho = 2$ and $\vartheta = 4$, the exact solution is given by [60, 61]

$$\varphi(\xi,\eta) = c + \frac{15}{19}\sqrt{\frac{11}{19}}\left(11\ \tanh^3\left(K(\phi)\right) - 9\tanh\left(K(\phi)\right)\right), \tag{4.55}$$

where $\phi = -c\eta + \xi - \xi_0$.

The initial and boundary conditions of GKS are given by

$$\varphi(\xi,0) = f(\xi), \quad 0 < \xi < 1, \tag{4.56}$$

$$\varphi(0,t) = h_1(\xi), \tag{4.57}$$

$$\varphi(1,t) = h_2(\xi), \tag{4.58}$$

$$\varphi_\xi(0,t) = h_3(\xi), \tag{4.59}$$

$$\varphi_\xi(1,t) = h_4(\xi). \tag{4.60}$$

We approximate the function $\varphi(\xi,\eta)$ as

$$\varphi_m(\xi,\eta) = \sum_{i=0}^{m} b_i(t)P_i(\xi), \tag{4.61}$$

and then we can solve Eq. (4.45) numerically as in Examples 1 and 2, using the Legendre collocation method with FDM. To transform (4.55) into the set of differential equations, we substitute (4.44) and (4.61) into Eq. (4.54) as

$$\begin{aligned}&\sum_{i=0}^{m}\frac{dc_i(t)}{dt}\bar{\zeta}_i(\xi) + \left(\sum_{i=0}^{m} d_i(t)\bar{\zeta}_i(\xi)\right)\left(\sum_{i=\lceil\theta\rceil}^{m}\sum_{j=\lceil\theta\rceil}^{i} d_i(t)H_{i,j}^{\theta}(\xi)\right)\\&+\alpha\left(\sum_{i=\lceil\rho\rceil}^{m}\sum_{j=\lceil\rho\rceil}^{i} d_i(t)H_{i,j}^{\rho}(\xi)\right) + \beta\left(\sum_{i=\lceil\vartheta\rceil}^{m}\sum_{j=\lceil\vartheta\rceil}^{i} d_i(t)H_{i,j}^{\vartheta}(\xi)\right) = 0.\end{aligned} \tag{4.62}$$

By following the same treatment as in Example 1, the following nonlinear algebraic equations can be obtained:

$$\sum_{i=0}^{m}\left(\frac{d_i^s-d_i^{s-1}}{\tau}\right)P_i(\xi_r)+\left(\sum_{i=0}^{m}d_i^s P_i(\xi_r)\right)\left(\sum_{i=\lceil\theta\rceil}^{m}\sum_{j=\lceil\theta\rceil}^{i}c_i^s H_{i,j}^{\theta}(\xi_r)\right)$$
$$+\alpha\left(\sum_{i=\lceil\rho\rceil}^{m}\sum_{j=\lceil\rho\rceil}^{i}d_i^s H_{i,j}^{\rho}(\xi_r)\right)+\beta\left(\sum_{i=\lceil\vartheta\rceil}^{m}\sum_{j=\lceil\vartheta\rceil}^{i}d_i^s H_{i,j}^{\vartheta}(\xi_r)\right)=0. \quad (4.63)$$

Finally, in this example, we follow the same technique as in the previous examples, using NRM, to solve the system (4.63) and finally find the approximate numerical solutions to Eq. (4.54).

4.4 Numerical Results and Discussion

In this section we present the numerical results of all the previous examples. The strategy followed is to compare the numerical and the exact solutions in the classical case, calculate the absolute error and represent it graphically. In the case of the fractional differential equations with non-integer order, we compute the residual error function, and represent it graphically. For Example 1 (a), we compare the numerical solution of (4.13) with the exact solution (4.18) for $\theta = 1$, $\rho = 2$, $\vartheta = 3$, $\epsilon = 3$, $m = 4$, and $\tau = 10^{-6}$ in Figure 4.1.

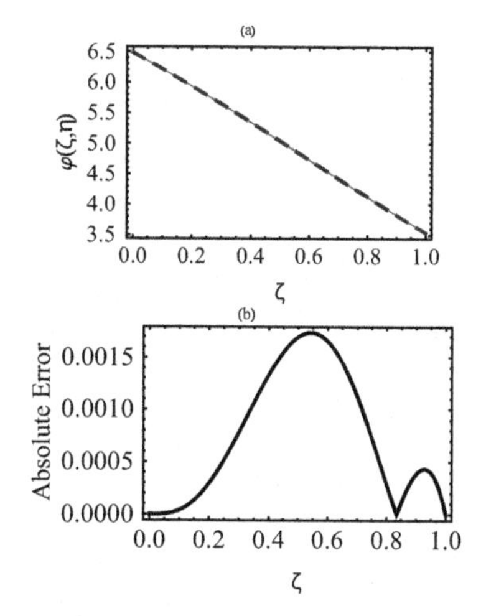

FIGURE 4.1
(a) Graph of the comparison between exact solution and numerical solution of (4.13) for $\theta = 1$, $\rho = 2$, $\vartheta = 3$, $\epsilon = 3$, $m = 4$, and $\tau = 10^{-6}$. (b) The absolute error between exact solution and numerical solutions for the same parameters. (Dashed blue line: numerical solution; solid red line: exact solution.)

The dashed blue line represents the numerical solution and the solid red line the exact solution. We adopt this procedure for most of the cases we discuss where results are compared in the classical case. In part (b) of the first example, the absolute error for the same parameters of (b) is calculated and presented graphically. In Figure 4.2, the residual error function of (4.13) is calculated for the values $\theta = 0.7$, $\rho = 1.7$, $\vartheta = 2.7$, $\epsilon = 0.1$, $m = 4$, and $\tau = 10^{-4}$. From these figures, we find that the amount of error is very small, and greater accuracy can be obtained when more terms are calculated. The error order ranges from 10^{-4} to 10^{-6}. In Figure 4.3(a)–(b), the exact solution (4.33) is

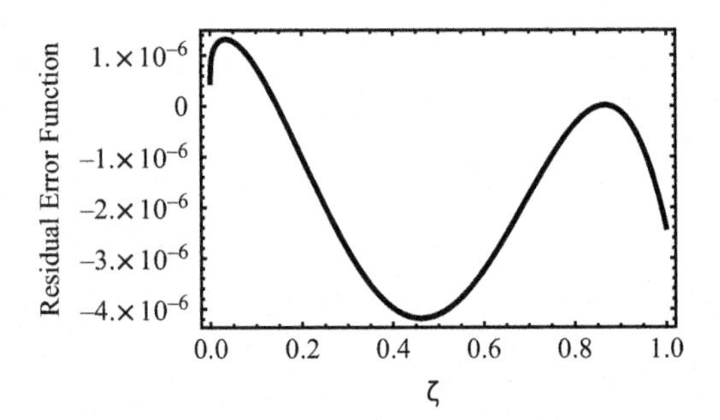

FIGURE 4.2
Graph of the residual error function of (4.13) for $\theta = 0.7$, $\rho = 1.7$, $\vartheta = 2.7$, $\epsilon = 0.1$, $m = 4$, and $\tau = 10^{-4}$.

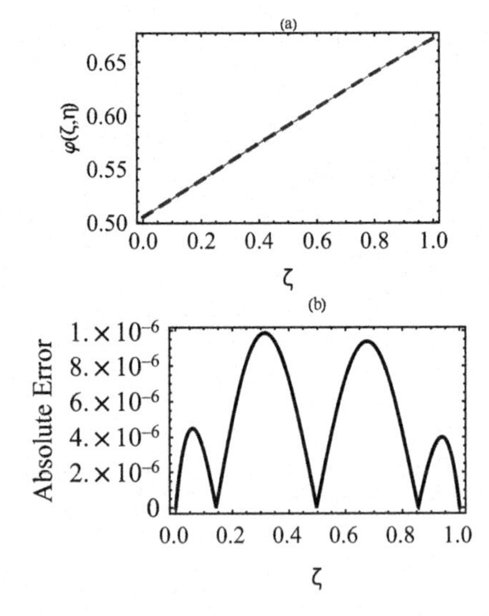

FIGURE 4.3
(a) Graph of the comparison between exact solution and numerical solution of (4.32) for $\nu = 2$, $\beta = 0.01$, $m = 4$, and $\tau = 10^{-6}$. (b) The absolute error between exact solution and the numerical solutions for the same parameters. (Dashed blue line: numerical solution; solid red line: exact solution.)

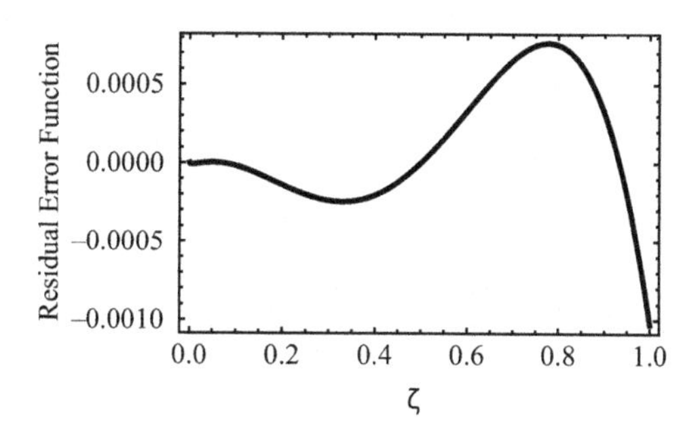

FIGURE 4.4
Graph of the residual error function of (4.32) for $\nu = 1.7$, $\beta = 0.3$, $m = 4$, and $\tau = 10^{-6}$.

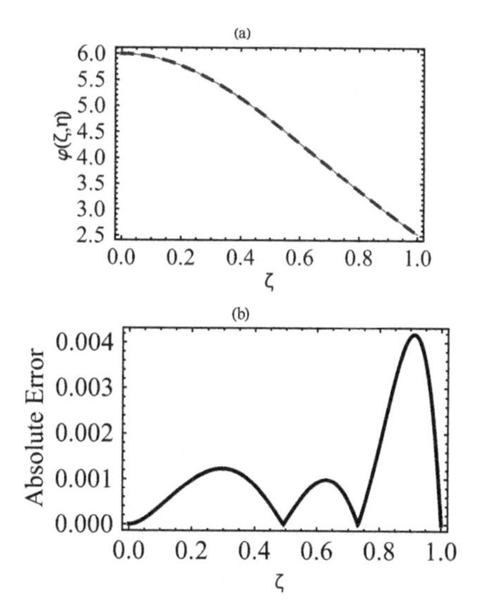

FIGURE 4.5
(a) Graph of the comparison between exact solution and numerical solution of (4.55) for $\theta = 1$, $\rho = 3$, $\sigma = 1$, $m = 5$, and $\tau = 10^{-5}$. (b) The absolute error between exact solution and the numerical solutions for the same parameters. (Dashed blue line: numerical solution; solid red line: exact solution.)

represented with an approximate solution of Eq. (4.32) for the parameters $\nu = 2$, $\beta = 0.01$, $m = 4$, and $\tau = 10^{-6}$. The absolute error is also calculated for the same parameters as in Figure 4.2 (a). In Figure 4.4, the residual error function is calculated for $\nu = 1.7$, $\beta = 0.3$, $m = 4$, and $\tau = 10^{-6}$. These discussions represent the numerical results in the presence of Chebyshev polynomials. The results are accurate and highly efficient.

Now we move on to the second part in the presence of Legendre polynomial functions. In Figure 4.5(a), the exact and numerical solutions of

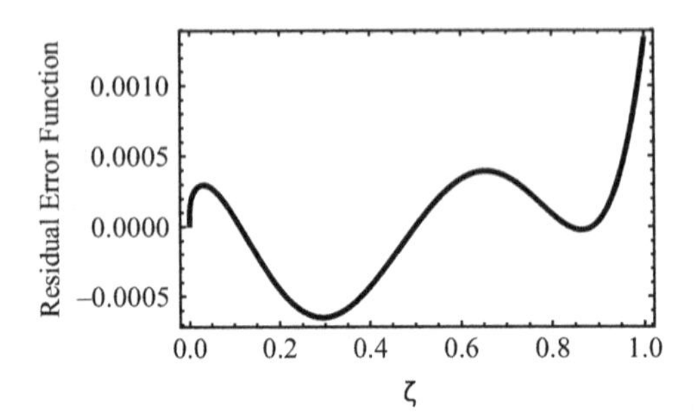

FIGURE 4.6
Graph of the residual error function of (4.55) for $\theta = 0.8$, $\rho = 2.8$, $\sigma = 0.25$, $m = 5$, and $\tau = 10^{-5}$.

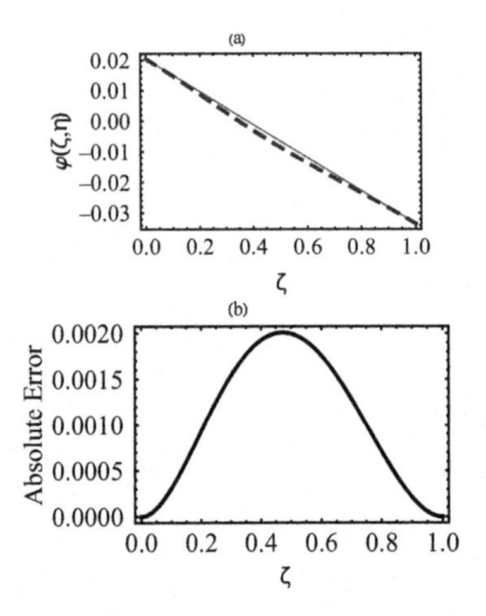

FIGURE 4.7
(a) Graph of the comparison between exact solution and numerical solution of (4.13) for $\theta = 1$, $\rho = 2$, $\vartheta = 3$, $\theta = 0.2$, $m = 4$, and $\tau = 10^{-6}$. (b) The absolute error between exact solution and the numerical solutions for the same parameters. (Dashed blue line: numerical solution; solid red line: exact solution.)

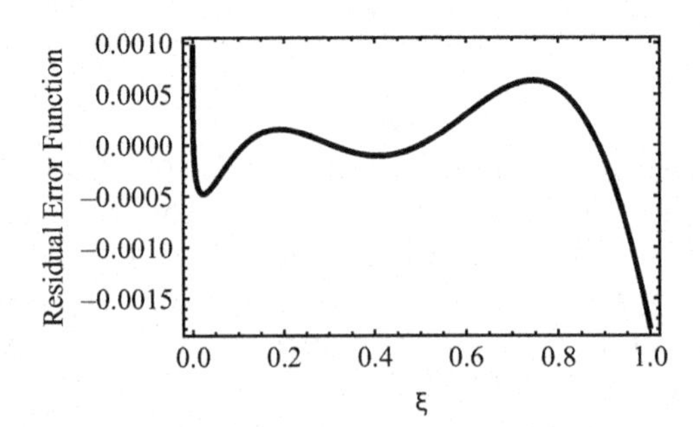

FIGURE 4.8
Graph of the residual error function of (4.13) for $\theta = 1$, $\rho = 2$, $\vartheta = 3$, $\theta = 0.2$, $m = 4$, and $\tau = 10^{-6}$.

Eq. (4.45) are compared for $\theta = 1$, $\rho = 2$, $\vartheta = 3$, $\theta = 0.2$, $m = 4$, and $\tau = 10^{-6}$. For the same parameters the absolute error between these solutions is represented in Figure 4.5(b). In Figure 4.6 the residual error function of Eq. (4.45) is represented for the values $\theta = 0.8$, $\rho = 2.8$, $\sigma = 0.25$, $m = 5$, and $\tau = 10^{-5}$. Also, here we find that the error order is 10^{-3}. In the last example, we follow the previous procedures. In Figure 4.7, the exact and approximate solutions of Eq. (4.54) are combined, and the absolute error between them is represented. In Figure 4.8, the residuals error function of the equation is represented for the parameters $\theta = 1$, $\rho = 2$, $\vartheta = 3$, $\theta = 0.2$, $m = 4$, and $\tau = 10^{-6}$.

4.5 Conclusion

In this chapter we presented a systematic study of the spectral collocation of models of fractional partial differential equations with the help of the Chebyshev and Legendre polynomials. We converted the models to involve a set of ordinary differential equations which we then solved by the finite difference method. Finally, we applied the Newton–Raphson method to find the numerical solutions for a set of nonlinear algebraic equations. In all examples, we verified the accuracy and effectiveness by comparing the exact and numerical solutions with integer order; for non-integer we computed the residual error function. In all cases, the errors are very small, and, by increasing the terms, we can reduce the error order.

In the future we can extend the study based on other special functions and different fractional operators (see, for details, [62] and [63]).

References

[1] K. B. Oldham, J. Spanier, *The Fractional Calculus*, Academic Press, New York, NY, 1974.

[2] I. Podlubny, *Fractional Differential Equations: An Introduction to Fractional Derivatives, Fractional Differential Equations, to Methods of Their Solution and Some of Their Applications. Mathematics in Science and Engineering*, Vol. 198, Academic Press, New York, London, Sydney, Tokyo and Toronto, 1999.

[3] A. A. Kilbas, H. M. Srivastava, J. J. Trujillo, *Theory and Applications of Fractional Differential Equations*. North-Holland Mathematical Studies, Vol. 204, Elsevier (North-Holland) Science Publishers, Amsterdam, London and New York, 2006.

[4] R. Jiwari, J. Yuan, A computational modeling of two dimensional reaction–diffusion Brusselator system arising in chemical processes, *Journal of Mathematical Chemistry* 52, (2014), 1535–1551.

[5] R. C. Mittal, R. Jiwari, Numerical solution of two-dimensional reaction–diffusion Brusselator system, *Applied Mathematics and Computation* 217(12), (2011), 5404–5415.

[6] R. Jiwari, A hybrid numerical scheme for the numerical solution of the Burgers' equation, *Computer Physics Communications* 188, (2015), 59–67.

[7] R. Jiwari, Haar wavelet quasilinearization approach for numerical simulation of Burgers' equation, *Computer Physics Communications* 183, (2012), 2413–2423.

[8] R. J. Iwari, R. C. Mittal, K. K. Sharma, A numerical scheme based on weighted average differential quadrature method for the numerical solution of Burgers' equation, *Applied Mathematics and Computation* 219, (2013), 6680–6691.

[9] H. M. Srivastava and M. M. Khader, Numerical simulation for the treatment of nonlinear predator-prey equations by using the finite element optimization method, *Fractal and Fractional* 5 (2021), Article ID 56, 1–9.

[10] O. P. Yadav, R. Jiwari, Finite element analysis and approximation of Burger's-Fisher equation, *Numerical Methods for Partial Differential Equations* 33(5), (2017), 1652–1677.

[11] J. Singh, D. Kumar, R. Swroop, S. Kumar, An Efficient computational approach for time-fractional Rosenau-Hyman equation, *Neural Computing and Applications* 30(6), (2017), 1–10.

[12] J. Singh, D. Kumar, D. Baleanu, S. Rathore, On the local fractional wave equation in fractal strings, *Mathematical Methods in the Applied Sciences* 42(5), (2019), 1588–1595.

[13] J. Singh, D. Kumar, D. Baleanu, S. Rathore, An efficient numerical algorithm for the fractional Drinfeld-Sokolov-Wilson equation, *Applied Mathematics and Computation* 335, (2018), 12–24.

[14] J. Singh, D. Kumar, D. Baleanu, On the analysis of fractional diabetes model with exponential law, *Adv. Difference Equ.* 2018, (2018), 231.

[15] M. M. Khader, K. M. Saad, A numerical approach for solving the problem of biological invasion (fractional Fisher equation) using Chebyshev spectral collocation method, *Chaos, Solitons & Fractals* 110, (2018), 169–177.

[16] M. M. Khader, K. M. Saad, On the numerical evaluation for studying the fractional KdV, KdV-Burger's, and Burger's equations, *The European Physical Journal Plus* 133(335), (2018), 1–13.

[17] M. M. Khader, K. M. Saad, A numerical study using Chebyshev collocation method for a problem of biological invasion: fractional Fisher equation, *International Journal of Biomathematics* 11(8), (2018), 1–15.

[18] K.M. Saad, E. H. F. Al-Shareef, A. K. Alomari, D. Baleanu, J. F. Gómez-Aguilar, On exact solutions for time-fractional Korteweg-de Vries and Korteweg-de Vries-Burgers equations using homotopy analysis transform method, *Chinese Journal of Physics* 63, (2020), 149–162.

[19] K. M. Saad, A reliable analytical algorithm for space-time fractional cubic isothermal autocatalytic chemical system, *Pramana-Journal of Physics* 91(4), (2018), 1–15.

[20] K. M. Saad, E. H. F. Al-Sharif, Comparative study of a cubic autocatalytic reaction via different analysis methods, *Discrete and Continuous Dynamical Systems-S* 12(3), (2019), 665–684.

[21] K. M. Saad, S. Deniz, D. Baleanu, On a new modified fractional analysis of Nagumo equation, *International Journal of Biomathematics* 12(3) (2019), 1950034.
[22] H. M. Srivastava, K. M. Saad, Some new and modified fractional analysis of the time-fractional Drinfeld-Sokolov-Wilson system, *Chaos: An Interdisciplinary Journal of Nonlinear Science* 30(11), (2020), Article ID 113104, 1–10.
[23] H. M. Srivastava, S. Deniz, K. M. Saad, An efficient semi-analytical method for solving the generalized regularized long wave equations with a new fractional derivative operator, *Journal of King Saud University Science* 33, (2021), Article ID 101345, 1–7.
[24] H. M. Srivastava, K. M. Saad, A comparative study of the fractional-order clock chemical model, *Mathematics* 8(9), (2020), Article ID 1436, 1–14.
[25] H. M. Srivastava, K. M. Saad, J. F. Gómez-Aguilar, A. A. Almadiy, Some new mathematical models of the fractional-order system of human immune against IAV infection, *Mathematical Biosciences and Engineering* 17(5), (2020), 4942–4969.
[26] H. M. Srivastava, K. M. Saad and M. M. Khader, An efficient spectral collocation method for the dynamic simulation of the fractional epidemiological model of the Ebola virus, *Chaos, Solitons & Fractals* 140, (2020), Article ID 110174, 1–7.
[27] H. Singh, H.M. Srivastava, D. Kumar, A reliable algorithm for the approximate solution of the nonlinear Lane-Emden type equations arising in astrophysics, *Numerical Methods for Partial Differential Equations* 34(5), (2018), 1524–1555.
[28] H. Singh, H.M. Srivastava, D. Kumar, A reliable numerical algorithm for the fractional vibration equation, *Chaos, Solitons and Fractals* 103, (2017), 131–138.
[29] H. Singh, H. M. Srivastava, Jacobi collocation method for the approximate solution of some fractional-order Riccati differential equations with variable coefficients, *Physica A* 523, (2019), 1130–1149.
[30] H. Singh, R. K. Pandey, H. M. Srivastava, Solving non-linear fractional variational problems using Jacobi polynomials, *Mathematics* 7(3), (2019), 224.
[31] H. Singh, H. M. Srivastava, Numerical simulation for fractional-order Bloch equation arising in nuclear magnetic resonance by using the Jacobi polynomials, *Applied Sciences* 10(8), (2020), 2850.
[32] H. Singh, H. M. Srivastava, Numerical investigation of the fractional-order Liénard and Duffing Equations arising in oscillating circuit theory, *Frontiers in Physics* 8, (2020), 120.
[33] K. M. Saad, Comparative study on fractional isothermal chemical model, *Alexandria Engineering Journal* 60, (2021), 3265–3274.
[34] C. Canuto, M. Y. Hussaini, A. Quarteroni, T. A. Zang, Spectral methods in fluid dynamics, Technical report, Springer, 1988.
[35] A. Doha, A. H. Bhrawy, S. S. Ezz-Eldien, Efficient Chebyshev spectral methods for solving multi-term fractional orders differential equations, *Applied Mathematical Modelling* 35(12), (2011), 5662–5672.
[36] S. Esmaeili, M. Shamsi. A pseudo-spectral scheme for the approximate solution of a family of fractional differential equations, *Communications in Nonlinear Science and Numerical Simulation* 16(9), (2011), 3646–3654.
[37] S. Kumar, R. K. Pandey, H. M. Srivastava, G. N. Singh, A convergent collocation approach for generalized fractional integro-differential equations using Jacobi poly-fractonomials, *Mathematics* 9(2021), Article ID 979, 1–17.

[38] H. M. Srivastava, Fractional-order derivatives and integrals: Introductory overview and recent developments, *Kyungpook Mathematical Journal* 60, (2020), 73–116.
[39] D. Baleanu, B. Shiri, H. M. Srivastava, M. Al Qurashi, A Chebyshev spectral method based on operational matrix for fractional differential equations involving non-singular Mittag-Leffler kernel, *Adv. Difference Equ.* 2018 (2018), Article ID 353, 1–23.
[40] M. A. Snyder, *Chebyshev Methods in Numerical Approximation*, Prentice-Hall, Inc., Englewood Cliffs, (1966).
[41] M. M. Khader, On the numerical solutions for the fractional diffusion equation, *Communications in Nonlinear Science and Numerical Simulation* 16, (2011), 2535–2542.
[42] M. M. Khader, M. M. Babatin, Numerical treatment for solving fractional SIRC model and influenza A, *Computational and Applied Mathematics* 33(3), (2014), 543–556.
[43] R. S. Johnson, A non-linear equation incorporating damping and dispersion, *Journal of Fluid Mechanics* 42, (1970), 49–60.
[44] W. L. Kath, N. F. Smyth, Interaction of soliton evolution and radiation loss for the Korteweg-de Vries equation, *Physical Review E* 51, (1995), 661–670.
[45] Z. Feng, Travelling wave solutions and proper solutions to the two-dimensional Burger's-Korteweg de Vries equation, *Journal of Physics A* 36, (2003), 8817–8827.
[46] Z. Feng, On travelling wave solutions of the KdV, *Nonlinearity* 20, (2007), 343–356.
[47] A. H. Bhrawy, M. A. Alghamdi, Approximate solutions of Fishers type equations with variable coefficients, *Abstract and Applied Analysis* 2013. doi: 10.1155/2013/176730, 2013.
[48] W. W. Bell, *Special Functions for Scientists and Engineers*, Great Britain, Butler and Tanner Ltd., Frome and London, 1968.
[49] N. N. Lebedev, *Special Functions and Their Applications* (Translated from the Russian by R. A. Silverman), Prentice-Hall Incorporated, Englewood Cliffs, New Jersey, 1965; Dover Publications, New York, 1972.
[50] M. M. Khader, A. S. Hendy, The approximate and exact solutions of the fractional-order delay differential equations using Legendre pseudo-spectral method, *Indian Journal of Pure and Applied Mathematics* 74, (2012), 287–297.
[51] M. M. Khader, K. M. Saad, On the numerical evaluation for studying the fractional KdV, KdV-Burger's, and Burger's equations, *The European Physical Journal Plus* 133, (2018), 1–13.
[52] N. J. Zabusky, M. D. Kruskal, Interaction of "Solitons" in a collisionless plasma and the recurrence of initial states, *Physical Review Letters* 15, (1965), 240–243.
[53] A. H. Khater, R. S. Temsah, Numerical solutions of the generalized Kuramoto–Sivashinsky equation by Chebyshev spectral collocation methods, *Computers & Mathematics with Applications* 56, (2008), 1465–1472.
[54] A. P. Hooper, R. Grimshaw, Persistent propagation of concentration waves in dissipative media far from thermal equilibrium, *Physics of Fluids* 28, (1985), 37–45.

[55] Y. Kuramoto, T. Tsuzuki, Persistent propagation of concentration waves in dissipative media far from thermal equilibrium, *Progress of Theoretical and Experimental Physics* 55, (1976), 356.

[56] J. Rademacher, R. Wattenberg, Viscous shocks in the destabilized Kuramoto–Sivashinsky, *Journal of Computational and Nonlinear Dynamics* 1, (2006) 336–347.

[57] G. I. Sivashinsky, Instabilities,pattern-formation,and turbulence in flames, *Annual Review of Fluid Mechanics* 55, (1983), 179–199.

[58] X. Liu, Gevrey class regularity and approximate inertial manifolds for the Kuramoto–Sivashinsky equation, *Physica D* 50, (1991), 135–151.

[59] R. Grimshaw, A. P. Hooper, The non-existence of a certain class of travelling wave solutions of the Kuramoto–Sivashinsky equation, *Physica D* 50, (1991), 231–238.

[60] D. Baldwin, U. Goktas, W. Hereman, L. Hong, R.S. Martino, J. C. Miller, Symbolic computation of exact solutions expressible in hyperbolic and elliptic functions for nonlinear PDEs, *Journal of Symbolic Computation* 37, (2004) 669–705.

[61] E. J. Parkes, B. R. Duffy, An automated tanh-function method for finding solitary wave solutions to non-linear evolution equations, *Computer Physics Communications* 98(3), (1996), 288–300.

[62] H. M. Srivastava, Some parametric and argument variations of the operators of fractional calculus and related special functions and integral transformations, *Journal of Nonlinear Convex Analysis* 22 (2021), 1501–1520.

[63] H. M. Srivastava, An introductory overview of fractional-calculus operators based upon the Fox-Wright and related higher transcendental functions, *Journal of Advanced Engineering and Computation* 5 (2021), 135–166.

5

On the Wave Properties of the Conformable Generalized Bogoyavlensky–Konopelchenko Equation

Haci Mehmet Baskonus
Harran University, Sanliurfa, Turkey

Mine Senel
Mugla Sitki Kocman University, Mugla,Turkey

Ajay Kumar
Bakhtiyarpur College of Engineering, Patna, India

Gulnur Yel
Final International University, Kyrenia, Turkey

Bilgin Senel
Mugla Sitki Kocman University, Mugla, Turkey

Wei Gao
Yunnan Normal University, Yunnan, China

CONTENTS

DOI: 10.1201/9781003263517-5

5.1 Introduction

Soliton theory is found in different fields of nonlinear science and engineering, as it arises in a wide range of real-world problems. With the development of computational programs, these problems are being investigated in depth by experts from all over the world, in terms of time and space. To solve such models, experts have developed various methods, including the auxiliary equation method, the Cole–Hopf transformation, the Hirota bilinear method, the simple equation method, the modified Kudryashov method, the first integral method, the variable separatedmethodand the exp-$(-\varphi(\eta))$ function method [1–8]. More recently, many operators investigating deeper properties of these problems have been introduced to the literature. One of them is the conformable operator, which overcomes some of the limitations of traditional concepts. The conformable derivative which satisfies the basic properties of classical calculus is proposed by Khalil et al. [9]. Some new properties of the conformable operator, such as the conformable gradient vector, the conformable divergence theorem, and Clairaut's theorem, are submitted in [9, 10]. Soliton solutions such as topological, compound singular of a nonlinear wave model in nonhomogeneous Murnaghan's rod, have been observed using the extended sinh-Gordon equation expansion method and the modified exp(*phi*))-expansion function method [11, 12]. The linear and nonlinear wave theories for the sine-Gordon equation are proposed by Yan in [13]. Tumor-immune system interaction with fractional order is considered and dynamical chaotic behaviors are observed in [14]. Optical and other solutions for the conformable space-time fractional Fokas–Lenells equation is proposed in [15]. (G'/G^2) expansion method is used to find the exact solution of some space-time fractional evolution equations by Yaslan et al. [16]. The (2+1)-dimensional Calogero–Bogoyavlenskii–Schi and the Kadomtsev–Petviashvili hierarchy equations are studied for new solitary wave solutions in [8], [17]. A fractional epidemiological model which describes computer viruses with any order having a non-singular kernel is considered using the Caputo–Fabrizio derivative as a numerical operator [18]. The (2+1)-dimensional generalized Bogoyavlensky–Konopelchenko equation for Hirota's bilinear form is proposed by Pouyanmehr et al. [19]. The Atangana–Baleanu derivative operatorwith the use of q-HATM describes a model for fatal diseases in pregnant women [20]. M-fractional solitons and periodic wave solutions of the Hirota–Maccari system are submitted in [21]. Some periodic solutions were submitted to the (3+1)-dimensional mKdV-ZK with a conformable sense in [22]. Some real-world problems with various operators are proposed in [23]. The existence of complex combined dark-bright soliton solutions of the conformable (2+1)-dimensional water wave equation with a perturbation parameter is shown in [24]. There are many relevant studies in the literature, one can see in [25–37].

We consider the conformable time-fractional (2+1)-generalized Bogoyavlensky–Konopelchenko equation (GBKE) defined as [19]

$$\frac{\partial^\theta}{\partial t^\theta}\left(\frac{\partial u}{\partial x}\right)+\alpha\left(6\frac{\partial u}{\partial x}\frac{\partial^2 u}{\partial x^2}+\frac{\partial^4 u}{\partial x^4}\right)+\beta\left(\frac{\partial^4 u}{\partial x^3\partial y}+3\frac{\partial u}{\partial x}\frac{\partial^2 u}{\partial x\partial y}+3\frac{\partial^2 u}{\partial x^2}\frac{\partial u}{\partial y}\right)$$
$$+\gamma_1\frac{\partial^2 u}{\partial x^2}+\gamma_2\frac{\partial^2 u}{\partial x\partial y}+\gamma_3\frac{\partial^2 u}{\partial y^2}=0, \tag{5.1}$$

where $\theta \in (0,1]$, $\alpha, \beta, \gamma_1, \gamma_2$ and γ_3 are constants. Eq. (5.1) is used to explain the process of phase dissociation in iron alloy, and is generally used in solidification and nucleation problems.This paper is organized as follows. Some definitions and properties of the conformable derivative are given in the second section. The main structure of the SGEM is described in the third section. We apply the SGEM to the GBKE in the fourth section. Figures illustrating the solutions and their physical meanings are shown in the fifth section fifth. The sixth and final section contains outcomes and results of the schemes under consideration.

5.2 Some Remarks on Conformable

Definition: ([9, 10]) Given a function $h : [0,\infty) \to R$. Then the conformable fractional derivative of h order α is defined by

$$(T_\alpha h)(h)=\lim_{\varepsilon\to 0}\frac{h\left(t+\varepsilon t^{1-\alpha}\right)-h(t)}{\varepsilon}\text{ for all } t>0, \alpha\in(0,1].$$

Definition: ([9, 10]) Let consider $h : [a,\infty) \to R$. Then the conformable (left) fractional derivative of h order α is defined by

$$(T_\alpha h)(h)=\lim_{\varepsilon\to 0}\frac{h(t+\varepsilon t^{1-\alpha})-h(t)}{\varepsilon}\text{ for all } t>0,\ \alpha\in(0,1]$$

Definition: ([9, 10]) Given a function $h : [a,\infty) \to R$. Let $n < \alpha \leq n + 1$ and $\beta = \alpha - n$. Then the conformable (left) fractional derivative of h order α, where $h^{(n)}(t)$ is defined by

$$\left(T_\alpha^a h\right)(h)=\left(T_\beta^a h^{(n)}\right)(t).$$

Theorem(Chain Rule): ([9, 10]) Let $h, g : (a, \infty) \to R$ be (left) α differentiable functions, where $0 < \alpha \le 1$. Let $k(t) = h(g(t))$. Then $k(t)$ is (left) α differentiable and for all t with $t \neq a$ and $g(t) \neq 0$ we have,

$$\left(T_\alpha^a k\right)(t) = \left(T_\alpha^a h\right)(g(t)).\left(T_\alpha^a g\right)(t).g(t)^{\alpha-1}.$$

Theorem: ([9, 10]) Let L_α be the derivative operator with order α and $\alpha \in (0,1)$ and h, k be α-differentiable at a point $t > 0$. We then have the following

i. $L_\alpha(ah + bk) = aL_\alpha(h) + bL_\alpha(k), \forall\, a, b \in R$.
ii. $L_\alpha(t^p) = pt^{p-\alpha}, \forall\, p \in R$.
iii. $L_\alpha(hk) = hL_\alpha(g) + kL_\alpha(f)$.
iv. $L_\alpha\left(\frac{h}{k}\right) = \frac{kL_\alpha(h) - hL_\alpha(k)}{k^2}$.
v. $L_\alpha(\lambda) = 0$, for all constant functions $h(t) = \lambda$.
vi. If h is differentiable then $L_\alpha(h)(t) = t^{1-\alpha}\frac{dh}{dt}(t)$.

5.3 General Properties of the SGEM

Considering the sine-Gordon equation

$$u_{xx} - u_{tt} = m^2 \sin(u), \tag{5.2}$$

where $u = u(x,t)$ and m is a constant. $\xi = \mu(x - ct)$ wave transform turns the above PDE into the following ODE

$$U'' = \frac{m^2}{\mu^2\left(1-c^2\right)}\sin(U), \tag{5.3}$$

where $U(\xi) = u(x,t)$ and ξ, and c are the amplitude and velocity of the traveling wave, respectively. Eq. (5.3) can be reduced into first order ODE

$$\left(\left(\frac{U}{2}\right)'\right)^2 = \frac{m^2}{\mu^2\left(1-c^2\right)}\sin^2\left(\frac{U}{2}\right) + C, \tag{5.4}$$

where C is the integration constant. If we substitute $C=0$, $\frac{U}{2}=w(\xi)$ and $a^2=\frac{m^2}{\mu^2(1-c^2)}$, in the above equation, it gives

$$w'=\frac{dw}{d\xi}=\sin(w) \tag{5.5}$$

with $w = w(\xi)$ *and* $a = 1$. By solving Eq. (5.5) by separation of variable and doing some simplification, two interesting relations can be verified:

$$\sin(w)=\sin\left[w(\xi)\right]=\left.\frac{2pe^{\xi}}{p^2e^{2\xi}+1}\right|_{p=1}=sech(\xi), \tag{5.6}$$

$$\cos(w)=\cos\left[w(\xi)\right]=\left.\frac{p^2e^{2\xi}-1}{p^2e^{2\xi}+1}\right|_{p=1}=\tanh(\xi), \tag{5.7}$$

where $p \neq 0$ is the integration constant.

Let the nonlinear partial differential equation for which we seek the solution be seen as the following form:

$$P(u,u_t,u_x,u_{tt},u_{tx},\ldots.)=0.$$

Applying $u=u(x,t)=U(\xi)$, $\xi=\mu(x-ct)$ produces nonlinear ordinary differential equation (NODE)

$$N(U,U',U'',U^2,\cdots)=0.$$

In this equation, we consider the solution form as,

$$U(\xi)=\sum_{i=1}^{n}\tanh^{i-1}(\xi)\left(B_i sech(\xi)+A_i\tanh(\xi)\right)+A_0. \tag{5.8}$$

Rearranging Eq. (5.8) according to Eqs (5.5)–(5.7), we have:

$$U(w)=\sum_{i=1}^{n}\cos^{i-1}(w)\left(B_i\sin(w)+A_i\cos(w)\right)+A_0. \tag{5.9}$$

We employ the balance principle to determine the value of n. Considering the summation of coefficients of $\sin^i(w)\cos^i(w)$ with the same power is zero, this gives a system of equations. After solving the system of equations to obtain the values A_i, B_i, μ and c, we substitute these coefficients into Eq. (5.8).

5.4 Investigation of SGEM to CGBKE

Here, we employ the SGEM [13, 17, 24, 38, 39] to the (2+1)-dimensional conformable time-fractional generalized Bogoyavlensky–Konopelchenko equation (CGBKE) [19].

First, we transform Eq. (5.1) to a NODE by the following travelling wave transformation:

$$u(x,y,t)=U(\xi),\xi=kx+ly+\lambda\frac{t^{\theta}}{\theta}. \tag{5.10}$$

Substituting the partial derivatives of Eq. (5.10) into Eq. (5.1), the following NODES is obtained:

$$\left(\alpha k^{4}+\beta k^{3}l\right)U''\ +\left(6\alpha k^{3}+6\beta k^{2}l\right)\left(U'\right)\left(U''\right)+\left(\gamma_{1}k^{2}+\gamma_{2}kl+\gamma_{3}l^{2}+k\lambda\right)U''=0 \tag{5.11}$$

When we integrate the above equation once with regard to ξ

$$\left(\alpha k^{4}+\beta k^{3}l\right)U''+\left(3\alpha k^{3}+3\beta k^{2}l\right)\left(\left(U'\right)^{2}\right)+\left(\gamma_{1}k^{2}+\gamma_{2}kl+\gamma_{3}l^{2}+k\lambda\right)U'=0, \tag{5.12}$$

Let

$$V=U', \tag{5.13}$$

all valued put Eq. (5.12)

$$\left(\alpha k^{4}+\beta k^{3}l\right)V''+\left(3\alpha k^{3}+3\beta k^{2}l\right)V^{2}+\left(\gamma_{1}k^{2}+\gamma_{2}kl+\gamma_{3}l^{2}+k\lambda\right)V=0. \tag{5.14}$$

By considering the principle of balance, then, $n=2$ is found and it produces the following format

$$V(w)=B_{1}\sin(w)+A_{1}\cos(w)+B_{2}\cos(w)\sin(w)+A_{2}\cos^{2}(w)+A_{0}. \tag{5.15}$$

Differentiating Eq. (5.15) twice yields

$$\begin{gathered}V''=B_{1}\cos^{2}(w)\sin(w)-B_{1}\sin^{3}(w)-2A_{1}\sin^{2}(w)\cos(w)+\\ B_{2}\cos^{3}\sin(w)-5B_{2}\sin^{3}(w)\cos(w)-4A_{2}\cos^{2}(w)\sin^{2}(w)+2A_{2}\sin^{4}(w).\end{gathered} \tag{5.16}$$

By using Eqs (5.15)–(5.16) into Eq. (5.14) and doing mathematical operations, we find an algebraic equations system. When the method is subsequently solved using software, we find the following situations.

Case 1

When we select coefficients

$$A_0 = k; A_1 = 0; A_2 = -k; B_1 = 0; B_2 = ik; \lambda = -\frac{k^4\alpha + k^3 l\beta + k^2\gamma_1 + kl\gamma_2 + l^2\gamma_3}{k}$$

$$u_1(x,y,t) = -iksech\left[kx + ly - t^{\theta}\left(\frac{k^4\alpha + k^3 l\beta + k^2\gamma_1 + kl\gamma_2 + l^2\gamma_3}{k\theta}\right)\right] + ktanh\left[kx + ly - t^{\theta}\left(\frac{k^4\alpha + k^3 l\beta + k^2\gamma_1 + kl\gamma_2 + l^2\gamma_3}{k\theta}\right)\right], \tag{5.17}$$

is obtained where $k, l, \alpha, \gamma_1, \gamma_2, \gamma_3$ are real constants with non-zero.

Case 2

Taking as

$A_1 = 0;\ A_2 = -2k;\ B_1 = 0; B_2 = 0; l = -\frac{k\alpha}{\beta}; \lambda = -k\gamma_1 + \frac{k\alpha(\beta\gamma_2 - \alpha\gamma_3)}{\beta^2}$, produces another dark soliton solution to Eq. (5.1) as follows:

$$u_2(x,y,t) = -2k\left(kx - \frac{ky\alpha}{\beta} + \frac{t^{\theta}\left(-k\gamma_1 + \frac{k\alpha(\beta\gamma_2 - \alpha\gamma_3)}{\beta^2}\right)}{\theta}\right) + A_0\left(kx - \frac{ky\alpha}{\beta} + \frac{t^{\theta}\left(-k\gamma_1 + \frac{k\alpha(\beta\gamma_2 - \alpha\gamma_3)}{\beta^2}\right)}{\theta}\right) + 2k\tanh\left[\left(kx - \frac{ky\alpha}{\beta} + \frac{t^{\theta}\left(-k\gamma_1 + \frac{k\alpha(\beta\gamma_2 - \alpha\gamma_3)}{\beta^2}\right)}{\theta}\right)\right]. \tag{5.18}$$

Case 3

If we get

$$A_0 = \frac{2k}{3}; A_1 = 0; A_2 = -2k; B_1 = 0; B_2 = 0; \lambda = 4k^2(k\alpha + l\beta) - k\gamma_1 - \frac{l(k\gamma_2 + l\gamma_3)}{k},$$

it results in another new dark solution to Eq. (5.1)

$$u_3(x,y,t) = -\frac{4}{3}k\left(kx + ly + \frac{t^\theta\left(4k^2(k\alpha + l\beta) - k\gamma_1 - \frac{l(k\gamma_2 + l\gamma_3)}{k}\right)}{\theta}\right)$$
$$+ 2k\tanh\left[kx + ly + \frac{t^\theta\left(4k^2(k\alpha + l\beta) - k\gamma_1 - \frac{l(k\gamma_2 + l\gamma_3)}{k}\right)}{\theta}\right]. \tag{5.19}$$

Case 4

If we take

$$A_0 = 2k; A_1 = 0; A_2 = -2k; B_1 = 0; B_2 = 0; \gamma_3 = -\frac{k\left(4k^2(k\alpha + l\beta) + \lambda + k\gamma_1 + l\gamma_2\right)}{l^2}$$

the following hyperbolic function solution to Eq. (5.1) is found

$$u_4(x,y,t) = 2k\tanh\left[kx + ly + \frac{t^\theta \lambda}{\theta}\right]. \tag{5.20}$$

Case 5

$$A_1 = 0; A_2 = \frac{\lambda + l\gamma_2 - \sqrt{(\lambda + l\gamma_2)^2 - 4l^2\gamma_1\gamma_3}}{\gamma_1}; B_1 = 0; B_2 = 0;$$

$$k = -\frac{\lambda + l\gamma_2 - \sqrt{(\lambda + l\gamma_2)^2 - 4l^2\gamma_1\gamma_3}}{2\gamma_1}; \beta = \frac{\alpha\left(\lambda + l\gamma_2 - \sqrt{(\lambda + l\gamma_2)^2 - 4l^2\gamma_1\gamma_3}\right)}{2l\gamma_1}$$

$$u_5(x,y,t) = A_0\left(ly + \frac{t^\theta\lambda}{\theta} - \frac{x\left(\lambda + l\gamma_2 - \sqrt{(\lambda + l\gamma_2)^2 - 4l^2\gamma_1\gamma_3}\right)}{2\gamma_1}\right)$$
$$-\frac{1}{\gamma_1}\left(-\lambda - l\gamma_2 + \sqrt{\lambda^2 + 2l\lambda\gamma_2 + l^2\gamma_2^2 - 4l^2\gamma_1\gamma_3}\right)$$
$$\left(ly + \frac{t^\theta\lambda}{\theta} - \frac{x\left(\lambda + l\gamma_2 - \sqrt{(\lambda + l\gamma_2)^2 - 4l^2\gamma_1\gamma_3}\right)}{2\gamma_1}\right) \tag{5.21}$$
$$-\tanh\left[ly + \frac{t^\theta\lambda}{\theta} - \frac{x\left(\lambda + l\gamma_2 - \sqrt{(\lambda + l\gamma_2)^2 - 4l^2\gamma_1\gamma_3}\right)}{2\gamma_1}\right]).$$

Case 6

$$A_0 = iB_2; A_1 = 0; A_2 = -iB_2; B_1 = 0; k = iB_2; \lambda = l\beta B_2^2 + i\alpha B_2^3 - iB_2\gamma_1 - l\gamma_2 + \frac{il^2\gamma_3}{B_2}$$

$$u_6(x,y,t) = -sech\left[ly + ixB_2 + \frac{t^\theta\left(l\beta B_2^2 + i\alpha B_2^3 - iB_2\gamma_1 - l\gamma_2 + \frac{il^2\gamma_3}{B_2}\right)}{\theta}\right]B_2$$
$$+iB_2\tanh\left[ly + ixB_2 + \frac{t^\theta\left(l\beta B_2^2 + i\alpha B_2^3 - iB_2\gamma_1 - l\gamma_2 + \frac{il^2\gamma_3}{B_2}\right)}{\theta}\right] \tag{5.22}$$

Case 7

$$A_0 = -A_2; A_1 = 0; B_1 = 0; B_2 = iA_2; k = -A_2;$$

$$\alpha = \frac{A_2\left(\lambda + l\beta A_2^2 - A_2\gamma_1 + l\gamma_2\right) - l^2\gamma_3}{A_2^4}$$

$$u_7\left(x,y,t\right) = -isech\left[ly + \frac{t^{\theta}\lambda}{\theta} - xA_2\right]A_2 - A_2\tanh\left[ly + \frac{t^{\theta}\lambda}{\theta} - xA_2\right] \quad (5.23)$$

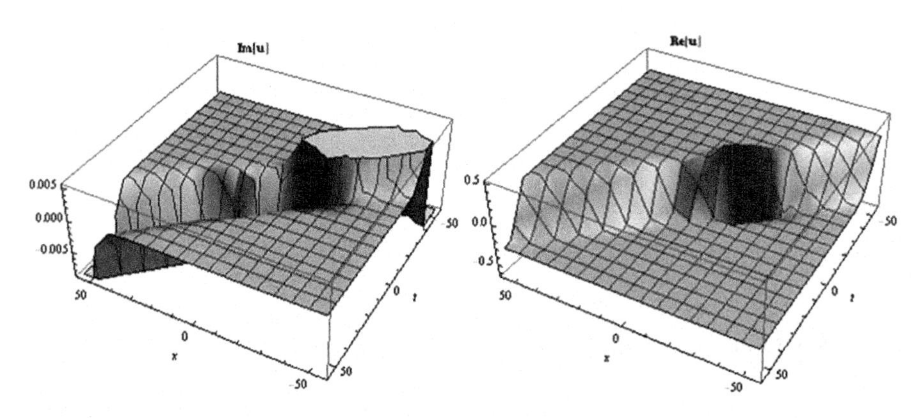

FIGURE 5.1
3D graphs to Eq. (5.17).

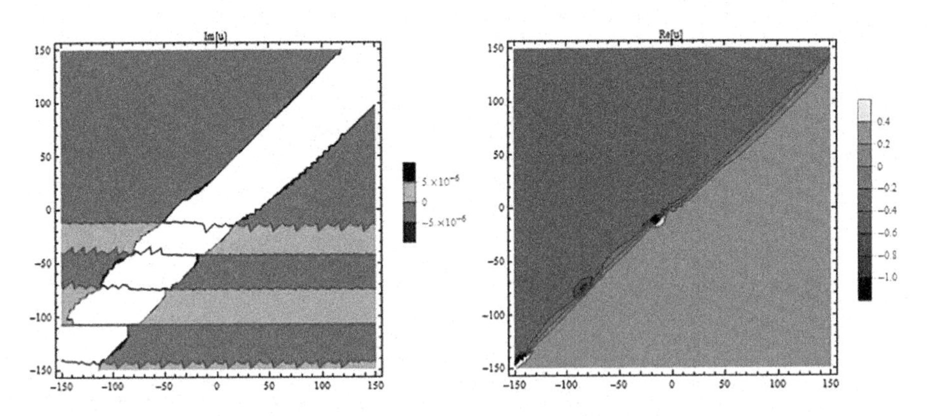

FIGURE 5.2
Contour surfaces to Eq. (5.17).

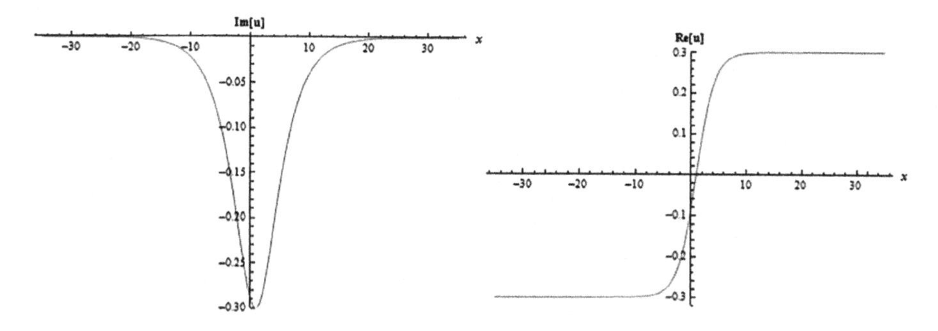

FIGURE 5.3
Two-dimensional imaginary and real plots to Eq. (5.17).

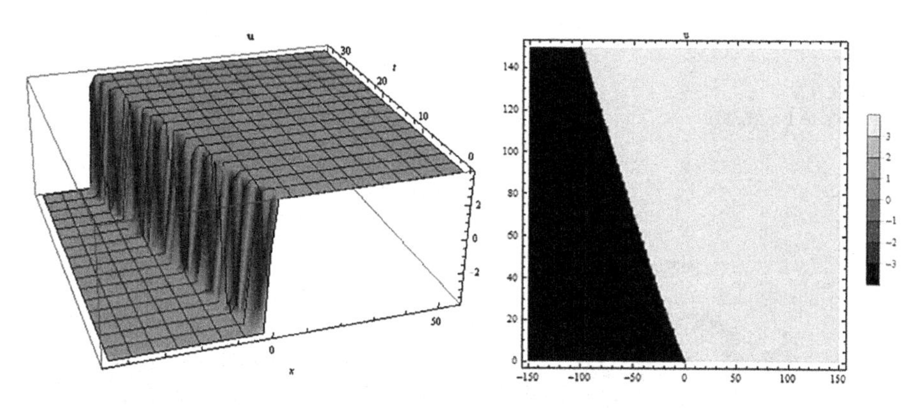

FIGURE 5.4
3D and contour graphs to Eq. (5.20).

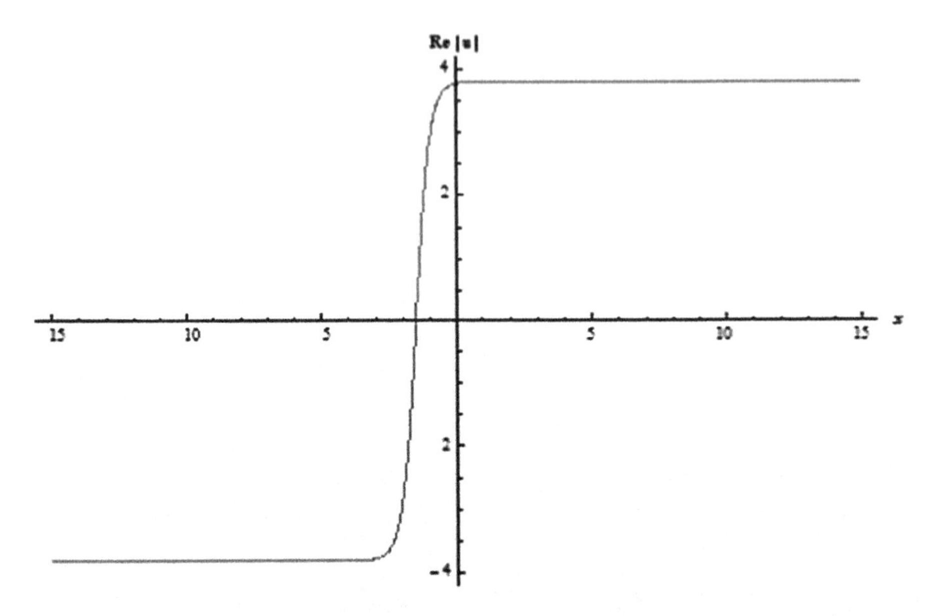

FIGURE 5.5
2D plots to Eq. (5.20).

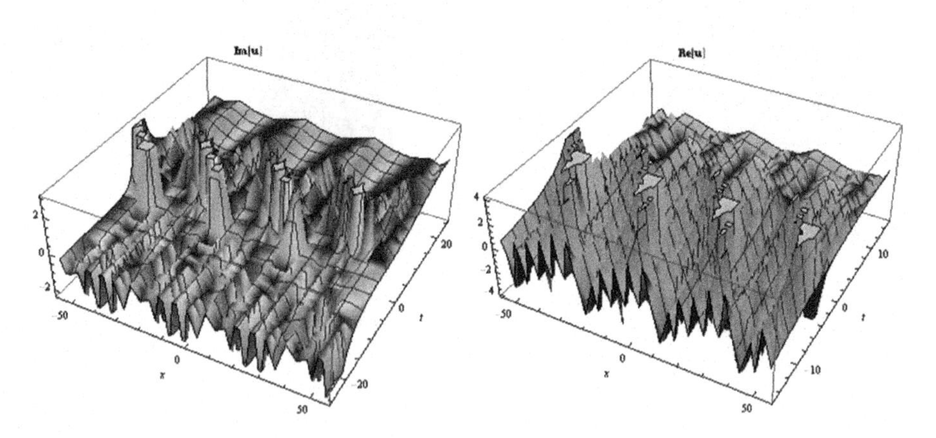

FIGURE 5.6
3D graphs to Eq. (5.22).

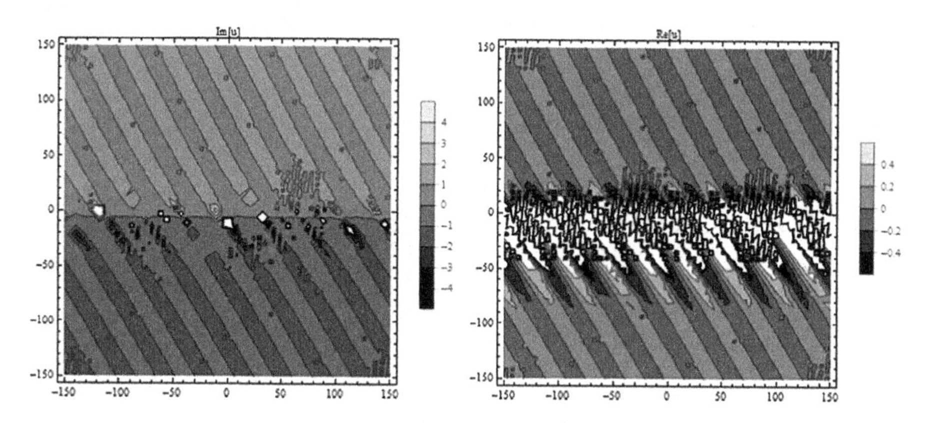

FIGURE 5.7
Contour graphs to Eq. (5.22).

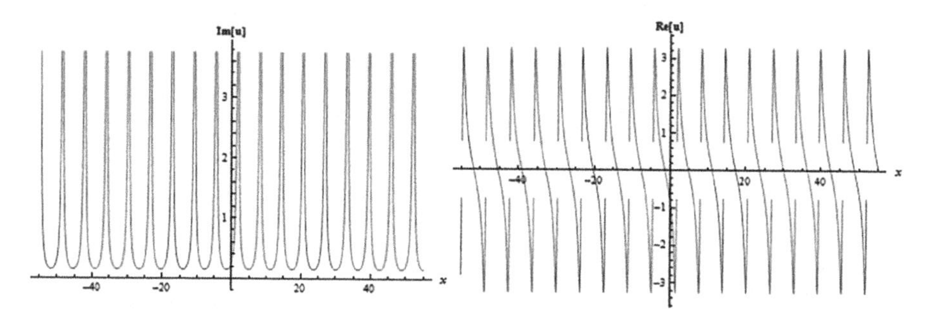

FIGURE 5.8
Two-dimensional imaginary and real plots to Eq. (5.22).

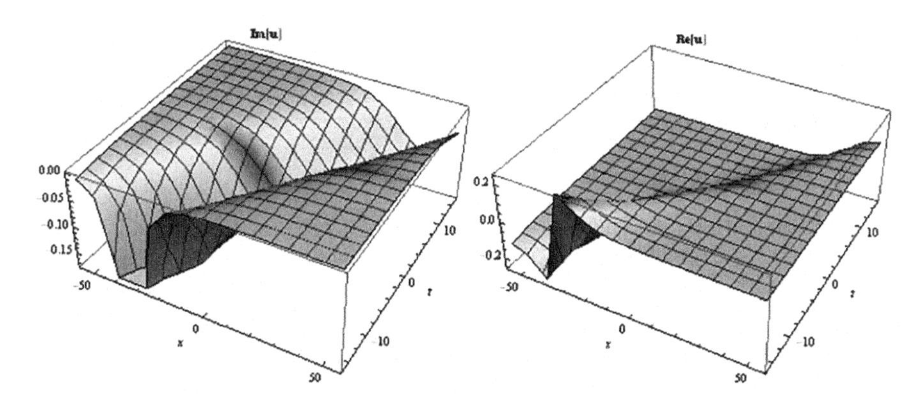

FIGURE 5.9
3D graphs to Eq. (5.23).

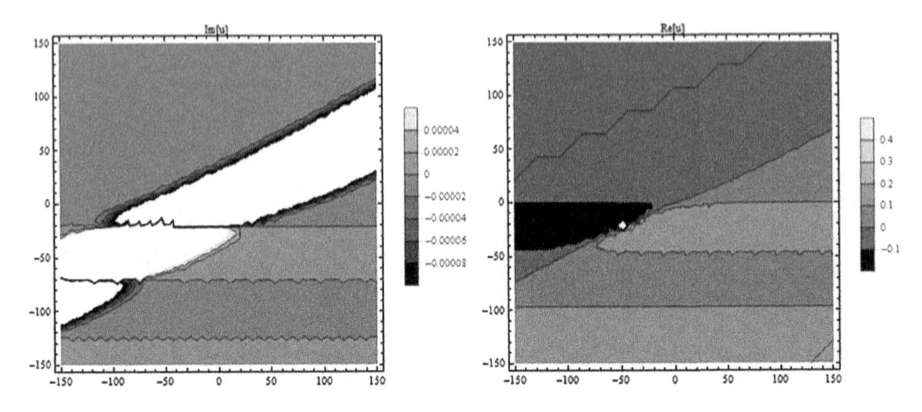

FIGURE 5.10
Contour graphs to Eq. (5.23).

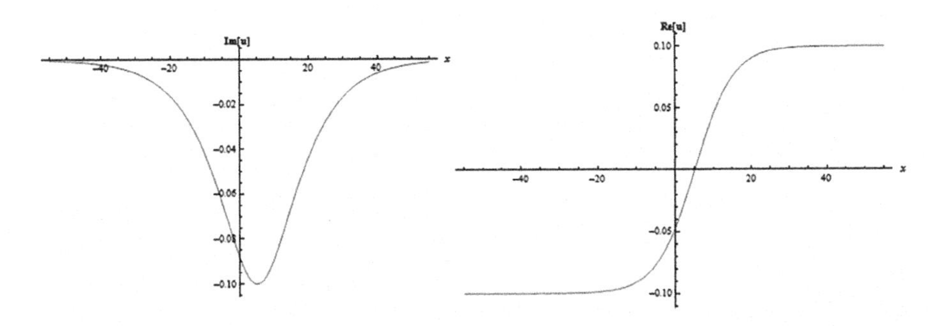

FIGURE 5.11
Imaginary and real plots to Eq. (5.23).

5.5 Results and Discussion

We plot the 3D, 2D and contour simulations to Eq. (5.17) with the values $k = 0.3, \alpha = 0.2, l = 0.3, \beta = 0.8, \gamma_1 = 0.4, \gamma_2 = 0.5, \gamma_3 = 0.6, y = 0.7, t = 0.9$. Figures 5.4 and 5.5 are plotted considering Eq. (5.20) with the values $k = 1.9$, $l = 1.9$, $\lambda = 1.9$, $y = 0.5$, $t = 0.9$. Figures 5.6–5.8 present valid solutions to Eq. (5.22) under the values $k = 1.3, \alpha = 0.2, l = 0.3, \beta = 0.8, \gamma_1 = 0.4, \gamma_2 = 0.5, \gamma_3 = 0.6, y = 0.7$, $B_2 = 1$, $t = 0.9$. Figures 5.9–5.11 represent 3D, 2D and contour surfaces to real and imaginary parts of Eq. (5.23) with $k = 0.3$, $\alpha = 0.2$, $l = 0.3$, $\beta = 0.8$, $A_2 = 0.1$, $\lambda = 0.3$, $y = 0.7$, $t = 0.9$. All figures are plotted while $\theta = 0.9$.

5.6 Conclusions

In this paper, we have successfully applied SGEM to find new analytical solutions to the (2+1)-dimensional generalized Bogoyavlensky–Konopelchenko equation by using conformable operator properties. We found many entirely new complex and dark soliton solutions to the governing model. It may be observed that all these results satisfied the governing model. Moreover, it can be observed from Figures 5.1–5.11 that these results symbolize the estimated wave behaviors of the main model by considering different values of parameters under strain conditions. It may be concluded that this is a very effective model for finding many new soliton solutions to nonlinear partial differential equations. Furthermore, it allows us to better understand behaviors of some physical phenomena. As a future direction, SGEM may be developed to apply to more complex and high nonlinear models. The original contribution of this paper is to investigate properties of the governing model in greater depth. The results show that the applications of conformable derivatives are simpler and very efficient.

Acknowledgments

This work was partially (not financial) supported by the Scientific Research Project Fund of Harran University with the project number HUBAP:21132.

References

[1] A. Ali, M.A. Iqbal, Q.M. Ul-Hassan, J. Ahmad, S.T. Mohyud Din, An efficient technique for higher order fractional differential equation, *Springer Plus*, 5(281), 1–10, 2016.

[2] A. Akbulut, M. Kaplan, Auxiliary equation method for time-fractional differential equations with conformable derivative, *Computers Mathematics with Applications*, 75(3), 876–882, 2018.

[3] A. Ali, R. Seadawy, D. Lu, Computational methods and traveling wave solutions for the fourth-order nonlinear Ablowitz-Kaup-Newell-Segur water wave dynamical equation via two methods and its applications, *Open Physics*, 16, 219–226, 2018.

[4] C. Chen, Y.-L. Jiang, Simplest equation method for some time-fractional partial differential equations with conformable derivative, *Computers Mathematics with Applications*, 75(8), 2978–2988, 2018.

[5] J.H. He, Exp-function Method for Fractional Differential Equations, *International Journal of Nonlinear Sciences and Numerical Simulation*, 14(6), 363–366, 2013.
[6] K. S. Al-Ghafri, H. Rezazadeh, Solitons and other solutions of (3+1)-dimensional space-time fractional modified KdV-Zakharov-Kuznetsov equation, *Applied Mathematics and Nonlinear Sciences*, 4(2), 289–304, 2019.
[7] M. Ekici, M. Mirzazadeh, M. Eslami, Q. Zhou, S.P. Moshokoa, A. Biswas, M. Belic, Optical soliton perturbation with fractional-temporal evolution by first integral method with conformable fractional derivatives, *Optik*, 127(22), 10659–10669, 2016.
[8] R. Hirota, Exact solution of the Korteweg-de Vries equation for multiple collisions of solitons, *Physical Review Letters*, 27, 1192–1194, 1971.
[9] R. Khalil, M. Al Horani, A. Yousef et al., A new definition of fractional derivative, *Journal of Computational and Applied Mathematics*, 264, 65–70, 2014.
[10] A. Atangana, D. Baleanu, A. Alsaedi, New properties of conformable derivative, *Open Mathematics*, 13, 889–898, 2015.
[11] C. Cattani, T.A. Sulaiman, H.M. Baskonus, H. Bulut, Solitons in an inhomogeneous Murnaghan's rod, *European Physical Journal Plus*, 133(228), 1–12, 2018.
[12] C. Cattani, T.A. Sulaiman, H. M. Baskonus, On the soliton solutions to the Nizhnik-Novikov-Veselov and the Drinfel'd-Sokolov systems, *Optical and Quantum Electronics*, 50(138), 1–10, 2018.
[13] C. Yan, A simple transformation for nonlinear waves, *Physics Letters A*, 224, 77–84, 1996.
[14] E. Balci, I. Ozturk, S. Kartal, Dynamical behaviour of fractional order tumor model with Caputo and conformable fractional derivative, *Chaos Solitons Fractals*, 123, 43–51, 2019.
[15] H. Bulut, T.A. Sulaiman, H.M. Baskonus, H. Rezazadeh, M. Eslami, M. Mirzazadeh, Optical solitons and other solutions to the conformable space-time fractional Fokas-Lenells equation, *Optik*, 172, 20–27, 2018.
[16] H.C. Yaslan, A. Girgin, New exact solutions for the conformable space-time fractional KdV, CDG, (2+1)-dimensional CBS and (2+1)- dimensional AKNS equations, *Journal of Taibah University for Science*, 13(1), 1–8, 2018.
[17] H.M. Baskonus, T.A. Sulaiman, H. Bulut, New Solitary Wave Solutions to the (2+1)-dimensional Calogero-Bogoyavlenskii-Schi and the Kadomtsev-Petviashvili Hierarchy Equations, *Indian Journal of Physics*, 91(10), 1237–1243, 2017.
[18] J. Singh, D. Kumar, Z. Hammouch, A. Atangana, A fractional epidemiological model for computer viruses pertaining to a new fractional derivative, *Applied Mathematics and Computation*, 316, 504–515, 2018.
[19] R. Pouyanmehr, K. Hosseini, R. Ansari, S.H. Alavi, Different Wave Structures to the (2+1)-Dimensional Generalized Bogoyavlensky-Konopelchenko Equation, *International Journal of Applied and Computational Mathematics*, 5(149), 2019.
[20] W. Gao, P. Veeresha, D.G. Prakasha, H.M. Baskonus, G. Yel, New approach for the model describing the deathly disease in pregnant women using Mittag-Leffler function, *Chaos Solitons and Fractals*, 134, 109696, 2020.
[21] T.A. Sulaiman, G. Yel, H. Bulut, M-fractional solitons and periodic wave solutions to the Hirota Maccari system, *Modern Physics Letters B*, 33(1950052), 1–20, 2019.
[22] G. Yel, T.A. Sulaiman, H.M. Baskonus, On the complex solutions to the (3+1)-dimensional conformable fractional modified KdV-Zakharov-Kuznetsov equation, *Modern Physics Letters B*, 34(5), 2050069, 2020.

[23] W. Gao, B. Ghanbari, H.M. Baskonus, New numerical simulations for some real world problems with Atangana-Baleanu fractional derivative, *Chaos Solitons Fractals*, 128, 34–43, 2019.
[24] W. Gao, G. Yel, H.M. Baskonus, C. Cattani, Complex solitons in the conformable (2+1)-dimensional Ablowitz-Kaup-Newell-Segur equation, *AIMS Mathematics*, 5(1), 507–521, 2019.
[25] X. Ma, S.T. Chen, Exact Solutions to a Generalized Bogoyavlensky Konopelchenko Equation via Maple Symbolic Computations, *Complexity*, 2019, 6, 2019.
[26] Y. Zhen-Ya, Z. Hong-Oing, F. En-Gui, New explicit and travelling wave solutions for a class of nonlinear evolution equations, *Acta Physica Sinica*, 48, 1–5, 1999.
[27] Z. Hammouch, T. Mekkaoui, Travelling-wave solutions for some fractional partial differential equation by means of generalized trigonometry functions, *International Journal of Applied Mathematical Research*, 1, 206–212, 2012.
[28] Z. Cheng, X. Hao, The periodic wave solutions for a (2+1)-dimensional AKNS equation, *Applied Mathematics and Computation*, 234, 118–126, 2014.
[29] Z. Yan, H. Zhang, New explicit and exact travelling wave solutions for a system of variant Boussinesq equations in mathematical physics, *Physic Letters A*, 252, 291–296, 1999.
[30] A. R. Seadawy, K. K. Ali, R.I. Nuruddeen, A variety of soliton solutions for the fractional Wazwaz-Benjamin-Bona-Mahony equations, *Results in Physics*, 12, 2234–2241, 2019.
[31] K.K. Ali, R.I. Nuruddeen, K.R. Raslan, New structures for the space time fractional simplified MCH and SRLW equations, *Chaos Solitons and Fractals*, 106, 304–309, 2018.
[32] P. Veeresha, D.G. Prakasha, Solution for fractional Kuramoto-Sivashinsky equation using novel computational technique, *International Journal of Applied and Computational Mathematics*, 7(33), 1–20, 2021.
[33] P. Veeresha, D.G. Prakasha, D. Baleanu, An efficient technique for fractional coupled system arisen in magneto thermoelasticity with rotation using Mittag-Leffler kernel, *Journal of Computational Nonlinear Dynamics*, 16(1). DOI: 10.1115/1.4048577, 2021.
[34] H. Singh, A.M. Wazwaz, Computational Method for Reaction Diffusion Model Arising in a Spherical Catalyst, *International Journal of Applied and Computational Mathematics*, 7(3), 65, 2021.
[35] H. Singh, Analysis for fractional dynamics of Ebola virus model, *Chaos Solitons Fractals*, 138, 109992, 2020.
[36] H. Singh, D. Kumar, D. Baleanu, *Methods of Mathematical Modelling: Fractional Differential Equations*, CRC Press Taylor and Francis, 2019.
[37] F. Dusunceli, New Exact Solutions for Generalized (3+1) Shallow Water Like (SWL) Equation, *Applied Mathematics and Nonlinear Sciences*, 4(2), 365–370, 2019.
[38] H.M. Baskonus, New acoustic wave behaviors to the Davey-Stewartson equation with power-law nonlinearity arising in fluid dynamics, *Nonlinear Dynamics*, 86(1), 177–183, 2016.
[39] H. Bulut, T.A. Sulaiman, H.M. Baskonus, New solitary and optical wave structures to the Korteweg–de Vries equation with dual-power law nonlinearity, *Optical and Quantum Electronics*, 12(564), 1–14, 2016.

6

Analytical Solution of a Time-Fractional Damped Gardner Equation Arising from a Collisional Effect on Dust-ion-acoustic Waves in a Dusty Plasma with Bi-Maxwellian Electrons

Naresh M. Chadha
DIT University, India

Santanu Raut
Mathabhanga College, Cooch Behar, India

Kajal Mondal
Cooch Behar Panchanan Barma University, India

Shruti Tomar
DIT University, Uttarakhand, India

CONTENTS

DOI: 10.1201/9781003263517-6

6.1 Introduction

Nonlinear evolution equations (NLEEs) have recently attracted considerable attention due to their wide applications in various branches of nonlinear sciences. For instance, NLEEs can be used to formulate various problems emerging in protein chemistry, such as ecological modelling, quantum mechanics, plasma physics, propagation of shallow water waves, and chemical kinetics. A search for their analytical solutions has attracted diverse groups of mathematicians and physicists. In particular, much attention is paid to variable coefficient nonlinear evolution equations, which can be used to formulate many interesting nonlinear phenomena more realistically than constant coefficients analogous equations. Consequently, many powerful methods have been developed in recent years to study NLEE analytical solutions, such as the inverse scattering method [1], Hirota's method [2], the Exp-function method [3], the sine-cosine method [4], the F-expansion method [5], the tanh-method [6, 7], the extended tanh-method [8], the classic G'/G–expansion method and its variants, [9–13], the Jacobi elliptic function method [14], and the homotopy perturbation method [15–17], to name just a few.

Another area closely related to the study of NLEEs is the propagation of ion acoustic waves (IAWs) in a dusty plasma, which has gained a lot of attention due to its ubiquitous applications in laboratory plasmas as well as in distinct astrophysical plasma studies related to the lower and upper mesosphere, cometary tails, planetary rings, interstellar media, etc. [18–21]. During the last few decades, a number of physicists and applied mathematicians have made extensive studies of nonlinear waves, especially DIAWs in a plasma environment, to analyze localized electrostatic perturbations in both laboratory and space plasma [20, 22].

The Gardner equation is an NLEE that has been used to describe various interesting real-world phenomena emerging in plasma physics, shallow-water fluid dynamics, quantum field theory, and so on. See for example [23–26]. The Gardner equation, in its conventional form, reads

$$\frac{\partial \phi}{\partial \tau} + A_1 \phi \frac{\partial \phi}{\partial \xi} + A_2 \phi^2 \frac{\partial \phi}{\partial \xi} + A_3 \frac{\partial^3 \phi}{\partial \xi^3} = 0, \tag{6.1}$$

where $\phi(\xi, \tau)$ is the amplitude of wave mode, ξ and τ are the space and time coordinates, respectively, A_1 is the coefficient of the nonlinear term, A_2 denotes the coefficient of higher-order cubic nonlinear term of the form $\phi^2 \frac{\partial \phi}{\partial \xi}$, and the linear dispersion $\frac{\partial^3 \phi}{\partial \xi^3}$ is controlled by the coefficient A_3. Equation 6.1 is also known as the combined KdV–mKdV equation. In a case where $A_2 = 0$, the Gardner equation reduces to the KdV equation given by

$$\frac{\partial \phi}{\partial \tau} + A_1 \phi \frac{\partial \phi}{\partial \xi} + A_3 \frac{\partial^3 \phi}{\partial \xi^3} = 0. \tag{6.2}$$

In the present study, we consider the governing equations for the nonlinear propagation of dust-ion-acoustic waves in a dusty plasma with bi-Maxwellian electrons. From these governing equations, the KdV equation is derived using the reductive perturbation technique. It is observed that the formation of solitons is not possible for parametric zones where the coefficient of nonlinear term, A_1, is far above or below a critical value; similar behavior has been reported in [27–29]. In such cases, different stretching variables are required for the formulation. This leads to the formation of the modified KdV (mKdV) and the Gardner equation. Then, using Euler–Lagrange equations, the Gardner equation is modified into a fractional time-derivative damped Gardner (TFDG) equation involving a Riesz fractional derivative operator [30]. It has been reported in the literature that fractional operators are non-local operators more suitable than conventional operators for dealing with experimental data emerging from plasma physics [31–33]. The analytical solution of the time-fractional damped Gardner equation is obtained using the extended G'/G–expansion method. It is found that Gardner's solitons exist beyond the KdV limit of the parameters. The analytical solution is analyzed to understand the impact of the variation of various parameters, in particular with respect to the parameter α (fractional order of time derivative) and the parameters guiding the damping effect.

It is important to note that the G'/G–expansion method offers certain advantages over other methods commonly used for solving NLEEs. For instance, the G'/G–expansion method yields more general solutions with some free parameters. A suitable choice of these parameters may return more than one solution to the problem under consideration. Another significant advantage of this method is that the method handles an NLEE directly as it has no dependence on initial trail function(s) and/or initial or boundary conditions. Many methods such as the Adomain decomposition method return the solution in the form of a series which necessitates analyzing the convergence of this series solution to the exact solution. In view of these advantages, the extended G'/G–expansion method is employed in the present study.

Topics covered in this chapter include:

- The effect of different physical parameters on wave propagation for the KdV, damped mKdV, and Gardner equations are shown from a numerical standpoint.
- In general, charging of dust particles, collision of inner particles, and other physical properties such as viscosity may introduce a

weak damping effect into the plasma system. This damping effect is neglected for the sake of simplicity in most of the studies available in the literature. The numerical expositions presented in this chapter show that this damping term has a significant effect on wave propagation in both mKdV and Gardner equations.

- To the best of our knowledge, the range of parameters using level curves and other means are obtained mathematically for the first time here, which is very helpful in defining the critical zones for them.
- Taking all practical considerations into account related to the critical zones of the parameters and the limitations of stretched variables used in KdV and damped mKdV equations, the time-fractional damped Gardner equation is derived, and the analytical solutions obtained, using the extended G'/G–expansion method for the first time. The effect of the parameters and the fractional order α is clearly shown in the numerical results. The results presented here may be useful for further studies in laboratory plasma as well as in space plasma.

6.2 Governing Equation and Formation of KdV Soliton

Here, we consider propagation properties of dust acoustic waves in an unmagnetized dusty plasma comprising finite temperature T_j inertial ions and Boltzmann distributed electrons in two different thermal states with different temperatures T_{e1} and T_{e2} (one is cold and the other is hot), respectively. We assume that $T_j << T_{e1}, T_{e2}$, and $T_{e1} > 0$. At equilibrium, we take $n_{j0} = n_{e10} + n_{e20} + Z_d n_{d0}$, where n_{e10} and n_{e20} denote densities of the cold and hot temperature electrons, respectively, Z_d denotes the number of electrons, and n_{j0} is the equilibrium density of the ion (dust). The plasma system is governed by

$$\frac{\partial n_j}{\partial t} + \frac{\partial (n_j v_j)}{\partial x} = 0, \tag{6.3a}$$

$$\frac{\partial v_j}{\partial t} + v_j \frac{\partial v_j}{\partial x} = -\frac{\partial \psi}{\partial x}, \tag{6.3b}$$

$$\frac{\partial^2 \psi}{\partial x^2} = -\rho = \mu + \mu_{e1} e^{\psi} + (1 - \mu_{e1} - \mu) e^{\sigma\psi} - n_j. \tag{6.3c}$$

Here n_j denotes the normalized number density of ions with respect to its equilibrium value n_{j0}; v_j represents the normalized ion fluid velocity compared to wave speed $C_j = \sqrt{k_B T_{e1}/m_j}$; ψ is normalized electrostatic wave potential with respect to $k_B T_{e1}/e$; ρ is the normalized surface charge density; k_B represents Boltzmann constant; e is the magnitude of the electron charge; $\sigma = T_{e1}/T_{e2}$, $\mu_{e1} = n_{e10}/n_{j0}$, $\mu_{e2} = n_{e20}/n_{j0}$, $\mu = Z_d n_{d0}/n_{j0} = 1 - \mu_{e1} - \mu_{e2}$; and x, t are affirmed as space and time coordinates, respectively and normalized by the Debye length $\lambda_{Dm} = \sqrt{k_B T_{e1}/4\pi n_{j0} e^2}$ and period of ion plasma to $\omega_{pi}^{-1} = \sqrt{m_i/4\pi n_{j0} e^2}$, respectively.

In order to derive KdV equation from the basic governing equation, the depending variables n_j, v_j, ψ and ρ are expanded in power series of ϵ as follows:

$$\begin{aligned} n_j &= 1 + \epsilon n_1 + \epsilon^2 n_2 + \epsilon^3 n_3 + \cdots \\ v_j &= 0 + \epsilon v_1 + \epsilon^2 v_2 + \epsilon^3 v_3 + \cdots \\ \psi &= 0 + \epsilon \psi_1 + \epsilon^2 \psi_2 + \epsilon^3 \psi_3 + \cdots \\ \rho &= 0 + \epsilon \rho_1 + \epsilon^2 \rho_2 + \epsilon^3 \rho_3 + \cdots \end{aligned} \tag{6.4}$$

Further, we introduce the following new stretched coordinates:

$$\xi = \epsilon^{\frac{1}{2}}\left(x - u_p t\right), \quad \tau = \epsilon^{\frac{3}{2}} t. \tag{6.5}$$

Here, $u_p = 1/\sqrt{\mu_{e1} + \mu_{e2}\sigma}$ denotes the phase speed of the perturbation mode and the small parameter ϵ helps to measure the weakness as well as the dispersion of the wave perturbation. Using standard perturbation technique, and comparing the coefficients of ϵ, we obtain the following KdV equation:

$$\frac{\partial \phi}{\partial \tau} + A\phi \frac{\partial \phi}{\partial \xi} + B\frac{\partial^3 \phi}{\partial \xi^3} = 0, \qquad \text{where } A = \frac{u_p^3}{2}\left(\frac{3}{u_p^4} - \mu_{e1} - \mu_{e2}\sigma^2\right), \; B = \frac{u_p^3}{2}. \tag{6.6}$$

The solitary wave solution of the KdV equation given by Eq. 6.6 is available in the literature (see for example [34]). The solution is of the form

$$\psi = \lambda_0 \operatorname{sech}^2\left(\frac{\xi - V_0 \tau}{w}\right), \tag{6.7}$$

where $\lambda_0 = 3u_0/A$ is the amplitude, $w = \sqrt{4B/V_0}$ is the width of the solitary waves, and V_0 is the constant normalized velocity.

The solution profiles of the KdV equation and the effect of various parameters are shown in Figures 6.1 and 6.2, and corresponding three-dimensional plots are shown in Figure 6.3.

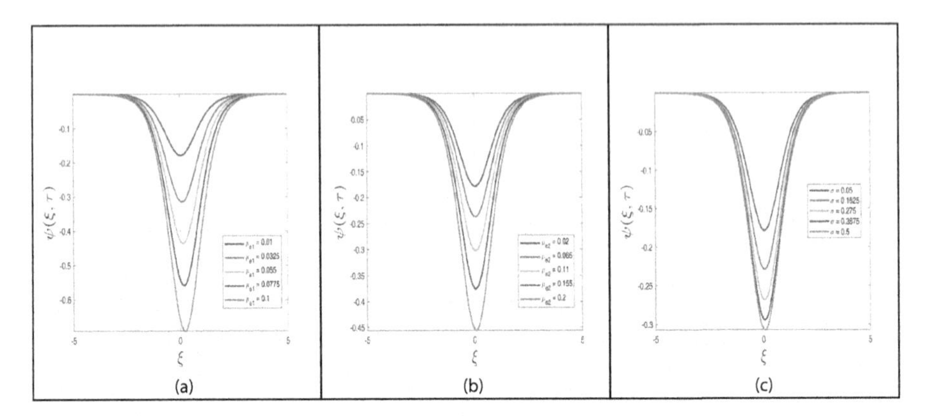

FIGURE 6.1
Profiles of solution of KdV equation for $\tau = 5$ and $V_0 = .5$; for three plots the following different values of the parameters are used: (a) $\mu_{e1} = p * 0.01, \mu_{e2} = 0.02, \sigma = 0.05$; (b) $\mu_{e1} = 0.01, \mu_{e2} = p * 0.02$, $\sigma = 0.05$; (c) $\mu_{e1} = 0.01, \mu_{e2} = 0.02, \sigma = p * 0.05$. Here p denotes five different values between 1 and 10.

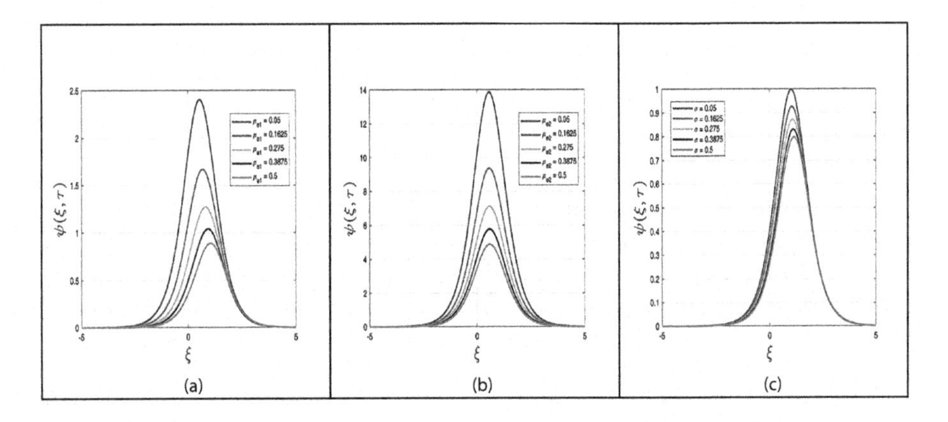

FIGURE 6.2
Profiles of solution of KdV equation for $\tau = 5$ and $V_0 = .5$; for three plots the following different values of the parameters are used: (a) $\mu_{e1} = p * 0.05, \mu_{e2} = 0.5, \sigma = 0.55$; (b) $\mu_{e1} = 0.35, \mu_{e2} = p *$ $0.05, \sigma = 0.05$; (c) $\mu_{e1} = 0.75, \mu_{e2} = 0.30, \sigma = p * 0.05$; here p is a multiplier which takes five equidistributed values between 1 and 10.

Figures 6.1–6.3 demonstrate that the solitons may completely change their nature from being compressive to rarefactive depending on the range of the parameters. Assuming the coefficient of dispersion term B remains positive throughout, the solution profiles are compressive or rarefactive depending on the sign of the nonlinear coefficient A (compressive for $A > 0$, and rarefactive for $A < 0$). It has been reported in the literature that in many practical applications the nonlinear coefficient A in Eq. 6.6 may vanish; see for

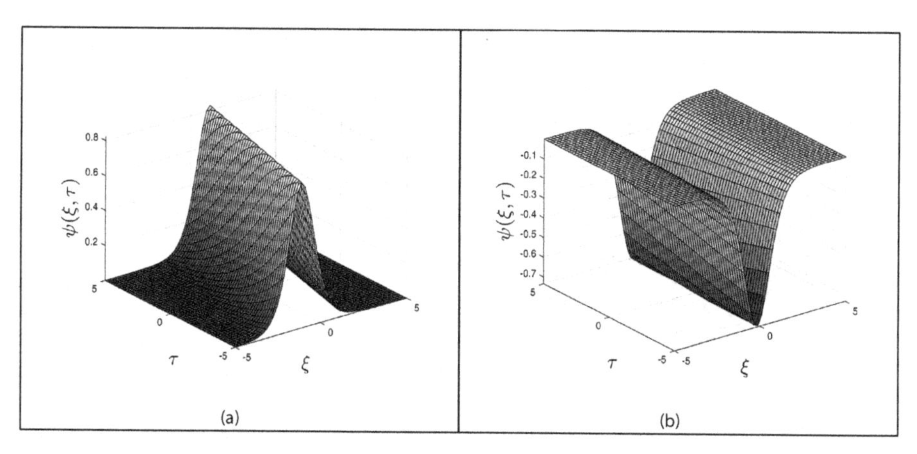

FIGURE 6.3
The solution profile of KdV equation changes from a hump to a kink for two different set of values of the parameters. The parameters are: (a) μ_{e1} = 0.14, μ_{e2} = 0.85, σ = 0.82, V_0 = .5; (b) μ_{e1} = 0.05, μ_{e2} = 0.09, σ = 0.25, V_0 = .5.

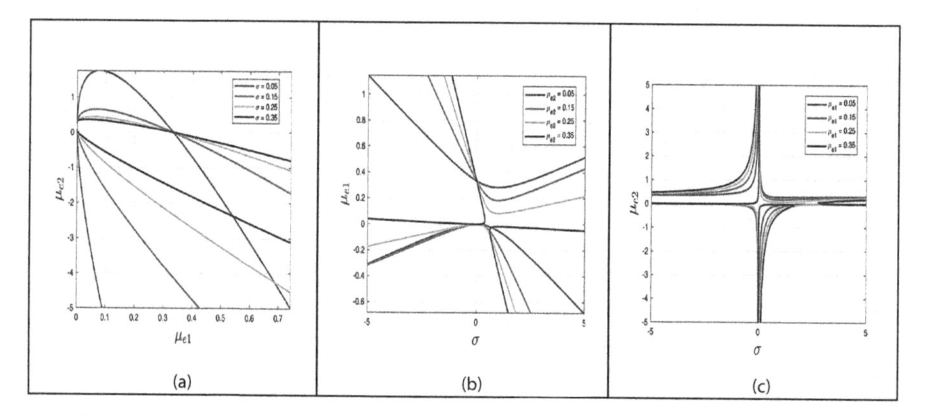

FIGURE 6.4
The trajectories for fixed values of a parameter along which the nonlinear coefficient $a = u_p^3 / 2\left(3/u_p^4 - \mu_{e1} - \mu_{e2}\sigma^2\right)$ is zero in the KdV equation given by Equation 6.6.

example [27–29]. For the values of the parameters for which $A = 0$, KdV solitons break down, which warrants seeking another appropriate equation to describe the evolution of the system. One of the ways to deal with this is to consider a new set of stretching coordinates.

In Figure 6.4, the trajectories for two-variable planes are shown for fixed values of a parameter along which the non-linear coefficient $A = u_p^3 / 2\left(3/u_p^4 - \mu_{e1} - \mu_{e2}\sigma^2\right)$ is zero. Figure 6.5 shows three-dimensional plots of the solution profiles of the KdV equation given by Eq.6.7.

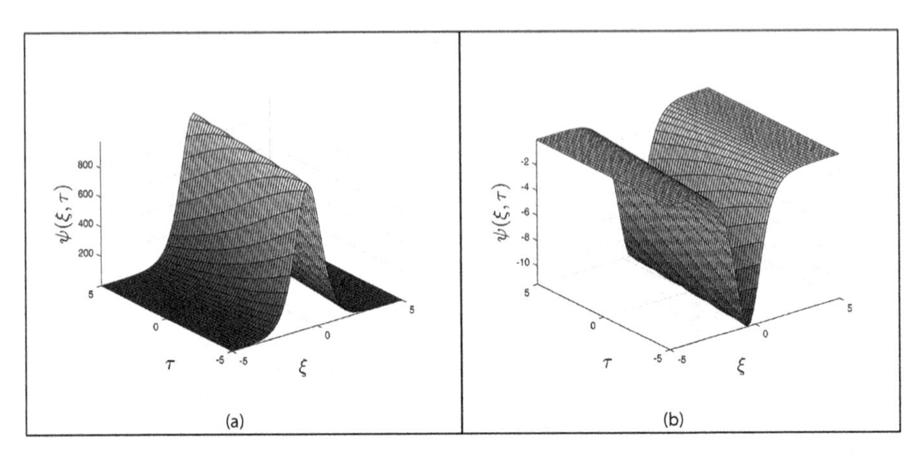

FIGURE 6.5
The solution profile of the KdV equation for the values of the parameters for which the coefficient of non-linearity is very close to zero. The parameters are: (a) $\mu_{e1} = 0.27, \mu_{e2} = 0.62, \sigma = 0.05, V0 = .5$ for these values of the parameters $a = 2.53e - 4$; (b) $\mu_{e1} = 0.20, \mu_{e2} = 0.35, \sigma = 0.15, V0 = .5$ for these parameters $a = -4.7e - 3$. Parameters are chosen on the trajectories shown in Figure 6.4(a) for which a is very close to zero.

The parameters for generating plots in Figure 6.5 are chosen such that the nonlinear coefficient A is very close to zero, and they lie on two different trajectories in Figure 6.4(a). For these values of the parameters, a sudden jump in the hump in Figure 6.5(a) (or kink in Figure 6.5(b)) is observed, which is unacceptable. Thus, finding a feasible range of the parameters for which the solitons exist or may exhibit certain important phenomena is of utmost interest. The plots shown in Figure 6.5 also indicate that the stretching coefficients used here to derive the KdV equation may not be appropriate in a case where the coefficient of non-linearity is very close to zero. This gives rise to mKdV, which is discussed in the next section. Other trajectories shown in Figure 6.4(b) and (c) may also be used for the same purpose.

6.3 Derivation of Damped mKdV Equation

Here, we consider weak damping due to the presence of collision effects in this dusty plasma environment, which leads the basic governing to be adapted as

$$\frac{\partial n_j}{\partial t} + \frac{\partial (n_j v_j)}{\partial x} = 0, \tag{6.8a}$$

$$\frac{\partial v_j}{\partial t} + v_j \frac{\partial v_j}{\partial x} = -\frac{\partial \psi}{\partial x} - \nu_{id0} v_j, \tag{6.8b}$$

$$\frac{\partial^2 \psi}{\partial x^2} = -\rho = \mu + \mu_{e1} e^{\psi} + (1 - \mu_{e1} - \mu) e^{\sigma \psi} - n_j, \tag{6.8c}$$

$$\nu_{jd} \approx \epsilon^3 \nu_{jd0}. \tag{6.8d}$$

In this case, we consider following stretching co-ordinates

$$\xi = \epsilon(x - u_p t), \tau = \epsilon^3 t. \tag{6.9}$$

Considering the third-order calculation for ϵ, we finally obtain the following mKdV equation with damping term:

$$\frac{\partial \phi}{\partial \tau} + A\phi^2 \frac{\partial \phi}{\partial \xi} + B \frac{\partial^3 \phi}{\partial \xi^3} + C\phi = 0, \tag{6.10}$$

where $A = a_1 a_2$, $a_1 = \left(\frac{15}{2u_p^6} - \frac{1}{2}\mu_{e1} - \frac{1}{2}\mu_{e2}\sigma^3 \right)$, $B = a_2 = \frac{u_p^3}{2}$, $C = \frac{\nu_{jd0}}{2}$.

6.4 Solution of Damped mKdV Equation

The analytical solution for an mKdV equation with a damping term such as Equation 6.10 is not available. In this section, we derive an analytical approximate solution for it. For the standard mKdV equation given by

$$\frac{\partial \phi_1}{\partial \tau} + A\phi_1^2 \frac{\partial \phi_1}{\partial \xi} + B \frac{\partial^3 \phi_1}{\partial \xi^3} = 0, \tag{6.11}$$

the solitary wave solution is available and it takes the following form [18]

$$\psi = \lambda_0 \operatorname{sech} \eta_0, \ \eta_0 = w(\xi - V_0 \tau), \tag{6.12}$$

$$\text{with} \quad w^2 = \frac{A\lambda_0^2}{6B}, \ V_0 = \frac{A\lambda_0^2}{6}, \tag{6.13}$$

where λ_0 is the constant amplitude. Motivated by the solution given by Equation 6.12, we consider a progressive wave solution for the evolution Equation 6.10 of the form

$$\psi = \lambda(\tau)\operatorname{sech}\eta,\ \eta = w(\tau)(\xi - V(\tau)), \tag{6.14}$$

$$\text{with} \quad w^2(\tau) = \frac{A\lambda^2(\tau)}{6B},\ V'(\tau) = \frac{A\lambda^2(\tau)}{6}, \tag{6.15}$$

where the prime denotes differentiation with respect to time τ. As a matter of fact, the solutions given in Equations 6.14–6.15 are formally the same as in Equations 6.12–6.13, except that in Equation 6.15, $\lambda(\tau)$ is a function of time τ and still undetermined. However, when the expressions given by Equation 6.14 and Equation 6.15 are inserted into the evolution Equation 6.10, it will not be satisfied identically; there will, rather, be a residue term $R(\xi,\tau)$ given by

$$R(\eta,\tau) = [\lambda' + C\lambda]\operatorname{sech}\eta - \frac{\lambda w'\eta}{w}\operatorname{sech}\eta\tanh\eta. \tag{6.16}$$

The residue term cannot be forced to be zero point by point. Here, we note that the residue term $R(\xi,\tau)$ is an even function of the variable η. To obtain a differential equation for $\lambda(\tau)$, we shall use the weighted residual method (WRM) as implemented in [35, 36]. In order to get strong restrictions on the coefficient $\lambda(\tau)$, we must select an even function as a weight function. For that purpose, we choose sechη as the weight function for this problem. Multiplying Eq. 6.16 by sechη, integrating it from $\eta = -\infty$ to $\eta = \infty$ and setting the result equal to zero, we obtain

$$\lambda' + C\lambda - \frac{w'\lambda}{2w} = 0. \tag{6.17}$$

Eliminating w between Equations 6.15 and 6.17, we get

$$\lambda' + 2C\lambda = 0. \tag{6.18}$$

The solution of this differential equation is given by

$$\lambda(\tau) = \frac{A\lambda_0^2}{6}e^{-2C\tau}, \tag{6.19}$$

where λ_0 is the wave amplitude defined in Eq. 6.12. Inserting Eq. 6.19 into Eq. 6.14, the other unknown quantities are obtained. They are as follows:

$$w(\tau)=\sqrt{\frac{A^3\lambda_0^2}{216B}}e^{-2C\tau},\ V(\tau)=\frac{A^3\lambda_0^4}{864C}\left(1-e^{-4C\tau}\right). \tag{6.20}$$

Thus, we obtain the general approximated solution for the damped mKdV equation which is given by

$$\psi=\frac{A\lambda_0^2}{6}e^{-2C\tau}\operatorname{sech}\eta, \tag{6.21}$$

where $$\eta=\sqrt{\frac{A^3\lambda_0^2}{216B}}e^{-2C\tau}\left(\xi-\frac{A^3\lambda_0^4}{864C}\left(1-e^{-4C\tau}\right)\right). \tag{6.22}$$

The effect of changing the parameters on the solution profiles of damped mKdV is shown in Figure 6.6. In Figure 6.7, we demonstrate the effect of the damping coefficient on the solution profile. It is evident from the plots that with an increase in the damping coefficient, the hump with positive potential tends to become a sharper, more spike-like profile.

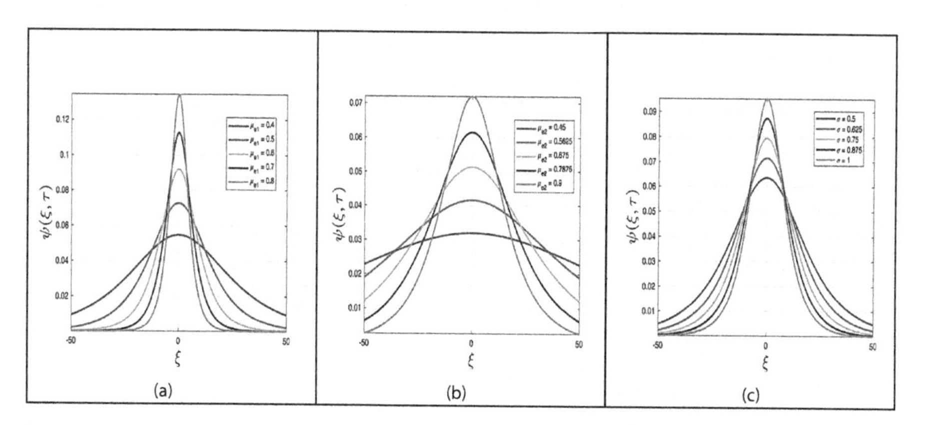

FIGURE 6.6
Profiles of solution of damped mKdV equation; the effect of variation of different parameters on the solution profiles is shown here. The values of the parameters are: (a) $\nu_{jd0}=0.01$, $\mu_{e1}=p*0.40$, $\mu_{e2}=0.35$, $\sigma=0.5$, $\lambda_0=0.5$; (b) $\nu_{jd0}=0.01$, $\mu_{e1}=0.20$, $\mu_{e2}=p*0.45$, $\sigma=0.5$; (c) $\nu_{jd0}=0.01$, $\mu_{e1}=0.45$; $\mu_{e1}=0.35$, $\sigma=p*0.5$, $\lambda_0=0.5$; p denotes a multiplier which takes five values between 1 and 2.

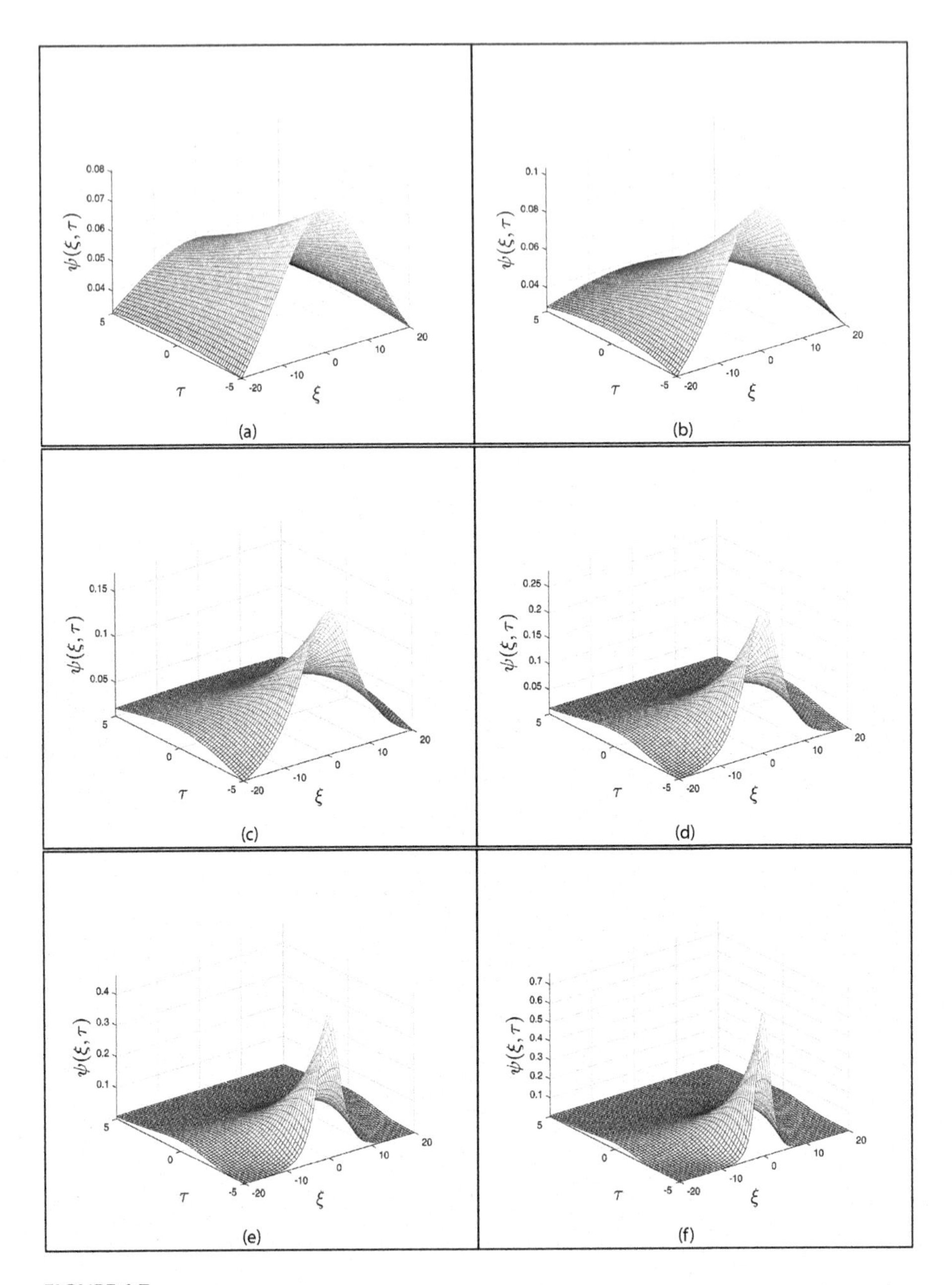

FIGURE 6.7
Profiles of damped mKdV equation in (ξ,τ) plane for different values of damping coefficients ν_{jd0}; the parameters are $\nu_{jd0} = p * 0.01, \mu_{e1} = 0.4, \mu_{e2} = 0.5, \sigma = .4, \lambda_0 = .5$; p denotes a multiplier which takes six different values given by $p = [5,10,20,30,40,50]$. It is clear that the hump with positive potential tends to become spikier with an increase in the damping coefficient.

6.5 Derivation of Damped Gardner Equation

As argued earlier, in a case where the coefficient of non-linearity tends to zero, there is a strong possibility that the formation of infinite amplitude solitons for both KdV and mKdV equations may need to be dealt with. To restrict these infinite amplitude solitons, we formulate the Gardner equation in this section.

It is obvious that the coefficient of non-linearity A in Eq. 6.13 is a function of μ_{e1}, μ_{e2}, and σ. So, to explore the parametric zones corresponding to $A = 0$, we express σ in terms of the other two parameters, namely, μ_{e1} and μ_{e2}). Thus, we have

$$\sigma_c = \frac{-3\mu_{e1}\mu_{e2} + \sqrt{\mu_{e1}\mu_{e2}\left(3\mu_{e1} + 3\mu_{e2} - 1\right)}}{\mu_{e2}\left(3\mu_{e2} - 1\right)} \tag{6.23}$$

where σ_c denotes the critical value of σ for which $A = 0$. In Figure 6.8 (a), we plot how σ varies with respect to the parameters μ_{e1} and μ_{e2}. In Figure 6.8 (b) level curves of a function $f\left(\mu_{e1}, \mu_{e2}, \sigma\right) \equiv A = \left(\frac{3}{u_p^4} - \mu_{e1} - \mu_{e2}\sigma^2\right) = 0$ are plotted. Note that $u_p = 1/\sqrt{\mu_{e1} + \mu_{e2}\sigma}$, hence a function of ($\mu_{e1}$, μ_{e2}, σ). These level curves are parallel to the (μ_{e1}, μ_{e2}) plane, and they may be useful to depict the range of the parameters for which the coefficient of non-linearity would tend to zero. So, in the neighborhood of the critical value (σ_c), $A = A_0$ can be expressed as

$$f\left(\mu_{e1}, \mu_{e2}, \sigma\right) \equiv A = \left(\frac{3}{u_p^4} - \mu_{e1} - \mu_{e2}\sigma^2\right) = 0$$

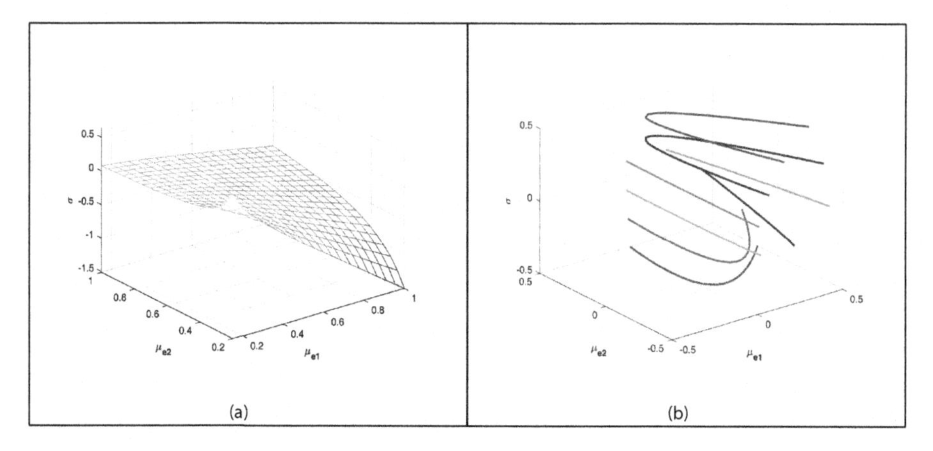

FIGURE 6.8
(a) Demonstrates how σ_c varies with respect to the parameters μ_{e1} and μ_{e2}, (b) shows the level curves of a function.

$$A_0 \approx s\left(\frac{\partial a}{\partial \sigma}\right)_{\sigma=\sigma_c}, \qquad |\sigma-\sigma_c| = c_1 s\epsilon, \tag{6.24}$$

where the constant c_1 depends on the parameters μ_{e1} and μ_{e2} and σ. $|\sigma - \sigma_c|$ is represented as a small dimensionless parameter which can also be considered as the expansion parameter ϵ, i.e., $|\sigma - \sigma_c| \approx \epsilon$, and $s = 1$ for $\sigma > \sigma_c$ and $s = -1$ for $\sigma < \sigma_c$. So, ρ_2 (in Eq. 6.5) can be expressed as

$$\epsilon^2 \rho_2 \approx -\epsilon^3 \frac{1}{2} c_1 s \phi^2, \tag{6.25}$$

which is included in the third-order Poisson's equation. To the next higher order in ϵ, we get the following equation:

$$\frac{\partial^2 \phi}{\partial \xi^2} = -\frac{1}{2} c_1 s \phi^2 + \frac{\left(\mu_{e1} + \mu_{e2}\sigma^3\right)}{6}\psi_1^3 + \left(\mu_{e1} + \mu_{e2}\sigma\right)\psi_3 + \left(\mu_{e1} + \mu_{e2}\sigma^2\right)\phi\psi_2 - n_3. \tag{6.26}$$

Differentiating Equation 6.26 with respect to ξ, we obtain

$$\begin{aligned}\frac{\partial^3 \phi}{\partial \xi^3} = &-c_1 s \phi \frac{\partial \phi}{\partial \xi} + \frac{\left(\mu_{e1} + \mu_{e2}\sigma^3\right)}{2}\psi_1^2 \frac{\partial \psi_1}{\partial \xi} + \left(\mu_{e1} + \mu_{e2}\sigma\right)\frac{\partial \psi_3}{\partial \xi} \\ &+ \left(\mu_{e1} + \mu_{e2}\sigma^2\right)\frac{\partial}{\partial \xi}\left(\phi\psi_2\right) - \frac{\partial n_3}{\partial \xi}.\end{aligned} \tag{6.27}$$

From Eqs 6.14 and 6.15, we have the higher $O(\epsilon)$ equation

$$\frac{\partial n_1}{\partial \tau} - u_p \frac{\partial n_3}{\partial \xi} + \frac{\partial}{\partial \xi}\left(n_1 v_2\right) + \frac{\partial}{\partial \xi}\left(n_2 v_1\right) + \frac{\partial v_3}{\partial \xi} = 0, \tag{6.28a}$$

$$\frac{\partial v_1}{\partial \tau} - u_p \frac{\partial v_3}{\partial \xi} + v_1 \frac{\partial v_2}{\partial \xi} + v_2 \frac{\partial v_1}{\partial \xi} + \frac{\partial \psi_3}{\partial \xi} + \nu_{jd0} v_1 = 0. \tag{6.28b}$$

Using Eqs 6.28a and 6.28b, in Eq. 6.27 we have

$$\frac{\partial \phi}{\partial \tau} + A_1 \phi \frac{\partial \phi}{\partial \xi} + A_2 \phi^2 \frac{\partial \phi}{\partial \xi} + A_3 \frac{\partial^3 \phi}{\partial \xi^3} + A_4 \phi = 0, \tag{6.29}$$

where $A_1 = \dfrac{c_1 s u_p^3}{2}$, $A_2 = \dfrac{u_p^3}{2}\left(\dfrac{15}{2u_p^6} - \dfrac{1}{2}\mu_{e1} - \dfrac{1}{2}\mu_{e2}\sigma^3\right)$, $A_3 = \dfrac{u_p^3}{2}$, $A_4 = \dfrac{\nu_{jd0}}{2}$.

Equation 6.29 is known as a damped Gardner equation. If we consider the damping term as zero, Eq. 6.29 reduces to a standard Gardner equation as

$$\frac{\partial \phi}{\partial \tau} + A_1 \phi \frac{\partial \phi}{\partial \xi} + A_2 \phi^2 \frac{\partial \phi}{\partial \xi} + A_3 \frac{\partial^3 \phi}{\partial \xi^3} = 0. \tag{6.30}$$

The reader is referred to [37, 38], where references for the analytical solution of the Gardner equation can be found. It is known that the solution of the Gardner Eq. 6.30 strongly depends on the sign of the coefficient A_2. For instance, for $A_2 < 0$, there exists one family of solitary waves, commonly referred to as solitons, and for $A_2 > 0$, two families of the solitons along with certain oscillating wave packets may exist; refer to [39, 40] for more details. In these papers, the authors have shown that the interaction of solitons in the Gardner equation is very different from that in the KdV equation. Due to the presence of a damping term in Eq. 6.26, the solitons may exhibit very distinctive features which may be a subject for further investigation. Furthermore, there is evidence in the literature that in many practical cases both non-linear coefficients A_1 and A_2 may change their signs depending on certain parameters; see for example references in [25]. The non-linear evolution equations emerging from plasma physics, for example Eq. 6.29 or Eq. 6.30, may also exhibit similar behavior which will make such problems more challenging.

6.6 The Time-Fractional Damped Gardner Equation (TFDGE)

Fractional differential equations have recently attracted much attention. It has been observed that many complex phenomena in the areas of science and engineering, such as electromagnetics, acoustics, electrochemistry, cosmology, and surface engineering can be well formulated using fractional differential equations; see, for example, references in [41–51]. For the numerical treatment of fractional differential equations, the reader is referred to [52–62]; see also [63, 64] for various applications of fractional differential equations and some recent advancements in related areas.

The main motivation for deriving the time-fractional Gardner equation is that in many real-world problems, information about historical states may also be required to describe the current or next state of a physical phenomenon. A fractional differential equation is more suitable for understanding various nonlinear physical phenomena that may appear due to interactions and collisions between dust plasma and ions, in particular in circumstances when the coefficients of non-linear terms may be expected to behave abruptly. In this section, we discuss the derivation of the time-fractional damped Gardner equation.

The Gardner Equation 6.30 can be converted into the time-fractional Gardner equation (TFGE) by employing the potential function $\phi(\xi,\tau)$ where $\phi(\xi,\tau) = U\xi(\xi,\tau)$. We consider the following Lagrangian equation as

$$L = -\frac{1}{2}U_\tau U_\xi - \frac{1}{6}A_1 U_\xi^3 - \frac{1}{12}A_2 U_\xi^4 + \frac{1}{2}A_3 U_{\xi\xi}. \tag{6.31}$$

The time-fractional Lagrangian equation for the Gardner equation can be written as

$$F\left({}_0D_\tau^\alpha U, U_\xi, U_{\xi\xi}\right) = -\frac{1}{2}{}_0D_\tau^\alpha U_\tau U_\xi - \frac{1}{6}A_1 U_\xi^3 - \frac{1}{12}A_2 U_\xi^4 + \frac{1}{2}A_3 U_{\xi\xi},\ 0 \le \alpha < 1, \tag{6.32}$$

where ${}_0D_\tau^\alpha$ is left Riemann–Liouville fractional derivative defined as follows [43, 45]

$${}_aD_\tau^\alpha = \frac{1}{\Gamma(M-\alpha)}\frac{d^M}{dt^M}\left(\int_a^t d\tau\left(t-\tau\right)^{M-\alpha-1} f\left(\tau\right)\right),\ M-1 \le \alpha \le M,\ t \in \left[a,b\right]. \tag{6.33}$$

Then, the functional of the TFGE can be written as

$$J\left(U\right) = \int_R d\xi \int_T d\tau F\left({}_0D_\tau^\alpha U, U_\xi, U_{\xi\xi}\right). \tag{6.34}$$

Considering the variational functional Eq. 6.34, and imposing the optimization constraints, i.e., $\delta U|_T = \delta U|_R = \delta U\xi|_R = 0$ with respect to $U(\xi,\tau)$ [65–67], leads to the following Euler–Lagrange equation

$${}_\tau D_{T_0}^\alpha\left(\frac{\partial F}{\partial_0 D_\tau^\alpha U}\right) - \frac{\partial}{\partial \xi}\left(\frac{\partial F}{\partial U_\xi}\right) + \frac{\partial^2}{\partial \xi^2}\left(\frac{\partial F}{\partial U_{\xi\xi}}\right) = 0. \tag{6.35}$$

Employing the Lagrangian of the TFGE Equation 6.32 in the Euler–Lagrange formula Eq.6.35, we get

$$-\frac{1}{2}{}_\tau D_{T_0}^\alpha U_\xi\left(\xi,\tau\right) + \frac{1}{2}{}_0D_\tau^\alpha U_\xi\left(\xi,\tau\right) + \left(A_1 U_\xi\left(\xi,\tau\right) + A_2 U_\xi^2\left(\xi,\tau\right)\right)U_{\xi\xi}\left(\xi,\tau\right) + A_3 U_{\xi\xi\xi\xi}\left(\xi,\tau\right) = 0. \tag{6.36}$$

Switching for the potential function $U\xi(\xi,\tau)=\phi(\xi,\tau)$ yields the TFGE for the state function $\psi(\xi,\tau)$ in the following form

$$-\frac{1}{2}{}_{\tau}D_{T_0}^{\alpha}\phi(\xi,\tau)+\frac{1}{2}{}_{0}D_{\tau}^{\alpha}\phi(\xi,\tau)+\left(A_1\phi(\xi,\tau)+A_2\phi^2(\xi,\tau)\right)\phi_{\xi}(\xi,\tau)+A_3\phi_{\xi\xi\xi}(\xi,\tau)=0. \tag{6.37}$$

The TFGE represented in Equation 6.37 can be rewritten as follows

$${}_{0}^{R}D_{\tau}^{\alpha}\phi(\xi,\tau)+\left(A_1\phi(\xi,\tau)+A_2\phi^2(\xi,\tau)\right)\phi_{\xi}(\xi,\tau)+A_3\phi_{\xi\xi\xi}(\xi,\tau)=0, \tag{6.38}$$

where the fractional operator ${}_{0}^{R}D_{\tau}^{\alpha}$ is a Riesz fractional derivative operator and can be represented as follows [30, 43]

$$\begin{aligned}{}_{0}^{R}D_{t}^{\alpha}f(t)&=\frac{1}{2}\left[{}_{0}D_{t}^{\alpha}f(t)+(-1)^{k}{}_{t}D_{T_0}^{\alpha}f(t)\right]\\&=\frac{1}{2}\frac{1}{\Gamma(k-\alpha)}\frac{d^k}{dt^k}\left(\int_a^t(t-\tau)^{k-\alpha-1}f(\tau)d\tau\right),\end{aligned} \tag{6.39}$$

where $k-1\le\alpha\le k,\ t\in[a,b]$, a and b are real. In the presence of a damping term in Eq. 6.38, the equation will be known as the time-fractional damped Gardner equation given by

$${}_{0}^{R}D_{\tau}^{\alpha}\phi(\xi,\tau)+\left(A_1\phi(\xi,\tau)+A_2\phi^2(\xi,\tau)\right)\phi_{\xi}(\xi,\tau)+A_3\phi_{\xi\xi\xi}(\xi,\tau)+A_4\phi=0. \tag{6.40}$$

In order to solve the time-fractional damped Gardner equation given by Eq. 6.40, we employ the extended $G'/G-$expansion method, which is briefly described in the next section.

6.7 Key Steps for Implementing the Extended $G'/G-$Expansion Method to Solve Time-Fractional Damped Gardner Equation

Let us take a nonlinear time fractional differential equation (FDE) as

$$\phi_{\tau}=\mathcal{N}\left(\phi,\phi_{\tau}^{\alpha},\phi_{\xi},\phi_{\xi\xi},\phi_{\xi\xi\xi},\phi_{\eta\eta}\cdots\right). \tag{6.41}$$

Here, $\phi = \phi(\xi, \tau)$ is an unknown function. To combine the real variables ξ and τ, we introduce a new variable η as,

$$\phi(\xi,\tau) = u(\eta), \quad \eta = k\left(\xi - \frac{V\tau^{\alpha}}{\Gamma(1+\alpha)}\right), \tag{6.42}$$

where V represents the speed of the travelling wave. Eq. 6.41 is transformed into the following ordinary differential equation

$$Q(u, -Vku', ku', k^2u'', k^3u''', \cdots) = 0, \tag{6.43}$$

where Q is a polynomial of $u = u(\eta)$ and its derivatives. The primes (′) denote the derivatives of u with respect to η. The key steps of the extended $G'/G-$expansion method are briefly described below.

Step 1. Time-fraction differential equation given by Eq. 6.41 is transformed into an ordinary differential equation given by Eq. 6.43 by employing the transformation defined in Eq. 6.42.

Step 2. Following the extended $G'/G-$expansion method [10, 13], the following form of solution $u(\eta)$ of Eq. 6.43 is chosen

$$u(\eta) = P_0 + \sum_{j=1}^{N} P_j \left(\frac{G'(\eta)}{G(\eta)}\right)^j + \sum_{j=1}^{N} Q_j \left(\frac{G'(\eta)}{G(\eta)}\right)^{-j}, \tag{6.44}$$

where $\left(G'(\eta)/G(\eta)\right)^j$ is a Cole–Hoph transformation and the constants P_j, ($j = 0 \cdots N$, N is a positive integer with $P_N \neq 0$. The multipliers P_j are to be determined and the function $G(\eta)$ will satisfy

$$G''(\eta) + \lambda G'(\eta) + \mu G(\eta) = 0, \tag{6.45}$$

where λ and μ are arbitrary constants to be determined. The general solution of Eq. 6.45 is given by

$$\frac{G'(\eta)}{G(\eta)} = \begin{cases} \dfrac{\sqrt{\lambda^2-4\mu}}{2}\left(\dfrac{r_1 \sinh\left(\dfrac{\sqrt{\lambda^2-4\mu}}{2}\eta\right) + r_2 \cosh\left(\dfrac{\sqrt{\lambda^2-4\mu}}{2}\eta\right)}{r_1 \cosh\left(\dfrac{\sqrt{\lambda^2-4\mu}}{2}\eta\right) + r_2 \sinh\left(\dfrac{\sqrt{\lambda^2-4\mu}}{2}\eta\right)}\right) - \dfrac{\lambda}{2}, & \lambda^2 - 4\mu > 0, \\[2ex] \dfrac{\sqrt{-\lambda^2+4\mu}}{2}\left(\dfrac{-r_1 \sin\left(\dfrac{\sqrt{-\lambda^2+4\mu}}{2}\eta\right) + r_2 \cos\left(\dfrac{\sqrt{-\lambda^2+4\mu}}{2}\eta\right)}{r_1 \cos\left(\dfrac{\sqrt{-\lambda^2+4\mu}}{2}\eta\right) + r_2 \sin\left(\dfrac{\sqrt{-\lambda^2+4\mu}}{2}\eta\right)}\right) - \dfrac{\lambda}{2}, & \lambda^2 - 4\mu < 0. \end{cases}$$

Step 3. The positive constant N can be determined by balancing the highest-order derivative and nonlinear terms in Eq. 6.43.

Step 4. Substitute the expansion given by Eq. 6.44 into Eq. 6.43, and using Eq. 6.45, the polynomial of (G'/G) can be obtained. Furthermore, collect all the coefficients of the same power terms of (G'/G), and then set them to zero. The system of algebraic equations thus obtained will yield the solution of the time-fractional differential equation with help of Maple symbolic software.

6.8 Solution of Time-fractional Damped Gardner Equation

We employ the extended G'/G–expansion method to obtain an analytical solution of TFDGE Eq. 6.40. The transformation is defined as $u(\eta) = \phi(\xi,\tau)$, $\eta = k\left(\xi - V\tau^{\alpha}/\Gamma(1+\alpha)\right)$, where V is the wave speed. This yields the following ordinary differential equation

$$-Vku' + A_1kuu' + A_2ku^2u' + A_3k^3u''' + A_4u = 0, \tag{6.46}$$

where prime denotes the derivative with respect to η. Now, we consider Eq. 6.44 as a solution of Eq. 6.46. Balancing the terms u^2u' and u''' in Eq. 6.46, we get $N = 1$. Thus, we have the solution of Eq. 6.46 in the form

$$u(\eta) = P_0 + P_1\left(\frac{G'}{G}\right) + Q_1\left(\frac{G'}{G}\right)^{-1}. \tag{6.47}$$

And in view of Eqs 6.47 and 6.45, we have the following derivatives:

$$u' = -P_1\mu - P_1\lambda\left(\frac{G'}{G}\right) - P_1\left(\frac{G'}{G}\right)^2 + Q_1 + Q_1\lambda\left(\frac{G'}{G}\right)^{-1} + Q_1\mu\left(\frac{G'}{G}\right)^{-2}, \tag{6.48}$$

$$\begin{aligned} u'' &= P_1\lambda\mu + P_1\left(\lambda^2 + 2\mu\right)\left(\frac{G'}{G}\right) + 3P_1\lambda\left(\frac{G'}{G}\right)^2 + 2P_1\left(\frac{G'}{G}\right)^3 \\ &\quad + Q_1\lambda + Q_1\left(\lambda^2 + 2\mu\right)\left(\frac{G'}{G}\right)^{-1} + 3Q_1\lambda\mu\left(\frac{G'}{G}\right)^{-2} + 2Q_1\mu^2\left(\frac{G'}{G}\right)^{-3}, \end{aligned} \tag{6.49}$$

$$u''' = -P_1\mu(\lambda^2+2\mu) - P_1\lambda(8\mu+\lambda^2)\left(\frac{G'}{G}\right) - P_1(8\mu+7\lambda^2)\left(\frac{G'}{G}\right)^2$$
$$-12P_1\lambda\left(\frac{G'}{G}\right)^3 - 6P_1\left(\frac{G'}{G}\right)^4 + Q_1(\lambda^2+2\mu) + Q_1\lambda(\lambda^2+8\mu)\left(\frac{G'}{G}\right)^{-1} \quad (6.50)$$
$$+Q_1\mu(8\mu+7\lambda^2)\left(\frac{G'}{G}\right)^{-2} + 12Q_1\lambda\mu^2\left(\frac{G'}{G}\right)^{-3} + 6Q_1\mu^3\left(\frac{G'}{G}\right)^{-4}.$$

Here, the prime denotes the derivative with respect to η. Substituting expression given by Eq. 6.47 in Eq. 6.46, and utilizing the polynomials in $(G'/G)^j$ $(j = 0, 1, \cdots, 4)$, we set the coefficients of $(G'/G)^j$ $(j = 0, 1, \cdots, 4)$ to zero and obtain the following system of algebraic equations for P_0, P_1, λ, μ and V:

$$\left(\frac{G'}{G}\right)^{-4}: 6k^3\mu^3A_3Q_1 + k\mu A_2Q_1^3 = 0,$$
$$\left(\frac{G'}{G}\right)^{-3}: 12A_3k^3Q_1\lambda\mu^2 + A_1kQ_1^2\mu + A_2kQ_1^3\lambda + 2A_2kP_0Q_1^2\mu = 0,$$
$$\left(\frac{G'}{G}\right)^{-2}: 2A_2kP_0Q_1^2\lambda + A_2kQ_1^3 + A_2kP_1Q_1^2\mu + 7A_3k^3Q_1\lambda^2\mu + A_2kP_0^2Q_1\mu - VkQ_1\mu$$
$$+A_1kQ_1\mu P_0 + A_1kQ_1^2\lambda + 8A_3k^3Q_1\mu^2 = 0,$$
$$\left(\frac{G'}{G}\right)^{-1}: A_2kP_0^2Q_1\lambda + A_1kQ_1^2 + 2A_2kP_0Q_1^2 + A_1kQ_1\lambda P_0 - VkQ_1\lambda + A_2kP_1Q_1^2\lambda + A_4Q_1$$
$$+8A_3k^3Q_1\lambda\mu + A_3k^3Q_1\lambda^3 = 0,$$
$$\left(\frac{G'}{G}\right)^{0}: -A_3k^3P_1\lambda^2\mu + A_1kQ_1P_0 + 2A_3k^3Q_1\mu + A_4P_0 - A_2kP_0^2P_1\mu - 2A_3k^3P_1\mu^2 - A_2kP_1^2Q_1\mu$$
$$+A_3k^3Q_1\lambda^2 + VkP_1\mu - VkQ_1 - A_1kP_1\mu P_0 + A_2kP_0^2Q_1 + A_2kP_1Q_1^2 = 0,$$

$$\left(\frac{G'}{G}\right)^{1}: -A_1kP_1\lambda P_0 + A_4P_1 - A_1kP_1^2\mu - A_2kP_0^2P_1\lambda - 8A_3k^3P_1\lambda\mu - A_3k^3P_1\lambda^3 - A_2kP_1^2Q_1\lambda$$
$$+VkP_1\lambda - 2A_2kP_0P_1^2\mu = 0,$$
$$\left(\frac{G'}{G}\right)^{2}: -A_1kP_1P_0 - A_1kP_1^2\lambda - A_2kP_1^3\mu - 7A_3k^3P_1\lambda^2 - A_2kP_1^2Q_1 - 2A_2kP_0P_1^2\lambda - A_2kP_0^2P_1$$
$$-8A_3k^3P_1\mu + VkP_1 = 0,$$
$$\left(\frac{G'}{G}\right)^{3}: -12k^3\lambda A_3P_1 - k\lambda A_2P_1^3 - 2kA_2P_0P_1^2 - kA_1P_1^2 = 0,$$
$$\left(\frac{G'}{G}\right)^{4}: -6k^3A_3P_1 - kA_2P_1^3 = 0.$$

Solving the above equations using the Maple symbolic system, we find the following results:

Solution Set 1

$$V = \frac{(-k^3\lambda^{11}A_3 + 10\mu k^3\lambda^9 A_3 - 36\mu^2 k^3\lambda^7 A_3 + 56\mu^3 k^3\lambda^5 A_3 - 38\mu^4 k^3\lambda^3 A_3)}{(k\lambda^3(-\lambda^2+2\mu)^3)} + \frac{(12\mu^5 k^3\lambda A_3 + \lambda^8 A_4 - 4\mu\lambda^6 A_4 + 6\mu^2\lambda^4 A_4 - 4\mu^3\lambda^2 A_4 + \mu^4 A_4)}{k\lambda^3(-\lambda^2+2\mu)^3},$$

$$P_1 = \frac{(-6k^3\lambda^7 A_3 + 36\mu k^3\lambda^5 A_3 - 60\mu^2 k^3\lambda^3 A_3 + 24\mu^3 k^3\lambda A_3 - \lambda^4 A_4 + 2\mu^2 A_4)}{k\lambda^2(-\lambda^2+2\mu)^2 A_1}, \tag{6.51}$$

$$P_0 = \frac{\mu P_1(-\lambda^2+\mu)}{(\lambda(-\lambda^2+2\mu))}, \; Q_1 = 0.$$

Substituting these relations from Eq. 6.51 into Eq. 6.47, we obtain the following analytical solutions.

Case 1

When $\Delta = \lambda^2 - 4\mu > 0$, the hyperbolic solution of Eq. 6.46 is as follows:

$$u_1(\eta) = P_0 + P_1\left(\frac{\sqrt{\Delta}}{2}\left(\frac{r_1\cosh\left(\frac{\sqrt{\Delta}}{2}\eta\right) + r_2\sinh\left(\frac{\sqrt{\Delta}}{2}\eta\right)}{r_1\sinh\left(\frac{\sqrt{\Delta}}{2}\eta\right) + r_2\cosh\left(\frac{\sqrt{\Delta}}{2}\eta\right)}\right) - \frac{\lambda}{2}\right), \tag{6.52}$$

where $\eta = k\left(\xi - \frac{V\tau^\alpha}{\Gamma(1+\alpha)}\right)$, and r_1, r_2 are arbitary constants.

Case 2

When $\Delta = \lambda^2 - 4\mu < 0$, the trigonometric from of solution of Eq. 6.46 is as follows:

$$u_2(\eta) = P_0 + P_1\left(\frac{\sqrt{-\Delta}}{2}\left(\frac{-r_1\sin\left(\frac{\sqrt{-\Delta}}{2}\eta\right) + r_2\cos\left(\frac{\sqrt{-\Delta}}{2}\eta\right)}{r_1\cos\left(\frac{\sqrt{-\Delta}}{2}\eta\right) + r_2\sin\left(\frac{\sqrt{-\Delta}}{2}\eta\right)}\right) - \frac{\lambda}{2}\right), \tag{6.53}$$

where $\eta = k\left(\xi - \frac{V\tau^\alpha}{\Gamma(1+\alpha)}\right)$, and r_1, r_2 are arbitary constants.

Solution Set 2

Another set of solutions which leads to a feasible solution is given by

$$V=\frac{\left(-k^3\lambda^{11}A_3+10\mu k^3\lambda^9A_3-36\mu^2k^3\lambda^7A_3+56\mu^3k^3\lambda^5A_3-38\mu^4k^3\lambda^3A_3\right)}{k\lambda^3\left(-\lambda^2+2\mu\right)^3}$$
$$+\frac{\left(12\mu^5k^3\lambda A_3-\lambda^8A_4+4\mu\lambda^6A_4-6\mu^2\lambda^4A_4+4\mu^3\lambda^2A_4-\mu^4A_4\right)}{k\lambda^3\left(-\lambda^2+2\mu\right)^3},$$
$$Q_1=\frac{\mu\left(-6k^3\lambda^7A_3+36\mu k^3\lambda^5A_3-60\mu^2k^3\lambda^3A_3+24\mu^3k^3\lambda A_3+\lambda^4A_4-2\mu^2A_4\right)}{k\lambda^2\left(-\lambda^2+2\mu\right)^2A_1},$$
$$P_0=\frac{Q_1\left(-\lambda^2+\mu\right)}{\left(\lambda\left(-\lambda^2+2\mu\right)\right)},\quad P_1=0. \tag{6.54}$$

Substituting these relations from Eq. 6.54 into Eq. 6.47, we obtain the following analytical solutions.

Case 1

When $\Delta=\lambda^2-4\mu>0$, the hyperbolic solution of Eq. 46 is as follows:

$$u_1(\eta)=P_0+P_1\left(\frac{\sqrt{\Delta}}{2}\left(\frac{r_1\cosh(\frac{\sqrt{\Delta}}{2}\eta)+r_2\sinh(\frac{\sqrt{\Delta}}{2}\eta)}{r_1\sinh(\frac{\sqrt{\Delta}}{2}\eta)+r_2\cosh(\frac{\sqrt{\Delta}}{2}\eta)}\right)-\frac{\lambda}{2}\right), \tag{6.55}$$

where $\eta=k\left(\xi-\frac{V\tau^{\alpha}}{\Gamma(1+\alpha)}\right)$, and r_1,r_2 are arbitrary constants.

Case 2

When $\Delta=\lambda^2-4\mu<0$, the trigonometric from of solution of Eq. 6.46 is as follows:

$$u_2(\eta)=P_0+P_1\left(\frac{\sqrt{-\Delta}}{2}\left(\frac{-r_1\sin(\frac{\sqrt{-\Delta}}{2}\eta)+r_2\cos(\frac{\sqrt{-\Delta}}{2}\eta)}{r_1\cos(\frac{\sqrt{-\Delta}}{2}\eta)+r_2\sin(\frac{\sqrt{-\Delta}}{2}\eta)}\right)-\frac{\lambda}{2}\right), \tag{6.56}$$

where $\eta = k\left(\xi - \frac{V\tau^{\alpha}}{\Gamma(1+\alpha)}\right)$, and r_1, r_2 are arbitrary constants.

For our numerical experiments, we consider the solution given by Eq. (6.52) with $r_1 = 0$. To be specific, we consider

$$u_1(\eta) = P_0 + P_1\left(\frac{\sqrt{\Delta}}{2}\tanh\left(\frac{\sqrt{\Delta}}{2}\eta\right) - \frac{\lambda}{2}\right), \quad \text{where } \eta = k\left(\xi - \frac{V\tau^{\alpha}}{\Gamma(1+\alpha)}\right).$$

The parameters P_0, P_1 and V are given in Equation 6.51.

In Figure 6.9, we demonstrate the effect of our model's two most important parameters α and ν_{id0} on the wave propagation. The parameters considered for Figure 6.9(a) are when $\mu_{e1} = 0.05$, $\mu_{e2} = 0.30$, $\sigma = 0.5$, $c_1 = 0.5$, $s = 0.5$, $\tau = 10$, $\nu_{id0} = 0.01$, $\lambda = 0.8$, $\mu = 0.01$, $k = .05$. For Figure 6.9(a), we consider following values of the parameters: $\alpha = 0.5$, $\mu_{e1} = .35$, $\mu_{e2} = 0.05$, $\sigma = 0.8$, $c_1 = 0.04$, $s = 0.4$, $\tau = 5$, $\lambda = 0.9$, $k = .05$. It is clear from the figure that a bigger value of the damping coefficient makes the wave steeper.

Three-dimensional wave profiles are drawn in Figure 6.10 to exhibit the impact of the fractional order alpha. Figure 6.10(a) and Figure 6.10(b) correspond to $\alpha = 1$ and $\alpha = 0.5$, respectively. The rest of the parameters remain the same for both the figures; they are: $\mu_{e1} = 0.05$, $\mu_{e2} = 0.6$, $\sigma = 0.8$, $c_1 = 1$, $s = 0.4$, $\nu_{id0} = 0.03$, $\mu = 0.01$, $\lambda = 0.8$, $k = 0.05$.

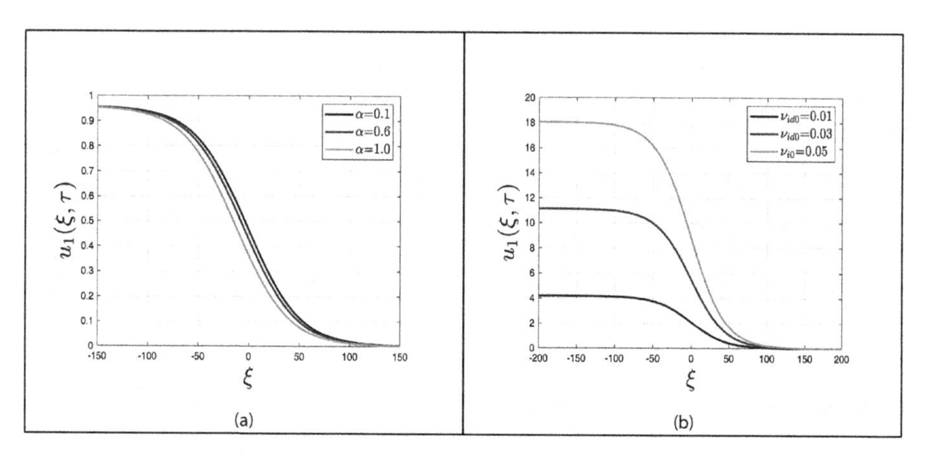

FIGURE 6.9
(a) The effect of different values of α on wave propagation with respect to the space variable ξ; profiles of Eq. (6.52), when $\mu_{e1} = 0.05$, $\mu_{e2} = 0.30$, $\sigma = 0.5$, $c_1 = 0.5$, $s = 0.5$, $\tau = 10$, $\nu_{id0} = 0.01$, $\lambda = 0.8$, $\mu = 0.01$, $k = .05$; (b) Various values of the damping factor are considered to observe its effect on the wave propagation with ξ taken on x–axis; profiles of Eq. (6.52), when $\alpha = 0.5$, $\mu_{e1} = .35$, $\mu_{e2} = 0.05$, $\sigma = 0.8$, $c_1 = 0.04$, $s = 0.4$, $\tau = 5$, $\lambda = 0.9$, $k = .05$.

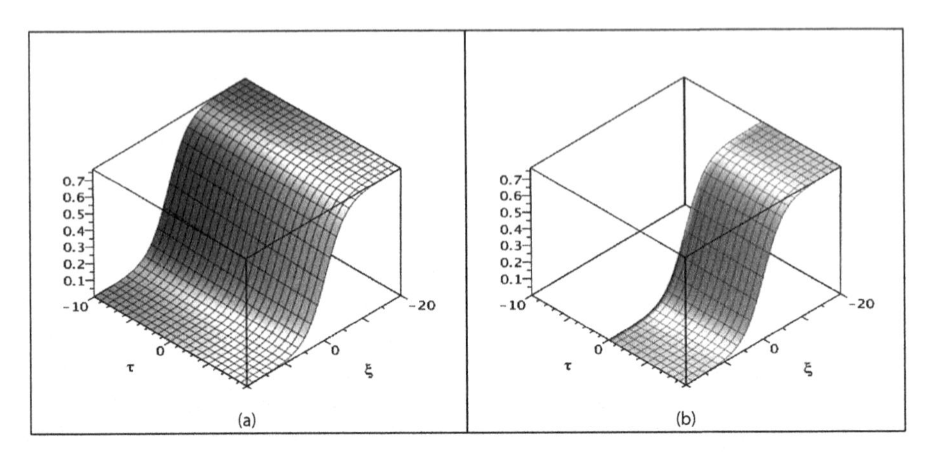

FIGURE 6.10
Profiles of Eq.(6.52), (a) when $\alpha = 1$, $\mu_{e1} = 0.05$, $\mu_{e2} = 0.6$, $\sigma = 0.8$, $c_1 = 1$, $s = 0.4$, $\nu_{id0} = 0.03$, $\mu = 0.01$, $\lambda = 0.8$, $k = 0.05$; (b) All other parameters remain the same except $\alpha = 0.5$.

6.9 Conclusion

In this chapter, using the reductive perturbation technique, we have derived KdV, damped MKdV and damped Gardner equations from the basic governing equation in a collisional dusty plasma. Finally, utilizing the Euler–Lagrange equation, a time-fractional damped Gardner equation is formulated to analyze the propagating characteristic of IAW in a fractional time–space domain. A new type of solution for this equation is formally derived by employing the extended G/G'–expansion method; $\alpha = 1$ would correspond to the solution of the standard damped Gardner equation. Furthermore, the solution profiles are plotted for different values of the parameters which show the impact of physical parameters on wave propagation from a numerical standpoint. It is also clear from the plots that with the change in the order of derivative (α), solution profiles change significantly.

6.10 Physical Significance

The classical KdV equation is a particular type of generic evolutionary PDE which is extensively utilized to model weakly nonlinear long waves. The related theories play a vital role in explaining various important features of unsteady internal waves in coastal oceans. From in-situ measurements and remote sensing, it is observed that long solitary-type waves are commonly seen in density stratified shallow water [68–70]. Recent investigations show that although the KdV framework is approved for a wide range of parameters, there are situations when the Gardner equation is preferred over the KdV equation for modelling purposes. In fact, the Gardner equation can be considered an extension of the KdV equation, considering dual nonlinearity in a classic KdV equation yields Gardner equation [70, 71].

In its classic form, in the Gardner equation, $\phi = \phi(x,t)$ signifies the amplitude of the wave mode, the nonlinear terms represent wave steepening, and the third-order derivative term denotes dispersive effects. In water wave phenomena, the coefficients of dual nonlinear terms and the dispersive term are respectively decided by the density of the oceanic background and stratification of water flow through the linear eigen mode of the internal waves. The Gardner equation is more suited to study a wider class of wave models in a density-stratified ocean in which internal gravity waves are also present. Apart from their applications in ocean engineering [72–79], the Gardner equation and its variants have been extensively used in diverse fields such as quantum field theory, plasma physics, solid-state physics, and so on [80, 81].

References

[1] M. J. Ablowitz, P. A. Clarkson, *Solitons, Nonlinear Evolution Equations and Inverse Scattering*. Cambridge University Press, New York, NY, 1991.

[2] R. Hirota, *The Direct Method in Soliton Theory*. Cambridge University Press, UK, 2004.

[3] Ji-Huan He, X. H. Wu, Exp-function method for nonlinear wave equations. *Chaos Soliton Fract.* 30(3): 700–708, 2006.

[4] A. M. Wazwaz, A sine-cosine method for handling nonlinear wave equations. *Math. Comput. Model.* 40(5–6): 499–508, 2004.

[5] J. L. Zhang, M. L. Wang, Y. M. Wang, Z. D. Fang, The improved F-expansion method and its applications. *Phys. Lett. A Gener. Atom. Solid State Phys.* 350(1–2): 103–109, 2006.

[6] W. Malfliet, W. Hereman, The tanh method: I. Exact solutions of nonlinear evolution and wave equations. *Phys. Scr.* 54: 563–568, 1996.

[7] A. M. Wazwaz, The tanh method for traveling wave solutions of nonlinear equations. *Appl. Math. Comput.* 154(3): 713–723, 2004.

[8] M. A. Abdou, The extended tanh method and its applications for solving nonlinear physical models. *Appl. Math. Comput.* 190(1): 988–996, 2007.

[9] R. Abazari, The (G'/G)-expansion method for Tzitzéica type nonlinear evolution equation. *Math. Comput. Model.* 52(9–10): 1834–1845, 2010.

[10] A. A. Al-Shawba, K. A. Gepreel, F. A. Abdullah, A. Azmi, Abundant closed form solutions of the conformable time fractional Sawada-Kotera-Ito equation using (G'/G)-expansion method. *Results Phys.* 9: 337–343, 2018.

[11] H. Naher, F. A. Abdullah. New approach of (G'/G)-expansion method and new approach of generalized (G'/G)-expansion method for nonlinear evolution equation. *AIP Adv.* 3(3): 032116, 2013.
[12] N. Taghizadeh, S. R. M. Noori, S. B. M. Noori, Application of the extended (G'/G)-expansion method to the improved Eckhaus equation. *Appl. Appl. Math. Int. J.* 9(1): 371–387, 2014.
[13] M. Wang, X. Li, J. Zhang, The G′/G expansion method and travelling wave solutions of nonlinear evolution equations in mathematical physics. *Phys. Lett. A*, 372(4): 417–423, 2008.
[14] S. Liu, Z. Fu, S. Liu, Q. Zhao, Jacobi elliptic function expansion method and periodic wave solutions of nonlinear wave equations. *Phys. Lett. A* 289(1–2): 69–74, 2001.
[15] J.-H. He, Homotopy perturbation method: a new nonlinear analytical technique. *Appl. Maths. Comput.* 135(1): 73–79, 2003.
[16] J.-H. He, Homotopy perturbation method for solving boundary value problems. *Phys. Lett. A* 350(1–2): 87–88, 2006.
[17] J.-H. He, Application of homotopy perturbation method to nonlinear wave equations. *Chaos Soliton. Fract.* 26(3): 695–700, 2005.
[18] S. Bansal, M. Aggarwal, T. S. Gill, Nonplanar ion acoustic waves in dusty plasma with two temperature electrons: Application to Saturn's E ring. *Phys. Plasmas* 27: 083704, 2020.
[19] C. K. Goertz, Dusty plasmas in the solar system. *Rev. Geophys.* 27(2): 271–292, 1989.
[20] P. K. Shukla, V. P. Silin, Dust ion-acoustic wave. *Phys. Scr.* 45: 508, 1992.
[21] P. K. Shukla, B. Eliasson, Colloquium: Fundamentals of dust-plasma interactions. *Rev. Mod. Phys.* 81: 25, 2009.
[22] A. Barkan, N. D'Angelo, R. L. Merlino, Experiments on ion-acoustic waves in dusty plasmas. *Planet. Space Sci.* 44: 239, 1996.
[23] J. A. Gear, R. Grimshaw, A second order theory for solitary waves in shallow fluids. *Phys. Fluids* 26: 14, 1983.
[24] P. E. Holloway, E. Pelinovsky, T. Talipova, B. Barnes, A Nonlinear Model of Internal Tide Transformation on the Australian North West Shelf. *J. Phys. Oceanogr.* 27: 871–896, 1997.
[25] P. E. Holloway, E. Pelinovsky, T. Talipova, A generalized Korteweg–de Vries model of internal tide transformation in the coastal zone. *J. Geophys. Res. C* 104(18): 333–350, 1999.
[26] S. Watanabe, Ion acoustic soliton in plasma with negative ion. *J. Phys. Soc. Jpn.* 53: 950–956, 1984.
[27] S. K. El-Labany, W. F. El-Taibany, Dust acoustic solitary waves and double layers in a dusty plasma with an arbitrary streaming ion beam. *Phys. Plasmas* 10: 989, 2003.
[28] M. M. Masud, M. Asaduzzaman, A. A. Mamun, Dust-ion-acoustic Gardner solitons in a dusty plasma with bi-Maxwellian electrons. *Phys. Plasmas* 19: 103706, 2012.
[29] S. G. Tagare, Dust-acoustic solitary waves and double layers in dusty plasma consisting of cold dust particles and two-temperature isothermal ions. *Phys. Plasmas* 4: 3167, 1997.

[30] O. P. Agrawal, Fractional variational calculus in terms of Riesz fractional derivatives. *J. Phys. A: Math. Theor.* 40: 6287, 2007.
[31] H. G. Abdelwahed, E. K. El-Shewy, A. A. Mahmoud, Cylindrical electron acoustic solitons for modified time- fractional nonlinear equation. *Phys. Plasmas* 24: 082107, 2017.
[32] H. G. Abdelwahed, E. K. El-Shewy, A. A. Mahmoud, Time fractional effect on ion acoustic shock waves in ion-pair plasma. *J. Exp. Theor. Phys.* 122(6): 1111–6, 2016.
[33] J. Sabatier, O. P. Agrawal, J. A. Machado Tenreiro, *Advances in Fractional Calculustheoretical Developments and Applications in Physics and Engineering*, Springer, Dordrecht, 2007.
[34] B. Sahu, D. Roy, Planar and nonplanar electron acoustic solitons in dissipative quantum plasma. *Citation*: *Phys. Plasmas*, 24: 112705, 2017.
[35] H. Demiray, An approximate wave solutions for perturbed KdV and dissipative NLS equations: Weighted residual method. *TWMS J. Appl. Eng. Math.*, 9: 786–791, 2019.
[36] H. Demiray, Analytical solution for nonplanar waves in a plasma with q-nonextensive nonthermal velocity distribution: Weighted residual method. *Chaos Soliton. Fract.*, 130: 109448, 2020.
[37] M. Ferdousi, S. Yasmin, S. Ashraf, A. A. Mamun, Nonlinear propagation of ion-acoustic waves in an electron-positron-ion plasma. *Astrophys. Space Sci.* 352(2): 579–584, 2014.
[38] S. A. Ema, M. Ferdousi, A. A. Mamun. Compressive and rarefactive dust-ion-acoustic Gardner solitons in a multi-component dusty plasma. *Phys. Plasmas* 22(4): 043702, 2015.
[39] A. V. Slyunyaev, E. N. Pelinovski, Dynamics of large-amplitude solitons. *JETP* 89: 173–181, 1999.
[40] A. V. Slyunyaev, Dynamics of localized waves with large amplitude in a weakly dispersive medium with a quadratic and positive cubic nonlinearity. *JETP* 92: 529–534, 2001.
[41] M. Cui, Compact finite difference method for the fractional diffusion equation. *J. Comput. Phys.* 228(20): 7792–7804, 2009.
[42] J.-H. He, Some applications of nonlinear fractional differential equations and their approximations. *Bull. Sci. Technol. Soc.* 15(2): 86–90, 1999.
[43] A. A. Kilbas, H. M. Srivastava, J. J. Trujillo, *Theory and Applications of Fractional Differential Equations*. Elsevier Science Inc., New York, 2006.
[44] R. L. Magin, Fractional calculus models of complex dynamics in biological tissues. *Comput. Math. Appl.* 59(5): 1586–1593, 2010.
[45] K. S. Miller, B. Ross, *An Introduction to the Fractional Calculus and Fractional Differential Equations*. John Wiley & Sons, New York, NY, 1993.
[46] Z. Odibat, S. Momani, The variational iteration method: an efficient scheme for handling fractional partial differential equations in fluid mechanics. *Comput. Math. Appl.* 58(11–12): 2199–2208, 2009.
[47] I. Podlubny, *Fractional Differential Equations*. Academic Press, San Diego, CA, 1999.
[48] S. G. Samko, A. A. Kilbas, O. I. Marichev, *Fractional Integrals and Derivatives: Theory and Applications*. Gordon and Breach, Yverdon, Switzerland, 1993.

[49] H. Singh, Analysis for fractional dynamics of Ebola virus model. *Chaos Soliton. Fract.* 138: 109992, 2020.

[50] H. Singh, Analysis of drug treatment of the fractional HIV infection model of CD4+ T-cell. *Chaos Soliton. Fract.* 146: 110868, 2021.

[51] B. J. West, M. Bologna, P. Grigolini, *Physics of Fractal Operators*. Springer, New York, 2003.

[52] K. Burrage, N. Hale, D. Kay, An efficient implicit FEM scheme for fractionalin-space reaction-diffusion equations. *SIAM J. Sci. Comput.* 34(4): A2145–A2172, 2012.

[53] K. Diethelm, An investigation of some nonclassical methods for the numerical approximation of Caputo-type fractional derivatives. *Numer. Algorithms* 47(4): 361–390, 2008.

[54] K. Diethelm, N. J. Ford, A. D. Freed, Y. Luchko, Algorithms for the fractional calculus: a selection of numerical methods. *Comput. Methods Appl. Mech. Engrg.* 194(6–8): 743–773, 2005.

[55] K. Diethelm, A. D. Freed, The FracPECE subroutine for the numerical solution of differential equations of fractional order. In S. Heinzel and T. Plesser (eds.) *Forschung und wissenschaftliches Rechnen 1998*, 1999, pp. 57–71.

[56] R. Garrappa, On some generalizations of the implicit Euler method for discontinuous fractional differential equations. *Math. Comput. Simulat.* 91: 213–228, 2012.

[57] A. Kadem, Y. Luchko, D. Baleanu, Spectral method for solution of the fractional transport equation. *Rep. Math. Phys.* 66(1): 103–115, 2010.

[58] C. Li, F. Zeng, The finite difference methods for fractional ordinary differential equations. *Numer. Funct. Anal. Optim.* 34(2): 149–179, 2013.

[59] C. Lubich, Discretized fractional calculus. *SIAM J. Math. Anal.* 17(3): 704–719, 1986.

[60] H. Singh, Jacobi collocation method for the fractional advection dispersion equation arising in porous media. *Numer. Meth. Part. Differ. Equ.* 37, 2020. DOI:10.1002/num.22674

[61] H. Singh, Numerical simulation for fractional delay differential equations. *Int. J. Dyn. Control* 9: 1–12, 2020.

[62] H. Singh, H. M. Srivastava, Numerical investigation of the fractional-order Liénard and Duffing equations arising in oscillating circuit theory. *Front. Phys.* 8: 120, 2020.

[63] H. Singh, D. Kumar, D. Baleanu, *Methods of Mathematical Modelling Fractional Differential Equations*, CRC Press, Taylor and Francis, 2019.

[64] H. Singh, J. Singh, S. D. Purohit, D. Kumar, *Advanced Numerical Methods for Differential Equations: Applications in Science and Engineering*, CRC Press, Taylor and Francis, 2021.

[65] O. P. Agrawal, Formulation of Euler-Lagrange equations for fractional variational problems. *J. Math. Anal. Appl.*, 272(1): 368–79, 2002.

[66] O. P. Agrawal, Fractional variational calculus in terms of Riesz fractional derivatives. *J. Phys. A Math. Theor.*, 40(24): 6287–303, 2007.

[67] O. P. Agrawal, A general finite element formulation for fractional variational problems. *J. Math. Anal. Appl.*, 337(1): 1–12, 2008.

[68] K. R. Helfrich, W.K. Melville, Long nonlinear internal waves. *Annu. Rev. Fluid Mech.* 38: 395–425, 2006.

[69] K. Lamb, L. Yan, The Evolution of Internal Wave Undular Bores: Comparisons of a Fully Nonlinear Numerical Model with Weakly Nonlinear Theory. *J. Phys. Oceanogr.* 26: 2712–2734, 1996.
[70] C. Y. Lee, R. C. Beardsley, The generation of long nonlinear internal waves in a weakly stratified shear flow. *J. Geophys. Res.* 79: 453–462, 1974.
[71] R. M. Miura, Korteweg-de Vries equation and generalizations I. A remarkable explicit nonlinear transformation. *J. Math. Phys.* 9: 1202–1204, 1968.
[72] A. K. Daoui, H. Triki, Solitary waves, shock waves and singular solitons of Gardner's equation for shallow water dynamics. *Acta Phys. Pol. B* 45(6):1135–1145, 2014.
[73] R. Grimshaw, Internal solitary waves. In: Grimshaw, R. (ed.) *Environmental Stratified Flows*. Kluwer, Dordcrecht, pp. 1–28, 2001.
[74] R. Grimshaw, E. Pelinovsky, X. Tian, Interaction of solitary wave with an external force. *Phys. D* 77: 405–433, 1994.
[75] R. Grimshaw, E. Pelinovsky, T. Talipova, Solitary wave transformation due to a change in polarity. *Stud. Appl. Math.* 101: 357–388, 1998.
[76] R. Grimshaw, D. Pelinovsky, E. Pelinovsky, A. Slunyaev, Generation of large-amplitude solitons in the extended Korteweg–de Vries equation. *Chaos* 12: 1070–1076, 2002a.
[77] R. Grimshaw, E. Pelinovsky, T. Talipova, Damping of large-amplitude solitary waves. *Wave Motion* 37: 351–364, 2003.
[78] R. Grimshaw, E. Pelinovsky, Y. Stepanyants, T. Talipova, Modelling internal solitary waves on the Australian NorthWest Shelf. *Mar. Freshw. R.es* 57: 265–272, 2006.
[79] P. Holloway, E. Pelinovsky, T. Talipova, Internal tide transformation and oceanic internal solitary waves. In: Grimshaw, R. (ed) *Environmental Stratified Flows*. Kluwer, Dordrecht, pp. 29–60, 2001.
[80] G. Betchewe, K. K. Victor, B. B. Thomas, K. T. Crepin, New solutions of the Gardner equation: Analytical and numerical analysis of its dynamical understanding. *Appl. Math. Comput.*, 223: 377–388, 2013.
[81] X. G. Xu, X. H. Meng, Y. T. Gao, X. Y. Wen, Analytic N-solitary-wave solution of a variable-coefficient Gardner equation from fluid dynamics and plasma physics. *Appl. Math. Comput.* 210(2): 313–320, 2009.

7

An Efficient Numerical Algorithm for Fractional Differential Equations

Ram K. Pandey
Dr. H.S. Gour Vishwavidyalaya, Sagar, India

Neelam Tiwari
Govt. P.M.R.S. College, GPM (C. G.), India

Harendra Singh
P. G. College, Ghazipur, India

CONTENTS

7.1 Introduction

A fractional derivative is a generalization of integer-order derivatives that consists of non-integer order, and it can be a rational, irrational, or even complex-valued number. Fractional differential equations have received considerable attention from mathematicians, physicists, and engineers, and have been widely used in many interdisciplinary applications, such as diffusion process, visco-elasticity, electro-chemistry, biological systems, and control theory [1–4]. In [5], the analytical fractional sub-equation method

DOI: 10.1201/9781003263517-7

is used to solve the historical Burgers–Huxley equation of fractional order. The Burgers–Huxley equation is widely used in many applications related to metallurgy, biology, chemistry, engineering, and mathematics. Several research papers have investigated the theory and solutions of fractional differential equations (see [6–9]).

It is well known that the integer-order differential operator is a local operator, whereas the fractional-order differential operator is non-local in the sense that the next state of the system depends not only upon its current state but also upon all of its preceding states. In the last decade, many authors have made notable contributions to both theory and application of fractional differential equations in areas as diverse as finance [10–12], physics [13–17], control theory [18], and hydrology [19]. Several published papers have shown the equivalence between transport equations using fractional-order derivatives and some heavy-tailed motions, thus extending the predictive capability of models built on the stochastic process of Brownian motion [20–21]. The motion can be heavy-tailed, implying extremely long-term correlation and fractional derivatives in time and/or space.

Finding the analytical solution of a fractional differential equation is not straightforward in general, due to the singular kernel appearing in the fractional derivative. Thus, it is necessary to develop accurate and efficient methods to solve fractional differential equations.

Until the early 1990s, no analytical method for such equations was available, even for linear fractional differential equations. In the 1990s powerful analytical methods, such as the homotopy analysis method (HAM) and variational iteration method (VIM), were proposed by Liao [22] and He [23] respectively. Several authors contributed research papers to find the numerical solution of initial value problems (IVPs) in fractional differential equations; ongoing research and relevant progress has been reported in the literature [24–27]. Other numerical methods proposed to find the numerical solution of FDEs include the spline collocation method [28], fractional Euler method, and modified trapezoidal rule using the generalized Taylor series expansion [29–30], fractional Adams method [31], Haar wavelet method [32], and operational matrix method [33]. Several other numerical algorithms to solve FDE are referenced in [34–39].

The family of RK formulae is one of the most widely used methods to find the numerical solution of IVPs in ordinary differential equations arising in various fields of applied mathematics and computational physics. An excellent book by J.C. Butcher [40] covers the development of Runge–Kutta methods and their applications. Several types of Runge–Kutta methods have been developed on the basis of stability properties and truncation error bounds. In the last two decades, several modifications to existing classical RK methods in the direction of new high-order more accurate RK methods have been envisaged. A detailed review of RK methods by Kalogiratou et al. [41] includes a short history of several modified RK methods.

The Runge–Kutta method initially derived to find the numerical solution of IVPs in first-order differential equations is given by

$$\frac{du}{dt} = f(t,y),\ u(t_0) = u_0,\ t \in [t_0, b]. \tag{7.1}$$

where $u = u(t)$ is unknown solution function and f: $R^2 \rightarrow R$, and it is assumed that f satisfies the Lipschitz condition so that a unique solution of the IVP (7.1) exists.

In the case of differential equations of integer order, the traditional RK family of formulae fails to solve IVPs having special characteristics in the solution, such as periodicity, energy conservation, oscillation, phase conservation, etc. To solve IVPs with periodic and oscillatory solutions, a first theoretical foundation was proposed by Gautschi [42] and Lyche [43] with a technique called functional fitting. The study of the exponentially fitted RK method is a modern development and is widely useful to solve equations with specific solution behavior. Much research has focused on construction of functionally fitted numerical methods. IVPs with exponential solutions can be solved exactly using an exponential fitting approach. Berghe et al. [44–46], Paternoster [47], and Simos [48] were the first to combine exponential fitting with existing RK methods, thus developing various types of exponentially fitted RK methods. Simos's approach differs from Berghe's in the sense that Berghe uses an additional parameter in the internal stages, which certainly improves the rate of convergence of the exponentially fitted Runge–Kutta method (ef-RKM), as shown in the numerical experiments given in Berghe et al. [44–46]. The principle of the fitting is based on the annihilation of the linear integral operator associated with the internal stages and external stage, provided the existing RK method has been converted into a functional form using Albrecht's approach [49]. Ixaru et al. [50] discovered an algorithm to compute the weights of exponentially fitted multi-step algorithms for ODEs. The development of exponential fitting of various numerical integrators is covered in an excellent monograph (see [51]). Ozawa [52] made the functional fitting of RKM with variable coefficients.

The main objective of the present paper is to construct a fractional Runge–Kutta method (FRKM) along with its exponential fitting to find the numerical solution of fractional linear and nonlinear differential equations. The fitting approach is based on the assumption of annihilation of internal and external integral operators associated with FRKM. Here, we assume that the internal and external integral operators associated with FRKM annihilate the set functions $\{e^{\pm \nu t}\}$ with unknown frequency $\nu \in \mathrm{R}$ (or $\nu \in i\mathrm{R}$). The optimum values of frequency ν are obtained by minimizing the term of local truncation error. Several exponential fitting techniques can be found in an excellent book by Ixaru and Berghe [51]. This theory can be generalized to solve the problem in fractional differential equations.

Here, we first construct an algorithm for solving fractional differential equations based on the Runge–Kutta method. We consider the following time-fractional differential equation:

$$\frac{d^{\alpha}u(t)}{dt^{\alpha}}=f\left(t,u(t)\right),\quad 0<\alpha\leq 1. \tag{7.2}$$

subject to the initial conditions:

$$u(0)=u_0,\, t\in\left[0,T\right] \tag{7.3}$$

The fractional derivatives are considered in the Caputo sense because many numerical solutions of the FDE derived from a fractional derivative based on the Riemann–Liouville or the Grünwald–Letnikov definitions [1, 53–54] may cause mass balance error as shown by Zhang et al. [55]. Several research papers on the development of interesting numerical algorithms to solve FDE based on Caputo derivatives can be seen in [56–58].

7.2 Fractional Calculus

In this section, we give some basic definitions and properties of fractional calculus [1, 36, 38, 53].

Definition 7.2.1

A real function $f(x)$, $x > 0$, is said to be in a space C_{μ}, $\mu \in R$ if there exists a real number $p(>\mu)$ such that $f(x) = x^{p}f_1(x)$ where $f_1(x) \in C[0,\infty)$, and it is said to be in the space C_{μ}^{m} if $f^{(m)} \in C_{\mu}$, $m \in N$.

Definition 7.2.2

The Riemann–Liouville fractional integral operator of order $\alpha \geq 0$, of a function $f \in C_{\mu}$, $\mu \geq -1$, is defined as

$$J^{\alpha}f(x)=\frac{1}{\Gamma(\alpha)}\int_{0}^{x}(x-t)^{\alpha-1}f(t)\,dt,\ \alpha>0, x>0, \tag{7.4}$$

$$J^{0}=f(x). \tag{7.5}$$

We need the following properties of the operator J^{α}:

$$J^{\alpha}(x-a)^{\gamma}=\frac{\Gamma(\gamma+1)}{\Gamma(1+\gamma+\alpha)}(x-a)^{\gamma+\alpha}$$

$$J^{\alpha}J^{\beta}f(x)=J^{\alpha+\beta}f(x)$$

$$J^{\alpha}J^{\beta}f(x)=J^{\beta}J^{\alpha}f(x).$$

Definition 7.2.3

For m to be the smallest integer that exceeds α, the Caputo time-fractional derivative operator of order $\alpha > 0$ is defined as

$$D_{\alpha}^{t}u(x,t)=\frac{\partial^{\alpha}u(x,t)}{\partial t^{\alpha}}=\begin{cases}\dfrac{1}{\Gamma(m-\alpha)}\displaystyle\int_{0}^{t}(t-\tau)^{m-\alpha-1}\frac{\partial^{m}u(x,\tau)}{\partial\tau^{m}}d\tau, & for \quad m-1<\alpha<m\\ \dfrac{\partial^{m}u(x,t)}{\partial t^{m}}, & for \quad \alpha=m\in N\end{cases}. \tag{7.6}$$

Definition 7.2.4

The fractional derivative of $f(x)$ in the Caputo sense is defined as

$$D^{\alpha}f(x)=J^{m-\alpha}D^{m}f(x)=\frac{1}{\Gamma(m-\alpha)}\int_{0}^{x}(x-t)^{m-\alpha-1}f^{m}(t)dt \tag{7.7}$$

for $m-1<\alpha\le m, m\in N, x>0, f\in C_{-1}^{m}$.
From (7.7), we get the following

$$D^{\alpha}(x-a)^{\gamma}=\begin{cases}\dfrac{\Gamma(\gamma+1)}{\Gamma(1+\gamma-\alpha)}(x-a)^{\gamma-\alpha}, & for \quad \alpha\le\gamma\\ 0, & for \quad \alpha>\gamma\end{cases} \tag{7.8}$$

Lemma 7.2.1

If $m-1<\alpha\le m$, $m\in N$ and $f\in C_{-1}^{m}$ and $\mu\ge -1$, then

$$D^{\alpha}J^{\alpha}f(x)=f(x), \tag{7.9}$$

$$J^{\alpha}D^{\alpha}f(x)=f(x)-\sum_{k=0}^{m-1}f^{(k)}(0^{+})\frac{(x-a)^{k}}{k!}, x>0. \tag{7.10}$$

Definition 7.2.5

Mittag–Leffler function

The Mittag-Leffler function is the generalization of the exponential function. It was first defined as a single-parameter function and is defined in the form of a series as [1, 36]

$$E_\alpha(z)=\sum_{k=0}^{\infty}\frac{z^k}{\Gamma(\alpha k+1)},\alpha>0,\alpha\in\mathbb{R},z\in\mathbb{C}. \tag{7.11}$$

Later on it is generalized to a two-parameter function as

$$E_{\alpha,\beta}(z)=\sum_{k=0}^{\infty}\frac{z^k}{\Gamma(\alpha k+\beta)},\alpha,\beta>0,\alpha,\beta\in\mathbb{R},z\in\mathbb{C}. \tag{7.12}$$

7.2.1 Fractional Taylor's Series Formula [59]

Let $f(x)$ be a real-valued function whose fractional derivatives of order up to $(n+1)$ exist and are continuous in $[a,b]$, i. e., $D_a^{k\alpha}f(x)\in C[a,b]$, for $k=0,1,2,\cdots,(n+1)$, then

$$f(x)=\sum_{i=0}^{n}\frac{(x-x_0)^{i\alpha}\left(D_a^{i\alpha}f\right)(x_0)}{\Gamma(i\alpha+1)}+\frac{(x-x_0)^{(n+1)\alpha}\left(D_a^{(n+1)\alpha}f\right)(\xi)}{\Gamma((n+1)\alpha+1)},\ x_0\le\xi\le x,\ \forall x\in[a,b]. \tag{7.13}$$

where $0<\alpha\le 1$, $D_a^{n\alpha}=D_a^{\alpha}.\,D_a^{\alpha}.\,D_a^{\alpha}\ldots D_a^{\alpha}$ (n - times).

If $\alpha=1$ then the generalized Taylor's formula (7.13) reduces to classic Taylor's formula.

$$f(x)=\sum_{i=0}^{n}\frac{(x-x_0)^{i}\left(D_a^{i}f\right)(x_0)}{i!}+\frac{(x-x_0)^{(n+1)}\left(D_a^{(n+1)}f\right)(\xi)}{(n+1)!},\ x_0\le\xi\le x,\ \forall x\in[a,b]. \tag{7.14}$$

7.3 Derivation of the Method

In this section, we first derive the fractional Runge–Kutta method (FRKM) for the numerical solution of IVPs of fractional-order ordinary differential equations (FODEs). The derivation is similar to that of the Runge–Kutta method for integer-order differential equations. Here we use the generalized Taylor's series formula instead of the classical formula. We also construct the exponential fitting (ef) of the FRKM. The motivation behind the fitting is that the ef version of FRKM can be used to solve FDEs having exponential/trigonometric solutions more efficiently. Where frequencies tend to zero, the exponentially fitted FRK method becomes the standard FRK method.

7.3.1 Fractional Runge–Kutta Method (FRKM)

In this subsection, we give the formulation of the FRK method proposed previously by several researchers [36]. In order to introduce FRKM for nonlinear FDE (7.2), we take $t_n = t_0 + nh$, $n = 0, 1, \cdots N - 1$ with $h = b - a/N$.

We denote u_n as the numerical approximation of the exact solution $u(t_n)$. A general s-stage FRKM, to solve the fractional IVP (7.2)–(7.3) is given by

$$u_{n+1} = u_n + \frac{h^\alpha}{\Gamma(1+\alpha)} \sum_{i=1}^{s} b_i k_i, \tag{7.15}$$

where

$$k_i = u_n + \frac{h^\alpha}{\Gamma(1+\alpha)} \sum_{j=1}^{s} a_{ij} f\left(t_n + c_j h^\alpha, k_j\right),\ i = 1,2,\cdots,s\left(n = 0,1,\cdots N-1\right). \tag{7.16}$$

The difference equation (7.15) is called the external (objective) stage, whereas (7.16) is the internal stage of the FRKM. The two-stage explicit FRKM may be rewritten (in functional form) as

$$u_{n+1} = u_n + \frac{h^\alpha}{\Gamma(1+\alpha)} \sum_{i=1}^{2} b_i f\left(t_n + c_i h^\alpha, U_i\right) \tag{7.17}$$

And

$$U_i = u_n + \frac{h^\alpha}{\Gamma(1+\alpha)} \sum_{j=1}^{i-1} a_{ij} f\left(t_n + c_j h^\alpha, U_j\right),\ i = 1, 2 \tag{7.18}$$

with $a_{1j} = 0, j = 1, 2$ (for explicit two-stage FRKM).

Applying the generalized Taylor's formula (7.13)–(7.14) in (7.17) and (7.18) we have the following order conditions:

$$b_1 + b_2 = \frac{1}{\Gamma(\alpha+1)}$$

$$b_2 c_2 = \frac{\Gamma(\alpha+1)}{\Gamma(2\alpha+1)}$$

$$b_2 a_{21} = \frac{\Gamma(\alpha+1)}{\Gamma(2\alpha+1)}.$$

Solving the above order conditions, we get $c_2 = a_{21}$ and
$b_2 = \frac{\Gamma(\alpha+1)}{c_2\Gamma(2\alpha+1)}, b_1 = \frac{1}{\Gamma(\alpha+1)} - \frac{\Gamma(\alpha+1)}{c_2\Gamma(2\alpha+1)}$ with free parameter as c_2 (node).

For $\alpha = 1$, FRKM is converted into the classical two-stage Runge–Kutta method.

7.3.2 Exponentially Fitted Fractional Runge–Kutta Method (ef-FRKM)

In this subsection, we construct the exponential fitting of FRKM given in (7.15) to (7.16). Here, we assume that the ef-FRKM integrates exactly the exponential functions $e^{\pm\nu t}$ ($\nu \in$ R or $\nu \in i$R). The ef-FRKM is therefore best suited to solving fractional initial value problems (FIVPs) with exponential/periodic solutions. In view of the approach given in Albrecht [49] to form the fitting process, we first rewrite the FRKM in its equivalent functional form of equations (7.15)–(7.16) ($n = 0, 1, 2, \ldots N-1$) as:

$$u_{n+1} = u_n + \frac{h^\alpha}{\Gamma(\alpha+1)}\sum_{i=1}^{s} b_i f\left(t_n + c_i h^\alpha, U_i\right) \tag{7.19}$$

$$U_i = u_n + \frac{h^\alpha}{\Gamma(\alpha+1)}\sum_{j=1}^{s} a_{ij} f\left(t_n + c_j h^\alpha, U_j\right),\ i = 1,2,\cdots,s\left(n = 0,1,\cdots N-1\right). \tag{7.20}$$

For the two-stage explicit ef-FRKM, we restrict $s = 2$ and $a_{1j} = 0$, $j = 1, 2$, $a_{21} = 0$.

The linear difference operator 'L' associated with the external stage (7.19) is defined as

$$L\left[u(t);h\right] \equiv u(t+h) - u(t) - \frac{h^\alpha}{\Gamma(1+\alpha)}\sum_{i=1}^{s} b_i f\left(t_n + c_i h^\alpha, U_i\right) \tag{7.21}$$

or

$$L\left[u(t);h\right] \equiv u(t+h) - u(t) - \frac{h^\alpha}{\Gamma(1+\alpha)}\sum_{i=1}^{s} b_i D^\alpha u\left(t + c_i h^\alpha\right) \tag{7.22}$$

And the linear operators 'L_i' associated with the internal stage (7.20) are defined as:

$$L_i\left[u(t);h\right] \equiv u\left(t + c_i h^\alpha\right) - u(t) - \frac{h^\alpha}{\Gamma(1+\alpha)}\sum_{j=1}^{s} b_i f\left(t_n + c_j h^\alpha, U_j\right) \tag{7.23}$$

or,

$$L_i\left[u(t);h\right] \equiv u\left(t+c_i h^{\alpha}\right)-u(t)-\frac{h^{\alpha}}{\Gamma(1+\alpha)}\sum_{j=1}^{s} b_i D^{\alpha}\left(t_n+c_j h^{\alpha}\right). \tag{7.24}$$

The operators L and L_i satisfy

$$L\left[1;h\right]=L_i\left[1;h\right]\equiv 0 \tag{7.25}$$

The operators L and L_i also annihilate the function $y(x) = x$ which ensures the consistency condition of ef-FRKM as:

$$c_i = \sum_{j=1}^{i-1} a_{ij}$$

under the limit $\alpha \to 1$.

For the two-stage explicit ef-FRKM the condition becomes $a_{21} = -\frac{\left(h^{\alpha}c_2\right)^{\alpha}\alpha(-1+\alpha)\pi \operatorname{cosec}(\pi\alpha)}{h^{\alpha}}$ and under the limit $\alpha \to 1$ it becomes $a_{21} = c_2$. That verifies the consistency condition for explicit two-stage RKM.

To compute the coefficients of ef-FRKM, we assume that the external and internal stage operators, given in equations (7.22) and (7.24) respectively, exactly integrate the two exponential functions $e^{\pm\nu t}$ ($\nu \in$ R or $\nu \in i$R).

So, taking $u(t) = e^{\pm\nu t}$ in equations (7.22) and (7.24), we can obtain the order conditions for exponential fitting of the two-stage explicit FRKM. Solving the order conditions, one can get the following coefficients for a two-stage explicit ef-FRKM under the limit $\alpha \to 1$

$$b_1 = \frac{-1+2c_2}{2c_2}-\frac{\left(1-4c_2+4c_2^2\right)\omega^2}{24c_2}+\frac{\left(-1+6c_2-10c_2^2+8c_2^4\right)\omega^4}{720c_2}+O(\omega)^5,$$

$$b_2 = \frac{1}{2c_2}-\frac{\left(-1+2c_2^2\right)\omega^2}{24c_2}+\frac{h^4\left(1-5c_2^2+7c_2^4\right)\omega^4}{720c_2}+O(\omega)^5,$$

$$a_{21} = c_2-\frac{1}{2}\left(c_2^2\omega\right)+\frac{1}{6}c_2^3\omega^2-\frac{1}{24}\left(c_2^4\omega^3\right)+\frac{1}{120}c_2^5\omega^4+O(\omega)^5.$$

where $\omega = \nu h$ with unknown frequency ν.

From the above expressions, it is inferred that under the limit $\omega \to 0$, ef-FRKM is converted into the standard FRKM with the following values of the coefficients:

$$b_1 = \frac{-1+2c_2}{2c_2},\ b_2 = \frac{1}{2c_2},\ a_{21} = c_2.$$

where node c_2 is the free parameter. The optimum value of node c_2 can be calculated by minimizing the truncation error (see Section 7.4).

7.4 Truncation Error

In order to find the local truncation error (LTE) for FRKM proposed in Section 7.3.1, we denote $u(t_{n+1})$ as an exact solution at the point t_{n+1}. We denote $e_{n+1} = u(t_{n+1}) - u_{n+1}$ as an error in the numerical solution u_{n+1}. We compute the LTE for a scalar IVP. In this computation, we use the generalized Taylor's series approximation for manipulation and simplification to find the expression of truncation error $e_{n+1} = u(t_{n+1}) - u_{n+1}$. The truncation error for two-stage explicit FRKM can be obtained in the following ways.

By generalized Taylor's series (7.13), we may express the exact solution $u(t_{n+1})$ as

$$u(t_{n+1}) = u(t_n + h) = u(t_n) + \frac{h^{\alpha}}{\Gamma(\alpha+1)} D_a^{\alpha} u(t_n) + \frac{h^{2\alpha}}{\Gamma(2\alpha+1)} D_a^{2\alpha} u(t_n) + \frac{h^{3\alpha}}{\Gamma(3\alpha+1)} D_a^{3\alpha} u(t_n) + \cdots \tag{7.26}$$

Following the approach given in [36], we use the values of various fractional derivatives as:

where

$$\begin{aligned} D_a^{\alpha} u(t_n) &= f(t_n, u_n), \\ D_a^{2\alpha} u(t_n) &= f_t(t_n, u_n) + f_y(t_n, u_n) D_a^{\alpha} u(t_n) = f_t(t_n, u_n) + f_y(t_n, u_n) f(t_n, u_n), \\ D_a^{3\alpha} u(t_n) &= f_{tt}(t_n, u_n) + 2D_a^{\alpha} u(t_n) f_{ty}(t_n, u_n) + \left(D_a^{\alpha} u(t_n)\right)^2 f_{yy}(t_n, u_n) + f_t(t_n, u_n) f_y(t_n, u_n) \\ &\quad + D_a^{\alpha} u(t_n) \left(f_y(t_n, u_n)\right)^2. \end{aligned}$$

On the other hand, the numerical solution u_{n+1} by the two-stage explicit FRKM is given in Section 7.3.1, represented by the equation (7.15), so the truncation error expression is given by:

$$e_{n+1} = h^{3\alpha} \left[\left(\frac{c_2}{2\Gamma(2\alpha+1)} - \frac{1}{\Gamma(3\alpha+1)} \right) D^{3\alpha} u(t_n) - \frac{c_2}{2\Gamma(2\alpha+1)} \begin{pmatrix} f_t(t_n, u_n) f_y(t_n, u_n) \\ + D^{\alpha} u(t_n) \left(f_y(t_n, u_n)\right)^2 \end{pmatrix} \right]. \tag{7.27}$$

From the expression of TE (7.27), the optimum value of the free parameter node c_2 can be obtained by minimizing it, and the optimum value of c_2 is given by

$c_2 = 2\Gamma(2\alpha + 1)/\Gamma(3\alpha + 1)$, and under the limit $\alpha \to 1$, $c_2 = 2/3$ (the optimum case of the classic two-stage Runge–Kutta method).

7.5 Numerical Experiments

In this section, we consider the two initial value problems in fractional-order differential equations and apply the FRKM and ef-RKM developed in Sections 7.3.1 and 7.3.2. The results obtained here are further compared in Tables 7.1 and 7.2, and the numerical solution profiles are depicted in Figures 7.1 and 7.2.

Example 1

First, we consider a fractional logistic growth model governed by the following fractional model [60–61]

$$\frac{d^\alpha N(t)}{dt^\alpha} = r^\alpha \frac{N(t)}{M}\left(1 - \frac{N(t)}{M}\right), \quad N(0) = N_0;\ 0 < \alpha \le 1. \tag{7.28}$$

where N_0 is the initial density of the population, r is the growth rate of the population, and M is maximum carrying capacity. In the asymptotic

TABLE 7.1

Numerical results of Example 1 for $\alpha = 1$ with step size $h = 0.001$.

t	N_{exact}	N_{apprx}.(by ef-FRKM)	Absolute Error $\lvert N_{exact} - N_{approx} \rvert$
0.0	0.750000	0.750000	0.0
0.1	0.768278	0.768278	5.1618 (−10)
0.2	0.785601	0.785601	9.0720 (−10)
0.3	0.801964	0.801964	1.19296 (−9)
0.4	0.817367	0.817367	1.39157 (−9)
0.5	0.831824	0.831824	1.51917 (−9)
0.6	0.845353	0.845353	1.58989 (−9)
0.7	0.857980	0.857980	1.61587 (−9)
0.8	0.869734	0.869734	1.60737 (−9)
0.9	0.880651	0.880651	1.57296 (−9)
1.0	0.890768	0.890768	1.51967 (−9)

Notation: $a(-b)$ means $a \times 10^{-b}$

TABLE 7.2

Comparison of absolute error for Example 2 for $\alpha = 1$ with varying step sizes h.

	Abs Error by ef-FRKM $\lvert N_{exact} - N_{approx^{\cdot}} \rvert$			
t	$h = 1/16$	$h = 1/32$	$h = 1/64$	$h = 1/128$
0.000	0.0	0.0	0.0	0.0
0.125	2.33 (−4)	5.79 (−4)	1.44 (−5)	3.59 (−6)
0.250	4.34 (−4)	1.08 (−4)	2.70 (−5)	6.75 (−6)
0.375	6.06 (−4)	1.52 (−4)	3.81 (−5)	9.53 (−6)
0.500	7.53 (−4)	1.90 (−4)	4.77 (−5)	1.19 (−5)
0.625	8.77 (−4)	2.23 (−4)	5.62 (−5)	1.14 (−5)
0.750	9.81 (−4)	2.51 (−4)	6.36 (−5)	1.60 (−5)
0.875	1.06 (−3)	2.75 (−4)	7.00 (−5)	1.76 (−5)
1.000	1.13 (−3)	2.96 (−4)	7.56 (−5)	1.91 (−5)

Notation: $a\,(-b)$ means $a \times 10^{-b}$

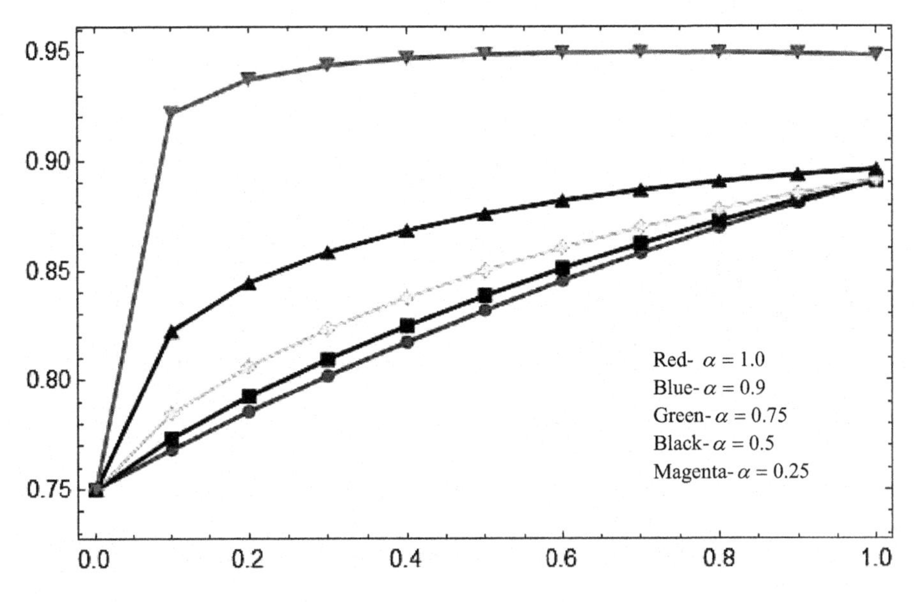

FIGURE 7.1
Comparison of numerical solutions to Example 1 for various values $\alpha = 1, 0.9, 0.75, 0.5, 0.25$ for $N(0) = 0.75$ with step size $h = 0.001$ by ef-FRKM

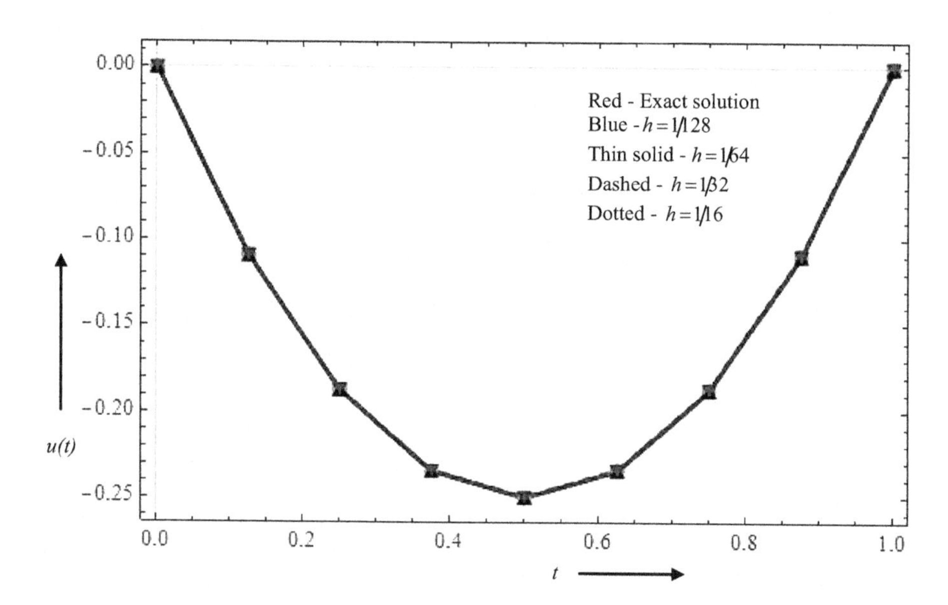

FIGURE 7.2
Comparison of solution profiles to Example 2 obtained by ef-FRKM for $\alpha = 1$ with various step sizes h

limit the normalized population N_0/M approaches to unity. The closed-form exact solution of FIVP (7.28) is given by [60]

$$N(t) = \frac{MN_0}{N_0 + (M - N)e^{-rt}}. \tag{7.29}$$

We have solved the fractional logistic growth model (7.28) using both our developed methods – FRKM (given in Section 7.3.1) and ef-FRKM (developed in Section 7.3.2). Results are reported in Figures 7.1 and 7.2 and Table 7.1. We denote $e_N = |y(t) - y_N|$ as absolute endpoint error and $e_n = |y_{exact}(t_n) - y_n|$ as pointwise absolute error.

Example 2

Next, we consider a fractional IVP governed by the following fractional model

$$\frac{d^\alpha u(t)}{dt^\alpha} = -\frac{t^{1-\alpha}}{\Gamma(2-\alpha)} + \frac{2t^{2-\alpha}}{\Gamma(3-\alpha)} + t^2 - t - u(t),\ u(0) = 0;\ 0 < \alpha \le 1,\ t \in [0,1]. \tag{7.30}$$

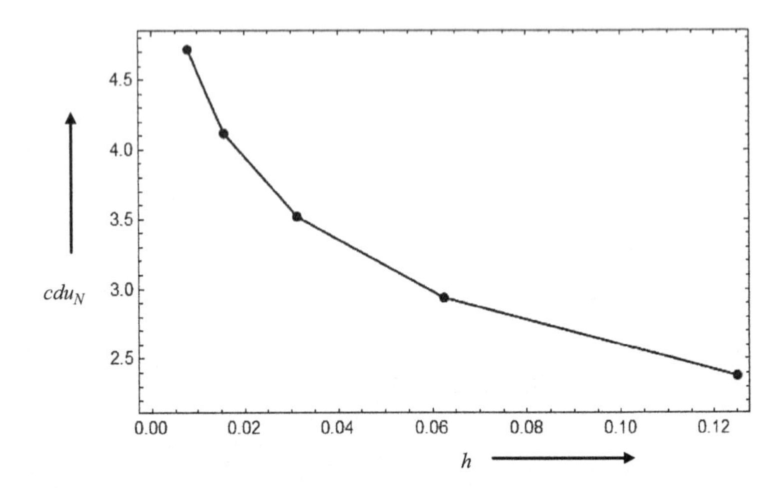

FIGURE 7.3
Comparison of correct digits of the solution ($t = 1$) versus step size h obtained by second-order ef-FRKM (Section 7.3.2) for Example 2 ($\alpha = 1$).

The closed-form exact solution of FIVP (7.30) is $u(t) = t^2 - t$.
We denote $e_N = |u(t) - u_N|$ as absolute endpoint error and
$e_n = |u(t_n) - u_n|$ as pointwise absolute error.

Also $cdu_N = -\log_{10}(|u(t) - u_N|)$ is denoted as correct digits in endpoint solution (a crude estimate of the error). We have implemented the ef-FRKM (Section 7.3.2) to solve numerically the FDE (7.30) for the step sizes $h = 1/N$ with $N = 2^i$, $i = 4, 5, 6, 7$, and results are reported in Table 7.2. The numerical solution profiles for $\alpha = 1$ with varying step sizes h are plotted in Figure 7.2. The comparison of the pointwise absolute error versus step size h is shown in Table 7.2. Figure 7.3 shows the relationship between step sizes h and correct digits in solution (logarithm of endpoint absolute error). From Figure 7.3, it is observed, we can achieve convergence by minimizing the step size h.

7.6 Conclusions

In this paper, we have successfully constructed a fractional two-stage Runge–Kutta method along with its exponential fitting for the numerical solution of IVPs in fractional differential equations. The proposed algorithms have tested two FIVPs that have physical importance. The ef-FRKM method exactly integrates two exponential functions: *exp* ($\pm vt$) with unknown frequency v. The computation of the local truncation error (LTE) is made and by minimizing this LTE the optimum value of free parameter c_2 can be calculated. Two fractional-order IVPs are solved using the methods given in Sections 7.3.1 and 7.3.2. Pointwise absolute errors are tabulated in Tables 7.1 and 7.2. From Tables 7.1–7.2 and Figures 7.1 and 7.2, it is observed that the errors incurred in two-stage explicit ef-FRKM (Section 7.3.2) are relatively small, ensuring the efficiency of the proposed method.

References

[1] I. Podlubny, *Fractional Differential Equations*, Academic Press, New York, 1999.

[2] A. A. Kilbas, H. M. Srivastava, and J. J. Trujillo, *Theory and Applications of fractional differential equations*. Elsevier, Amsterdam, 2006.

[3] A. Carpinteri and F. Mainardi, *Fractals and Fractional Calculus in Continuum Mechanics*. Vol. 378, Springer, Berlin, Germany, 2014.

[4] C. Drapaca, Fractional calculus in neuronal electromechanics. *J. Mech. Mat. Struct.*, 12(1) (2017) 35–55.

[5] N. K. Tripathi, Analytical solution of two dimensional nonlinear space-time fractional Burgers–Huxley equation using fractional sub-equation method. *Natl. Acad. Sci. Lett.*, 41(5) (2018) 295–299.

[6] K. Diethelm, *The Analysis of Fractional Differential Equations*. Springer, Berlin, 2004.

[7] V. Lakshmikantham, Theory of fractional functional differential equations. *Nonlinear Anal.*, 69 (2008) 3337–3343.

[8] P. Pakdaman, A. Ahmadian, S. Effati, S. Salahshour, and D. Baleanu, Solving differential equations of fractional order using an optimization technique based on training artificial neural network. *Appl. Math. Comput.*, 293 (2017) 81–95.

[9] S. R. Balachandar, K. Krishnaveni, K. Kannan, and S. G. Venkatesh, Analytical solution for fractional gas dynamics equation. *Natl. Acad. Sci. Lett.*, 42(1) (2019) 51–57.

[10] R. Gorenflo, F. Mainardi, E Scalas, and M. Roberto, Fractional calculus and continuos time finance, III, The diffusion limit. in: Math, Finance, Konstanz, in: *Trends Math*, Birkhauser, Basel, 2001, 171–180.

[11] M. Roberto, E. Scalas, and F. Mainardi, Waiting-times and returns in high frequency financial data: An empirical study. *Physica A*, 314 (2002) 749–755.

[12] L. Sabatelli, S. Keating, J. Dudley, and P. Richmond, Waiting time distributions in financial markets. *Eur. Phys. J. B.*, 27 (2002) 273–275.

[13] M. Meerschaaert, D. Benson, H. P. Schefflera, and B. Baeumer, Stochastic solution of space-time fractional diffusion equation. *Phys. Rev. E.*, 65 (2002) 1103–1106.

[14] B. West, M. Bologna, and P. Grigolini, *Physics of Fractal Operators*. Springer, New York, 2003.

[15] G. Zaslavsky, *Hamiltonian Choas and Fractional Dynamics*. Oxford University Press, Oxford, 2005.

[16] T. M. Atanacković, S. Pilipović, B. Stanković, and D. Zorica, *Fractional Calculus with Applications in Mechanics: Vibrations and Diffusion Processes*. John Wiley & Sons, Hoboken, 2014.

[17] R. Hilfer, *Application of Fractional Calculus in Physics*. World Scientific, Singapore, 2000.

[18] J. T. Mechado, Discrete time fractional-order controllers. *Frac. Cal. Appl. Anal.*, 4 (2001) 47–66.

[19] R. Schumer, D. A. Benson, M. M. Meerschaert, and B. Baeumer, Multiscaling fractional advection-dipersion equations and their solutions. *Water Resource Res.*, 39 (2003) 1022–1032.

[20] R. Metzler and J. Klafter, The restaurant at the end of random walk: Recent developments in the description of anomalous transport by fractional dynamics. *J. Phys. A*, 37 (2004) R161–R208.

[21] R. Metzler and J. Klafter, The random walk's guide to anomalous diffusion: A fractional dynamics approach. *Phys. Rep.*, 339 (2000) 1–77.

[22] S. J. Liao, *Beyond Perturbation: Introduction to Homotopy Analysis Method*. Chapman & Hall/CRC Press, Boca Raton, 2003.

[23] J. H. He, Approximate analytical solution for seepage flow with fractional derivatives in porous media. *Comput. Methods Appl. Mech. Eng.*, 167 (1998) 57–68.

[24] K. Diethelm and N. J. Ford, Analysis of fractional differential equations. *J. Math. Anal. Appl.*, 265 (2002) 229–248.

[25] K. Diethelm and A. D. Freed, The FracPECE subroutine for the numerical solution of differential equations of fractional order. In: *Forschung und wissenschaftliches Rechnen: Beiträge zum Heinz-Billing-Preis*, Gesellschaft für wissenschaftliche Datenverarbeitung, Göttingen 1, 1998, 57–71.

[26] R. Garrappa, Numerical solution of fractional differential equations: A survey and a software tutorial. *Mathematics*, 6 (2018) 1–23.

[27] G. E. Karniadakis, *Numerical Methods, Handbook of Fractional Calculus with Applications* (3rd ed.) DeGruyter, Berlin/Boston, 2019.

[28] A. Pedas and E. Tamme, Numerical solution of nonlinear fractional differential equations by spline collocation methods. *J. Comput. Appl. Math.*, 255 (2014) 216–230.

[29] Z. Odibat and S. Momani, An algorithm for numerical solution of differential equations of fractional order. *J. Appl. Math. Inform.*, 26 (2008) 15–27.

[30] Z. Odibat, Approximations of fractional integrals and Caputo fractional derivatives. *Appl. Math. Comput.*, 178 (2006) 527–533.

[31] K. Diethelm, N. J. Ford, and A. D. Freed, Detailed error analysis for a fractional Adams method, *Numer. Algorithms*, 36 (2004) 31–52.

[32] Ü. Lepik, Solving fractional integral equations by Haar wavelet method. *Appl. Math. Comput.*, 214 (2009) 468–478.

[33] E. H. Doha, A. H. Bhrawy, and S. S. Ezz-Eldien, A Chebyshev spectral method based on operational matrix for initial and boundary value problems of fractional order. *Comput. Math. Appl.*, 62 (2011) 2364–2373.

[34] C. Li and F. Zeng, *Numerical methods for fractional calculus*. Lecture notes in electrical engineering. Chapman and Hall/CRC, Boca Raton, 2015.

[35] C. Milici, J. T. Machado, and G. Draganescu, On the fractional Cornu spirals. *Commun. Nonlinear Sci. Numer. Simul.*, 67 (2019), 315–320.

[36] C. Milici, G. Draganescu, and J. T. Machado, *Introduction to fractional differential equations*. Springer, Cham, Switzerland, 2018.
[37] K. Diethelm, *The Analysis of Fractional Differential Equations* in: Lecture Notes in Mathematics, vol. 2004, Springer, Berlin, 2010.
[38] A. A. Kilbas, H. M. Srivastava, and J. J. Trujillo, *Theory and Applications of Fractional Differential Equations*. North-Holland Mathematics Studies, vol. 204, Elsevier, Amsterdam, 2006.
[39] M. H. Hamarsheh and E. A. Rawashdeh, A numerical method for solution of semi differential equations. *Mat. Vesnik*, 62 (2010) 117–126.
[40] J. C. Butcher, *Numerical Methods for Ordinary Differential Equations*. John Wiley & Sons, Ltd., UK, 2016.
[41] Z. Kalogiratou, T. Monovasilis, G. Psihoyios, and T. E. Simos, Runge-Kutta type methods with special properties for the numerical integration of ordinary differential equations. *Phys. Report.*, 536 (3) (2014) 75–146.
[42] W. Gautschi, Numerical integration of ordinary differential equations based on trigonometric polynomials. *Num. Math.*, 3 (1961) 381–397.
[43] T. Lyche, Chebyshevian multistep methods for ordinary differential equations. *Num. Math.*, 19 (1972) 65–75.
[44] G. V. Berghe, H. De Meyer, M. Van Daele, and T. Van Hecke, Exponentially-fitted explicit Runge-Kutta methods. *Comput. Phys. Comm.*, 123 (1999) 7–15.
[45] G. V. Berghe, H. De Meyer, M. Van Daele, and T. Van Hecke, Exponentially-fitted Runge-Kutta methods. *J. Comput. Appl. Math.*, 125 (2000) 107–115.
[46] G. V. Berghe, L.G. Ixaru, and M. V. Daele, Optimal implicit exponentially-fitted Runge-Kutta methods. *Comp. Phys. Comm.*, 140 (2001) 346–357.
[47] B. Paternoster, Runge-Kutta(-Nyström) methods for ODEs with periodic solutions based on trigonometric polynomials. *Appl. Num. Math.*, 28 (1998) 401–412.
[48] T. E. Simos, An exponential-fitted Runge-Kutta method for the numerical integration of initial value problems with periodic or oscillating solutions. *Comput. Phys. Comm.*, 115 (1998) 1–8.
[49] P. Albrecht, A new theoretical approach to Runge-Kutta methods. *SIAM J. Numerical Anal.*, 24 (2) (1987) 391–406.
[50] L. G. Ixaru, G. V. Berghe, and H. De Meyer, Frequency evaluation in exponential fitting multistep algorithms for ODEs. *J. Compt. Appl. Math.*, 140 (2002) 423–434.
[51] L. G. Ixaru and V. G. Berghe, *Exponential fitting*. Kluwer Academic Publishers, Dordrecht, 2004.
[52] K. Ozawa, A functional fitting Runge-Kutta-Nyström method with variable coefficients. *Jpn. J. Ind. Appl. Math.*, 18 (2001) 105–130.
[53] K. S. Miller and B. Ross, *An Introduction to the Fractional, Calculus and Fractional Differential Equation*. Wiley, New York, 1993.
[54] N. Heymans and I. Podlubny, Physical interpretation of initial conditions for fractional differential equations with Riemann-Liouville fractional derivatives. *Rheol Acta.*, (2006) 765–771.
[55] X. Zhang, M. Lv, J. W. Crawford, and I. M. Young, The impact of boundary on the fractional advection-dispersion equation for solute transport in soil: Defining the fractional dispersive flux with the Caputo derivatives. *Adv. Water Resour.*, 30(5) (2007) 1205–1217.

[56] H. Singh, H.M. Srivastava, and D. Kumar, A reliable numerical algorithm for the fractional vibration equation. *Chaos Solit. Fract.*, 103 (2017) 131–138.
[57] H. Singh, H.M. Srivastava, and D. Kumar, A reliable algorithm for the approximate solution of the nonlinear Lane-Emden type equations arising in astrophysics. *Numer. Methods Partial Differ. Equ.*, 34 (5) (2018) 1524–1555.
[58] H. Singh, R. K. Pandey, and H.M. Srivastava, Solving non-linear fractional variational problems using Jacobi polynomials. *Mathematics*, 7 (3) (2019) 224.
[59] Z. Odibat and N. T. Shawagfeh, Generalized Taylor's formula. *Appl. Math. Comput.*, 186 (2007) 286–293.
[60] B. J. West, Exact solution to fractional logistic equation. *Phys. A Stat. Mech. Appl.*, 429 (2015) 103–108.
[61] M. D'Ovidio, P. Loreti, and S. S. Ahrabi, Modified fractional logistic equation. *Phys. A Stat. Mech. Appl.*, 505 (2018) 818–824.

8

Generalization of Fractional Kinetic Equations Containing Incomplete I-Functions*

Kamlesh Jangid
Rajasthan Technical University, Kota, India

Sapna Meena and Sanjay Bhatter
Malaviya National Institute of Technology, Jaipur, India

S. D. Purohit
Rajasthan Technical University, Kota, India

CONTENTS

8.1 Introduction and Mathematical Preliminaries

Mathematical computation is a valuable tool for gaining knowledge of physical processes. A mathematical model clearly enables researchers to create, optimize, and forecast the future success of various treatment schemes. Fractional-order mathematical models are a step forward from traditional models using fractional calculus. Fractional calculus is sometimes useful in the analysis of these models and their solutions, which are then applied to a wide range of scientific and engineering subjects. Diffusion, reaction-diffusion, fluid flow, turbulence, oscillation, electric networks, control systems, chemical physics, electrochemistry, epidemiology, and other essential

* 2010 *Mathematics Subject Classification*: 26A33, 44A10, 33B20, 33C45, 33C60.

DOI: 10.1201/9781003263517-8

applications may be shown. As illustrated in [1, 4, 11, 22–27, 30, 31] a variety of mathematical models relevant to real-world problems have already revealed several implications of fractional calculus. For a detailed description of fractional calculus operators, including their characteristics and implications in numerous fields, see [28, 29].

Fractional kinetic equations are widely used in astrophysics, control systems, and mathematical physics; therefore the solution of fractional kinetic equations has attracted the interest of many researchers. As a result, a large number of recent research articles (see [2, 3, 5, 7, 12, 14, 16, 19–21, 35]) focus on the solution of these equations, including generalized Mittag–Leffler function, Bessel's function, Struve function, *G*-function, *H*-function, and Aleph-function.

Haubold and Mathai [5] defined the fractional differential equation for the production and destruction of species. If the rate of change of reaction, $\mathcal{N} = \mathcal{N}(t)$, the rate of destruction, $\delta(\mathcal{N}_t)$, and the rate of growth, $p(\mathcal{N}_t)$, then:

$$\frac{d\mathcal{N}}{dt} = -\delta(\mathcal{N}_t) + p(\mathcal{N}_t), \tag{8.1}$$

where, $\mathcal{N}_t$ is given by $\mathcal{N}_t(t^*) = \mathcal{N}(t - t^*)$, $t^* > 0$.

In addition, Haubold and Mathai [5] gave the limiting case of (8.1) when $\mathcal{N}(t)$ in the quantity of spatial fluctuations or homogeneities is ignored and given as

$$\frac{d\mathcal{N}_j}{dt} = -c_j \mathcal{N}_j(t), \tag{8.2}$$

where $\mathcal{N}_j(t = 0) = \mathcal{N}_0$ is the amount of density of species j at time $t = 0$, $c_j > 0$. If we drop the index j and integrate the typical kinetic equation (8.2), we receive

$$\mathcal{N}(t) - \mathcal{N}_0 = -c_0 D_t^{-1} \mathcal{N}(t), \tag{8.3}$$

where ${}_0D_t^{-1}$ is the specialized case of Riemann–Liouville integral operator ${}_0D_t^{-\nu}$ laid out as

$${}_0D_t^{-\nu} f(t) = \frac{1}{\Gamma(\nu)} \int_0^t (t-u)^{\nu-1} f(u)\,du, \quad (t > 0, \Re(\nu) > 0). \tag{8.4}$$

Haubold and Mathai [5] gave a fractional direction to the classical kinetic equation by considering fractional derivative rather than total derivative in (8.2)

$$\mathcal{N}(t) - \mathcal{N}_0 = -c^{\nu}{}_0D_t^{-\nu} \mathcal{N}(t), \tag{8.5}$$

then the solution for $\mathcal{N}(\mathrm{t})$ is a Mittag–Leffler function $\mathrm{E}_\nu(.)$

$$\mathcal{N}(\mathrm{t}) = \mathcal{N}_0 \sum_{r=0}^{\infty} \frac{(-1)^r (c\mathrm{t})^{\nu r}}{\Gamma(\nu r+1)} = \mathcal{N}_0 \mathrm{E}_\nu\left(-c^\nu \mathrm{t}^\nu\right). \tag{8.6}$$

In addition, Saxena and Kalla [19] thought about subsequent fractional kinetic equations.

$$\mathcal{N}(\mathrm{t}) - \mathcal{N}_0 f(\mathrm{t}) = -c^\nu {}_0D_\mathrm{t}^{-\nu} \mathcal{N}(\mathrm{t}), \tag{8.7}$$

where $f(\mathrm{t}) \in \mathrm{L}(0,\infty)$.

The Laplace transformation of the Riemann–Liouville fractional integration (8.4) is specified

$$\mathrm{L}\left[{}_0D_\mathrm{t}^{-\nu} f(\mathrm{t});\omega\right] = \omega^{-\nu} F(\omega), \left(\mathrm{t}>0, \Re(\nu)>0, \Re(\omega)>0\right), \tag{8.8}$$

where $F(\omega)$ is the Laplace transform of the function $f(\mathrm{t})$.

The usual incomplete Gamma functions $\gamma(s,x)$ and $\Gamma(s,x)$ represented by

$$\gamma(s,x) := \int_0^x \mathrm{t}^{s-1}\, e^{-\mathrm{t}}\, dt \qquad \left(\Re(s)>0; x \geqq 0\right) \tag{8.9}$$

and

$$\Gamma(s,x) := \int_x^\infty \mathrm{t}^{s-1}\, e^{-\mathrm{t}}\, dt \qquad \left(x \geqq 0; \Re(s)>0 \quad \text{if} \quad x=0\right), \tag{8.10}$$

satisfy the following rule of decomposition:

$$\gamma(s,x) + \Gamma(s,x) := \Gamma(s) \qquad \left(\Re(s)>0\right). \tag{8.11}$$

Throughout this chapter, $\mathbb{N}$, Z^- and C stand for positive integer sets, negative integer sets, and complex numbers, resepctively.

$$\mathbb{N}_0 := \mathbb{N} \cup \{0\} \qquad \text{and} \qquad \mathbb{Z}_0^- := \mathbb{Z}^- \cup \{0\}.$$

In addition, the $x \geqq 0$ parameter dealt with in (8.9) and (8.10), as well as elsewhere in the present chapter, is independent of $R(z)$ of the complex number $z \in \mathrm{C}$.

Recently, Srivastava et al. [33] introduced incomplete pochhammer symbols and incomplete hypergeometric functions in terms of incomplete Gamma functions and explored importance of these functions in the field of communication theory, probability theory and groundwater pumping modelling.

Srivastava et al. [34] recently presented a pair of Mellin–Barnes contour integral representations of incomplete H-functions $\gamma_{p,q}^{m,n}(z)$ and $\Gamma_{p,q}^{m,n}(z)$, and incomplete $\bar{H}$-functions $\bar{\gamma}_{p,q}^{m,n}(z)$ and $\bar{\Gamma}_{p,q}^{m,n}(z)$, in view of the $\gamma(s,x)$ and $\Gamma(s,x)$ represented by (8.9) and (8.10), respectively,

$$\begin{aligned}\gamma_{p,q}^{m,n}(z) &= \gamma_{p,q}^{m,n}\left[z\left|\begin{matrix}(\mathfrak{g}_1,\mathfrak{G}_1,x),(\mathfrak{g}_i,\mathfrak{G}_i)_{2,p}\\(\mathfrak{h}_j,\mathfrak{H}_j)_{1,q}\end{matrix}\right.\right]\\&= \gamma_{p,q}^{m,n}\left[z\left|\begin{matrix}(\mathfrak{g}_1,\mathfrak{G}_1,x),(\mathfrak{g}_2,\mathfrak{G}_2),\cdots,(\mathfrak{g}_p,\mathfrak{G}_p)\\(\mathfrak{h}_1,\mathfrak{H}_1),(\mathfrak{h}_2,\mathfrak{H}_2),\cdots,(\mathfrak{h}_q,\mathfrak{H}_q)\end{matrix}\right.\right]\\&= \frac{1}{2\pi i}\int_{\mathcal{L}}\psi(s,x)\,z^{-s}\,ds,\end{aligned}\tag{8.12}$$

where

$$\psi(s,x)=\frac{\gamma(1-\mathfrak{g}_1-\mathfrak{G}_1 s,x)\prod_{j=1}^{m}\Gamma(\mathfrak{h}_j+\mathfrak{H}_j s)\prod_{j=2}^{n}\Gamma(1-\mathfrak{g}_j-\mathfrak{G}_j s)}{\prod_{j=m+1}^{q}\Gamma(1-\mathfrak{h}_j-\mathfrak{H}_j s)\prod_{j=n+1}^{p}\Gamma(\mathfrak{g}_j+\mathfrak{G}_j s)},\tag{8.13}$$

and

$$\begin{aligned}\Gamma_{p,q}^{m,n}(z) &= \Gamma_{p,q}^{m,n}\left[z\left|\begin{matrix}(\mathfrak{g}_1,\mathfrak{G}_1,x),(\mathfrak{g}_i,\mathfrak{G}_i)_{2,p}\\(\mathfrak{h}_j,\mathfrak{H}_j)_{1,q}\end{matrix}\right.\right]\\&= \gamma_{p,q}^{m,n}\left[z\left|\begin{matrix}(\mathfrak{g}_1,\mathfrak{G}_1,x),(\mathfrak{g}_2,\mathfrak{G}_2),\cdots,(\mathfrak{g}_p,\mathfrak{G}_p)\\(\mathfrak{h}_1,\mathfrak{H}_1),(\mathfrak{h}_2,\mathfrak{H}_2),\cdots,(\mathfrak{h}_q,\mathfrak{H}_q)\end{matrix}\right.\right]\\&= \frac{1}{2\pi i}\int_{\mathcal{L}}\Psi(s,x)\,z^{-s}\,ds,\end{aligned}\tag{8.14}$$

where

$$\Psi(s,x)=\frac{\Gamma(1-\mathfrak{g}_1-\mathfrak{G}_1 s,x)\prod_{j=1}^{m}\Gamma(\mathfrak{h}_j+\mathfrak{H}_j s)\prod_{j=2}^{n}\Gamma(1-\mathfrak{g}_j-\mathfrak{G}_j s)}{\prod_{j=m+1}^{q}\Gamma(1-\mathfrak{h}_j-\mathfrak{H}_j s)\prod_{j=n+1}^{p}\Gamma(\mathfrak{g}_j+\mathfrak{G}_j s)}, \tag{8.15}$$

The incomplete H-functions $\gamma_{p,q}^{m,n}(z)$ and $\Gamma_{p,q}^{m,n}(z)$ in (8.12) and (8.14) exist for all $x \geq 0$ within similar contours and conditions as stated in Mathai and Saxena [13]. The denotations (8.12) and (8.14) readily yield the succeeding division formula:

$$\gamma_{p,q}^{m,n}(z)+\Gamma_{p,q}^{m,n}(z)=H_{p,q}^{m,n}(z), \tag{8.16}$$

for the familiar H-function.

Jangid et al. [6] introduced a family of incomplete I-functions ${}^{\gamma}I_{p,q}^{m,n}(z)$ and ${}^{\Gamma}I_{p,q}^{m,n}(z)$, which leads to a natural generalization of a variety of I-functions:

$$\begin{aligned}{}^{\gamma}I_{p,q}^{m,n}(z) &= {}^{\gamma}I_{p,q}^{m,n}\left[z\left|\begin{matrix}(\mathfrak{g}_1,\alpha_1;\mathfrak{G}_1:x),(\mathfrak{g}_2,\alpha_2;\mathfrak{G}_2),\cdots,(\mathfrak{g}_p,\alpha_p;\mathfrak{G}_p)\\(\mathfrak{h}_1,\beta_1;\mathfrak{H}_1),(\mathfrak{h}_2,\beta_2;\mathfrak{H}_2),\cdots,(\mathfrak{h}_q,\beta_q;\mathfrak{H}_q)\end{matrix}\right.\right]\\ &= \frac{1}{2\pi i}\int_{\pounds}\phi(s,x)\; z^s\; ds,\end{aligned} \tag{8.17}$$

and

$$\begin{aligned}{}^{\Gamma}I_{p,q}^{m,n}(z) &= {}^{\Gamma}I_{p,q}^{m,n}\left[z\left|\begin{matrix}(\mathfrak{g}_1,\alpha_1;\mathfrak{G}_1:x),(\mathfrak{g}_2,\alpha_2;\mathfrak{G}_2),\cdots,(\mathfrak{g}_p,\alpha_p;\mathfrak{G}_p)\\(\mathfrak{h}_1,\beta_1;\mathfrak{H}_1),(\mathfrak{h}_2,\beta_2;\mathfrak{H}_2),\cdots,(\mathfrak{h}_q,\beta_q;\mathfrak{H}_q)\end{matrix}\right.\right]\\ &= \frac{1}{2\pi i}\int_{\pounds}\Phi(s,x)\; z^s\; ds,\end{aligned} \tag{8.18}$$

$\forall\ z \neq 0$, provided

$$\phi(s,x)=\frac{\{\gamma(1-\mathfrak{g}_1+\alpha_1 s,x)\}^{\mathfrak{G}_1}\prod_{j=1}^{m}\{\Gamma(\mathfrak{h}_j-\beta_j s)\}^{\mathfrak{H}_j}\prod_{j=2}^{n}\{\Gamma(1-\mathfrak{g}_j+\alpha_j s)\}^{\mathfrak{G}_j}}{\prod_{j=n+1}^{p}\{\Gamma(\mathfrak{g}_j-\alpha_j s)\}^{\mathfrak{G}_j}\prod_{j=m+1}^{q}\{\Gamma(1-\mathfrak{h}_j+\beta_j s)\}^{\mathfrak{H}_j}}, \tag{8.19}$$

and

$$\Phi(s,x)=\frac{\{\Gamma(1-\mathfrak{g}_1+\alpha_1 s,x)\}^{\mathfrak{G}_1}\prod_{j=1}^{m}\{\Gamma(\mathfrak{h}_j-\beta_j s)\}^{\mathfrak{H}_j}\prod_{j=2}^{n}\{\Gamma(1-\mathfrak{g}_j+\alpha_j s)\}^{\mathfrak{G}_j}}{\prod_{j=n+1}^{p}\{\Gamma(\mathfrak{g}_j-\alpha_j s)\}^{\mathfrak{G}_j}\prod_{j=m+1}^{q}\{\Gamma(1-\mathfrak{h}_j+\beta_j s)\}^{\mathfrak{H}_j}}, \tag{8.20}$$

The incomplete I-functions ${}^{\gamma}I_{p,q}^{m,n}(z)$ and ${}^{\Gamma}I_{p,q}^{m,n}(z)$ in (8.17) and (8.18) exist for all $x \geq 0$ within similar contours and conditions as stated in Rathie [18].

Further, if we set $\mathfrak{H}_1=\cdots=\mathfrak{H}_m=1$ and $\mathfrak{G}_{n+1}=\cdots=\mathfrak{G}_p=1$, we define the following new incomplete $\bar{I}$-functions:

$$ {}^{\gamma}\bar{I}_{p,q}^{m,n}(z)={}^{\gamma}\bar{I}_{p,q}^{m,n}\left[z\left|\begin{matrix}(\mathfrak{g}_1,\alpha_1;\mathfrak{G}_1:x),\cdots,(\mathfrak{g}_n,\alpha_n;\mathfrak{G}_n),\\(\mathfrak{h}_1,\beta_1;1),\cdots,(\mathfrak{h}_m,\beta_m;1),\end{matrix}\right.\right.$$
$$\left.\begin{matrix}(\mathfrak{g}_{n+1},\alpha_{n+1};1),\cdots,(\mathfrak{g}_p,\alpha_p;1)\\(\mathfrak{h}_{m+1},\beta_{m+1};\mathfrak{H}_{m+1}),\cdots,(\mathfrak{h}_q,\beta_q;\mathfrak{H}_q)\end{matrix}\right], \tag{8.21}$$

and

$$ {}^{\Gamma}\bar{I}_{p,q}^{m,n}(z)={}^{\Gamma}\bar{I}_{p,q}^{m,n}\left[z\left|\begin{matrix}(\mathfrak{g}_1,\alpha_1;\mathfrak{G}_1:x),\cdots,(\mathfrak{g}_n,\alpha_n;\mathfrak{G}_n),\\(\mathfrak{h}_1,\beta_1;1),\cdots,(\mathfrak{h}_m,\beta_m;1),\end{matrix}\right.\right.$$
$$\left.\begin{matrix}(\mathfrak{g}_{n+1},\alpha_{n+1};1),\cdots,(\mathfrak{g}_p,\alpha_p;1)\\(\mathfrak{h}_{m+1},\beta_{m+1};\mathfrak{H}_{m+1}),\cdots,(\mathfrak{h}_q,\beta_q;\mathfrak{H}_q)\end{matrix}\right]. \tag{8.22}$$

The incomplete I-functions ${}^{\gamma}I_{p,q}^{m,n}(z)$ and ${}^{\Gamma}I_{p,q}^{m,n}(z)$ defined in (8.17) and (8.18) exist for $x \geq 0$, within the set of circumstances outlined by Rathie [18], with

$$\Delta>0, |\arg(z)|<\Delta\pi/2,$$

where

$$\Delta=\sum_{j=1}^{m}\mathfrak{H}_j\beta_j-\sum_{j=m+1}^{q}\mathfrak{H}_j\beta_j+\sum_{j=1}^{n}\mathfrak{G}_j\alpha_j-\sum_{j=n+1}^{p}\mathfrak{G}_j\alpha_j.$$

For more details on developments in incomplete functions and their applications, see recent research [6, 7, 9, 10, 15, 17].

Srivastava [32] introduced the general class of polynomials with dimension n, ($n = 0,1,2,\cdots$) as follows:

$$S_n^m(x) = \sum_{s=0}^{[n/m]} \frac{(-n)_{ms}}{s!} A_{n,s} x^s. \tag{8.23}$$

$A_{n,s} \in$ R (or C) are unrestricted positive constants, while m is a positive integer. Pochhammer symbol and largest integer function are denoted by the correspondences $(-n)_m$ and "[.]", respectively. Upon adequately specializing the coefficient $A_{n,s}$, Srivastava's polynomials yield a set of relevant polynomials as special instances.

In this chapter, we propose an extended model of the fractional kinetic equation and possible solutions. Together with the well-known Laplace transform technique, we provide another approach for deriving solutions to kinetic equations that incorporates a class of polynomials and the incomplete I-functions and incomplete $\bar{I}$-functions.

8.2 Solution of Generalized Fractional Kinetic Equations

Theorem 1

Assume that ζ, η, $\mathfrak{g}$, $\mathfrak{h} > 0$ *and* $\mu > 0$, *so the solution of*

$$\mathcal{N}(\mathrm{t}) - \mathcal{N}_0 \mathrm{t}^{\mu-1} S_n^m\left[\mathfrak{g}\mathrm{t}^{\zeta}\right]^{\Gamma} I_{u,v}^{r,s}\left[\mathfrak{h}\mathrm{t}^{\eta}\right] = -c^{\nu}{}_0D_{\mathrm{t}}^{-\nu}\mathcal{N}(\mathrm{t}), \tag{8.24}$$

is provided as

$$\mathcal{N}(\mathrm{t}) = \mathcal{N}_0 \mathrm{t}^{\mu-1} \sum_{i=0}^{\infty} \left(-c^{\nu}\mathrm{t}^{\nu}\right)^i \sum_{k=0}^{[n/m]} \frac{(-n)_{mk} A_{n,k}}{k!} \left(\mathfrak{g}\mathrm{t}^{\zeta}\right)^k$$
$$\times {}^{\Gamma}I_{u+1,v+1}^{r,s+1}\left[\mathfrak{h}\mathrm{t}^{\eta} \left| \begin{matrix} (\mathfrak{g}_1,\alpha_1;\mathfrak{G}_1 : x),(1-\mu-\zeta k,\eta;1),(\mathfrak{g}_j,\alpha_j;\mathfrak{G}_j)_{2,u} \\ (\mathfrak{h}_j,\beta_j;\mathfrak{H}_j)_{1,v},(1-\mu-\zeta k-\nu i,\eta;1) \end{matrix} \right.\right]. \tag{8.25}$$

Proof. The Laplace transform approach is used to establish the result. Take the Laplace transform of (8.24), and using (8.18), (8.23) and (8.8), after a little simplification we obtain

$$\left[1+c^{\nu}\omega^{-\nu}\right]\mathcal{N}(\omega) = \mathcal{N}_0 \sum_{k=0}^{[n/m]} \frac{(-n)_{mk} A_{n,k}\mathfrak{g}^k}{k!} \frac{1}{2\pi i} \int_{\mathcal{L}} \Phi(s,x)\mathfrak{h}^s \frac{\Gamma(\mu+\zeta k+\eta s)}{\omega^{\mu+\zeta k+\eta s}} ds, \tag{8.26}$$

where $\mathcal{N}(\omega)=\mathrm{L}\{\mathcal{N}(\mathrm{t});\omega\}$ and $\Phi(s,x)$ is defined in (8.20). Since $(1+x)^{-1}=\sum_{r=0}^{\infty}(-1)^r x^r$, therefore (8.26) implies that

$$\mathcal{N}(\omega)=\mathcal{N}_0\sum_{k=0}^{[n/m]}\frac{(-n)_{mk}A_{n,k}\mathfrak{g}^k}{k!}\frac{1}{2\pi i}\int_{\mathcal{L}}\Phi(s,x)\mathfrak{h}^s\Gamma(\mu+\zeta k+\eta s)\,ds \times\sum_{i=0}^{\infty}\left(-c^{\nu}\right)^i\omega^{-(\mu+\zeta k+\eta s+\nu i)}. \tag{8.27}$$

Taking the inverse Laplace transform of (8.27) gives us

$$\mathcal{N}(\mathrm{t})=\mathcal{N}_0\sum_{k=0}^{[n/m]}\frac{(-n)_{mk}A_{n,k}\mathfrak{g}^k}{k!}\times\frac{1}{2\pi i}\int_{\mathcal{L}}\Phi(s,x)\mathfrak{h}^s\;\Gamma(\mu+\zeta k+\eta s)\,ds \times\sum_{i=0}^{\infty}\left(-c^{\nu}\right)^i\frac{\mathrm{t}^{(\mu+\zeta k+\eta s+\nu i-1)}}{\Gamma(\mu+\zeta k+\eta s+\nu i)}. \tag{8.28}$$

Finally, using (8.18), we achieve the desired outcome (8.25).

Theorem 2

Assume that ζ, η, $\mathfrak{g}$, $\mathfrak{h}>0$ and $\mu>0$, then the solution of

$$\mathcal{N}(\mathrm{t})-\mathcal{N}_0\mathrm{t}^{\mu-1}S_n^m\left[\mathfrak{g}\mathrm{t}^{\zeta}\right]^{\gamma}I_{u,v}^{r,s}\left[\mathfrak{h}\mathrm{t}^{\eta}\right]=-c^{\nu}{}_0D_t^{-\nu}\mathcal{N}(\mathrm{t}), \tag{8.29}$$

is provided as

$$\mathcal{N}(\mathrm{t})=\mathcal{N}_0\mathrm{t}^{\mu-1}\sum_{i=0}^{\infty}\left(-c^{\nu}\mathrm{t}^{\nu}\right)^i\sum_{k=0}^{[n/m]}\frac{(-n)_{mk}A_{n,k}}{k!}\left(\mathfrak{g}\mathrm{t}^{\zeta}\right)^k \times{}^{\gamma}I_{u+1,v+1}^{r,s+1}\left[\mathfrak{h}\mathrm{t}^{\eta}\left|\begin{matrix}(\mathfrak{g}_1,\alpha_1;\mathfrak{G}_1:x),(1-\mu-\zeta k,\eta;1),(\mathfrak{g}_j,\alpha_j;\mathfrak{G}_j)_{2,u}\\(\mathfrak{h}_j,\beta_j;\mathfrak{H}_j)_{1,v},(1-\mu-\zeta k-\nu i,\eta;1)\end{matrix}\right.\right]. \tag{8.30}$$

Proof. The proof is the immediate consequence of definitions (8.17), (8.23) and parallel to Theorem 1, and hence we skip the proof.

With the following relationships, incomplete I-functions are connected to incomplete $\overline{I}$-functions:

$$^{\Gamma}\overline{I}_{u,v}^{r,s}(z)={}^{\Gamma}I_{u,v}^{r,s}\left[z\left|\begin{matrix}(\mathfrak{g}_1,\alpha_1;\mathfrak{G}_1,x),(\mathfrak{g}_j,\alpha_j;\mathfrak{G}_j)_{2,u}\\(\mathfrak{h}_j,\beta_j;1)_{1,r},(\mathfrak{h}_j,\beta_j;\mathfrak{H}_j)_{r+1,v}\end{matrix}\right.\right] \tag{8.31}$$

and

$$ {}^{\gamma}\bar{I}_{u,\,v}^{r,\,s}(z) = {}^{\gamma}I_{u,\,v}^{r,\,s}\left[z \,\middle|\, \begin{matrix} (\mathfrak{g}_1,\alpha_1;\mathfrak{G}_1,x),(\mathfrak{g}_j,\alpha_j;\mathfrak{G}_j)_{2,u} \\ (\mathfrak{h}_j,\beta_j;1)_{1,r},(\mathfrak{h}_j,\beta_j;\mathfrak{H}_j)_{r+1,v} \end{matrix}\right]. \tag{8.32} $$

If we give specific values to the parameters, such as $\mathfrak{H}_j = 1$ $(j = 1,\cdots,r)$ in (8.24), (8.25), (8.29) and (8.30) and utilizing the connection (8.31) and (8.32), we get the following corollaries:

Corollary 1

Assume that $\zeta,\ \eta,\ \mathfrak{g},\ \mathfrak{h} > 0$ and $\mu > 0$, so the solution of

$$ \mathcal{N}(\mathrm{t}) - \mathcal{N}_0 \mathrm{t}^{\mu-1} S_n^m\left[\mathfrak{g}\mathrm{t}^{\zeta}\right]^{\Gamma} \bar{I}_{u,v}^{r,s}\left[\mathfrak{h}\mathrm{t}^{\eta}\right] = -c^{\nu}{}_0D_{\mathrm{t}}^{-\nu}\mathcal{N}(\mathrm{t}), \tag{8.33} $$

is provided as

$$ \begin{aligned} \mathcal{N}(\mathrm{t}) = \mathcal{N}_0 \mathrm{t}^{\mu-1} \sum_{i=0}^{\infty} \left(-c^{\nu}\mathrm{t}^{\nu}\right)^i \sum_{k=0}^{[n/m]} \frac{(-n)_{mk} A_{n,k}}{k!} \left(\mathfrak{g}\mathrm{t}^{\zeta}\right)^k \\ \times {}^{\Gamma}\bar{I}_{u+1,v+1}^{r,s+1}\left[\mathfrak{h}\mathrm{t}^{\eta} \,\middle|\, \begin{matrix} (\mathfrak{g}_1,\alpha_1;\mathfrak{G}_1 : x),(1-\mu-\zeta k,\eta;1),(\mathfrak{g}_j,\alpha_j;\mathfrak{G}_j)_{2,u} \\ (\mathfrak{h}_j,\beta_j;1)_{1,r},(\mathfrak{h}_j,\beta_j;\mathfrak{H}_j)_{r+1,v},(1-\mu-\zeta k-\nu i,\eta;1) \end{matrix}\right]. \end{aligned} \tag{8.34} $$

Corollary 2

Assume that $\zeta,\ \eta,\ \mathfrak{g},\ \mathfrak{h} > 0$ and $\mu > 0$, then the solution of

$$ \mathcal{N}(\mathrm{t}) - \mathcal{N}_0 \mathrm{t}^{\mu-1} S_n^m\left[\mathfrak{g}\mathrm{t}^{\zeta}\right]^{\gamma} \bar{I}_{u,v}^{r,s}\left[\mathfrak{h}\mathrm{t}^{\eta}\right] = -c^{\nu}{}_0D_{\mathrm{t}}^{-\nu}\mathcal{N}(\mathrm{t}), \tag{8.35} $$

is provided as

$$ \begin{aligned} \mathcal{N}(\mathrm{t}) = \mathcal{N}_0 \mathrm{t}^{\mu-1} \sum_{i=0}^{\infty} \left(-c^{\nu}\mathrm{t}^{\nu}\right)^i \sum_{k=0}^{[n/m]} \frac{(-n)_{mk} A_{n,k}}{k!} \left(\mathfrak{g}\mathrm{t}^{\zeta}\right)^k \\ \times {}^{\gamma}\bar{I}_{u+1,v+1}^{r,s+1}\left[\mathfrak{h}\mathrm{t}^{\eta} \,\middle|\, \begin{matrix} (\mathfrak{g}_1,\alpha_1;\mathfrak{G}_1 : x),(1-\mu-\zeta k,\eta;1),(\mathfrak{g}_j,\alpha_j;\mathfrak{G}_j)_{2,u} \\ (\mathfrak{h}_j,\beta_j;1)_{1,r},(\mathfrak{h}_j,\beta_j;\mathfrak{H}_j)_{r+1,v},(1-\mu-\zeta k-\nu i,\eta;1) \end{matrix}\right]. \end{aligned} \tag{8.36} $$

The incomplete I-functions are related to the incomplete $\bar{H}$-functions, as in the relation below (see [34]):

$$\begin{aligned}\bar{\Gamma}_{u,\,v}^{r,\,s}(z) &= {}^{\Gamma}I_{u,\,v}^{r,\,s}\left[z\left|\begin{matrix}(\mathfrak{g}_1,\alpha_1;\mathfrak{G}_1:x),(\mathfrak{g}_j,\alpha_j;\mathfrak{G}_j)_{2,s},(\mathfrak{g}_j,\alpha_j;1)_{s+1,u}\\(\mathfrak{h}_j,\beta_j;1)_{1,r},(\mathfrak{h}_j,\beta_j;\mathfrak{H}_j)_{r+1,v}\end{matrix}\right.\right]\\ &= \bar{\Gamma}_{u,\,v}^{r,\,s}\left[z\left|\begin{matrix}(\mathfrak{g}_1,\alpha_1;\mathfrak{G}_1:x),(\mathfrak{g}_j,\alpha_j;\mathfrak{G}_j)_{2,s},(\mathfrak{g}_j,\alpha_j)_{s+1,u}\\(\mathfrak{h}_j,\beta_j)_{1,r},(\mathfrak{h}_j,\beta_j;\mathfrak{H}_j)_{r+1,v}\end{matrix}\right.\right],\end{aligned} \tag{8.37}$$

and

$$\begin{aligned}\bar{\gamma}_{u,\,v}^{r,\,s}(z) &= {}^{\gamma}I_{u,\,v}^{r,\,s}\left[z\left|\begin{matrix}(\mathfrak{g}_1,\alpha_1;\mathfrak{G}_1:x),(\mathfrak{g}_j,\alpha_j;\mathfrak{G}_j)_{2,s},(\mathfrak{g}_j,\alpha_j;1)_{s+1,u}\\(\mathfrak{h}_j,\beta_j;1)_{1,r},(\mathfrak{h}_j,\beta_j;\mathfrak{H}_j)_{r+1,v}\end{matrix}\right.\right]\\ &= \bar{\gamma}_{u,\,v}^{r,\,s}\left[z\left|\begin{matrix}(\mathfrak{g}_1,\alpha_1;\mathfrak{G}_1:x),(\mathfrak{g}_j,\alpha_j;\mathfrak{G}_j)_{2,s},(\mathfrak{g}_j,\alpha_j)_{s+1,u}\\(\mathfrak{h}_j,\beta_j)_{1,r},(\mathfrak{h}_j,\beta_j;\mathfrak{H}_j)_{r+1,v}\end{matrix}\right.\right].\end{aligned} \tag{8.38}$$

If we give specific values to the parameters, such as $\mathfrak{H}_j(j = 1,\cdots,r) = 1$ and $\mathfrak{G}_j(j = s+1,\cdots,u) = 1$ in (8.24), (8.25), (8.29) and (8.30), and utilizing the relation (8.37) and (8.38), we get the preceding corollaries:

Corollary 3

Suppose that $\zeta,\ \eta,\ \mathfrak{g},\ \mathfrak{h} > 0$ *and* $\mu > 0$, *therefore the solution of*

$$\mathcal{N}(\mathrm{t}) - \mathcal{N}_0\mathrm{t}^{\mu-1}S_n^m\left[\mathfrak{g}\mathrm{t}^{\zeta}\right]\bar{\Gamma}_{u,v}^{r,s}\left[\mathfrak{h}\mathrm{t}^{\eta}\right] = -c^{\nu}{}_0D_{\mathrm{t}}^{-\nu}\mathcal{N}(\mathrm{t}), \tag{8.39}$$

is provided as

$$\begin{aligned}\mathcal{N}(\mathrm{t}) = \mathcal{N}_0\mathrm{t}^{\mu-1}\sum_{i=0}^{\infty}\left(-c^{\nu}\mathrm{t}^{\nu}\right)^i\sum_{k=0}^{[n/m]}\frac{(-n)_{mk}A_{n,k}}{k!}\left(\mathfrak{g}\mathrm{t}^{\zeta}\right)^k\\ \times\bar{\Gamma}_{u+1,v+1}^{r,\,s+1}\left[\mathfrak{h}\mathrm{t}^{\eta}\left|\begin{matrix}(\mathfrak{g}_1,\alpha_1;\mathfrak{G}_1:x),(1-\mu-\zeta k,\eta;1),(\mathfrak{g}_j,\alpha_j;\mathfrak{G}_j)_{2,s},(\mathfrak{g}_j,\alpha_j)_{s+1,u}\\(\mathfrak{h}_j,\beta_j)_{1,r},(\mathfrak{h}_j,\beta_j;\mathfrak{H}_j)_{r+1,v},(1-\mu-\zeta k-\nu i,\eta;1)\end{matrix}\right.\right].\end{aligned} \tag{8.40}$$

Corollary 4

Assume that $\zeta,\ \eta,\ \mathfrak{g},\ \mathfrak{h}>0$ *and* $\mu>0$, *then the solution of*

$$\mathcal{N}(\mathrm{t})-\mathcal{N}_0\mathrm{t}^{\mu-1}S_n^m\left[\mathfrak{g}\mathrm{t}^{\zeta}\right]\bar{\gamma}_{u,v}^{r,s}\left[\mathfrak{h}\mathrm{t}^{\eta}\right]=-c^{\nu}{}_0D_{\mathrm{t}}^{-\nu}\mathcal{N}(\mathrm{t}), \tag{8.41}$$

is provided as

$$\mathcal{N}(\mathrm{t})=\mathcal{N}_0\mathrm{t}^{\mu-1}\sum_{i=0}^{\infty}\left(-c^{\nu}\mathrm{t}^{\nu}\right)^i\sum_{k=0}^{[n/m]}\frac{(-n)_{mk}A_{n,k}}{k!}\left(\mathfrak{g}\mathrm{t}^{\zeta}\right)^k$$
$$\times\,\bar{\gamma}_{u+1,v+1}^{r,s+1}\left[\mathfrak{h}\mathrm{t}^{\eta}\left|\begin{matrix}(\mathfrak{g}_1,\alpha_1;\mathfrak{G}_1:x),(1-\mu-\zeta k,\eta;1),(\mathfrak{g}_j,\alpha_j;\mathfrak{G}_j)_{2,s},(\mathfrak{g}_j,\alpha_j)_{s+1,u}\\(\mathfrak{h}_j,\beta_j)_{1,r},(\mathfrak{h}_j,\beta_j;\mathfrak{H}_j)_{r+1,v},(1-\mu-\zeta k-\nu i,\eta;1)\end{matrix}\right.\right]. \tag{8.42}$$

The incomplete $\bar{H}$-functions are related to the incomplete H-functions $\Gamma_{u,v}^{r,s}$ and $\gamma_{u,v}^{r,s}$ (see [34]). In consequence of (8.37) and (8.38), the incomplete I-functions are related to the incomplete H-functions as below:

$$\begin{aligned}\Gamma_{u,v}^{r,s}(z)&={}^{\Gamma}I_{u,v}^{r,s}\left[z\left|\begin{matrix}(\mathfrak{g}_1,\alpha_1;1:x),(\mathfrak{h}_j,\alpha_j;1)_{2,u}\\(\mathfrak{h}_j,\beta_j;1)_{1,v}\end{matrix}\right.\right]\\&=\Gamma_{u,v}^{r,s}\left[z\left|\begin{matrix}(\mathfrak{g}_1,\alpha_1:x),(\mathfrak{g}_j,\alpha_j)_{2,u}\\(\mathfrak{h}_j,\beta_j)_{1,v}\end{matrix}\right.\right],\end{aligned} \tag{8.43}$$

and

$$\begin{aligned}\gamma_{u,v}^{r,s}(z)&={}^{\gamma}I_{u,v}^{r,s}\left[z\left|\begin{matrix}(\mathfrak{g}_1,\alpha_1;1:x),(\mathfrak{g}_j,\alpha_j;1)_{2,u}\\(\mathfrak{h}_j,\beta_j;1)_{1,v}\end{matrix}\right.\right]\\&=\gamma_{u,v}^{r,s}\left[z\left|\begin{matrix}(\mathfrak{g}_1,\alpha_1:x),(\mathfrak{g}_j,\alpha_j)_{2,u}\\(\mathfrak{h}_j,\beta_j)_{1,v}\end{matrix}\right.\right].\end{aligned} \tag{8.44}$$

If we put $\mathfrak{H}_j(j=1,\cdots,v)=1$ and $\mathfrak{G}_j(j=1,\cdots,u)=1$ in (8.24), (8.25), (8.29) and (8.30), and making use of connection (8.43) and (8.44), then we get the known results due to [7]:

Corollary 5

Assume that $\zeta,\ \eta,\ \mathfrak{g},\ \mathfrak{h} > 0$ *and* $\mu > 0$, *then the solution of*

$$\mathcal{N}(\mathrm{t}) - \mathcal{N}_0 \mathrm{t}^{\mu-1} S_n^m\left[\mathfrak{g}\mathrm{t}^{\zeta}\right]\Gamma_{u,v}^{r,s}\left[\mathfrak{h}\mathrm{t}^{\eta}\right] = -c^{\nu}\,{}_0D_{\mathrm{t}}^{-\nu}\mathcal{N}(\mathrm{t}), \tag{8.45}$$

is provided as

$$\begin{aligned}\mathcal{N}(\mathrm{t}) = \mathcal{N}_0 \mathrm{t}^{\mu-1}\sum_{i=0}^{\infty}\left(-c^{\nu}\mathrm{t}^{\nu}\right)^i \sum_{k=0}^{[n/m]}\frac{(-n)_{mk}A_{n,k}}{k!}\left(\mathfrak{g}\mathrm{t}^{\zeta}\right)^k \\ \times \Gamma_{u+1,v+1}^{r,s+1}\left[\mathfrak{h}\mathrm{t}^{\eta}\left|\begin{matrix}(\mathfrak{g}_1,\alpha_1,x),(1-\mu-\zeta k,\eta),(\mathfrak{g}_j,\alpha_j)_{2,u}\\ (\mathfrak{h}_j,\beta_j)_{1,v},(1-\mu-\zeta k-\nu i,\eta)\end{matrix}\right.\right].\end{aligned} \tag{8.46}$$

Corollary 6

Assume that $\zeta,\ \eta,\ \mathfrak{g},\ \mathfrak{h} > 0$ *and* $\mu > 0$, *then the solution of*

$$\mathcal{N}(\mathrm{t}) - \mathcal{N}_0 \mathrm{t}^{\mu-1} S_n^m\left[\mathfrak{g}\mathrm{t}^{\zeta}\right]\gamma_{u,v}^{r,s}\left[\mathfrak{h}\mathrm{t}^{\eta}\right] = -c^{\nu}\,{}_0D_{t}^{-\nu}\mathcal{N}(\mathrm{t}), \tag{8.47}$$

is provided as

$$\begin{aligned}\mathcal{N}(\mathrm{t}) = \mathcal{N}_0 \mathrm{t}^{\mu-1}\sum_{i=0}^{\infty}\left(-c^{\nu}\mathrm{t}^{\nu}\right)^i \sum_{k=0}^{[n/m]}\frac{(-n)_{mk}A_{n,k}}{k!}\left(\mathfrak{g}\mathrm{t}^{\zeta}\right)^k \\ \times \gamma_{u+1,v+1}^{r,s+1}\left[\mathfrak{h}\mathrm{t}^{\eta}\left|\begin{matrix}(\mathfrak{g}_1,\alpha_1,x),(1-\mu-\zeta k,\eta),(\mathfrak{g}_j,\alpha_j)_{2,u}\\ (\mathfrak{h}_j,\beta_j)_{1,v},(1-\mu-\zeta k-\nu i,\eta)\end{matrix}\right.\right].\end{aligned} \tag{8.48}$$

8.3 Applications

In this section, we look at some implications and applications of the preceding findings. Particular examples of derived results can be generated by appropriately specializing the coefficient $A_{n,s}$ to obtain a wide number of known polynomial spectrums.

Example 1

Show that the solution of:

(i) $\mathcal{N}(t) - \mathcal{N}_0 t^{\mu-1} {}^{\Gamma}I_{u,v}^{r,s}\left[\mathfrak{h}t^{\eta}\right] = -c^{\nu}{}_0D_t^{-\nu}\mathcal{N}(t),$ (8.49)

is provided as

$$\mathcal{N}(t) = \mathcal{N}_0 t^{\mu-1} \sum_{i=0}^{\infty} \left(-c^{\nu} t^{\nu}\right)^i \times {}^{\Gamma}I_{u+1,v+1}^{r,s+1}\left[\mathfrak{h}t^{\eta} \left| \begin{matrix} (\mathfrak{g}_1, \alpha_1; \mathfrak{G}_1 : x), (1-\mu, \eta; 1), (\mathfrak{g}_j, \alpha_j; \mathfrak{G}_j)_{2,u} \\ (\mathfrak{h}_j, \beta_j; \mathfrak{H}_j)_{1,v}, (1-\mu-\nu i, \eta; 1) \end{matrix} \right.\right]. \quad (8.50)$$

(ii) $\mathcal{N}(t) - \mathcal{N}_0 t^{\mu-1} {}^{\Gamma}\overline{I}_{u,v}^{r,s}\left[\mathfrak{h}t^{\eta}\right] = -c^{\nu}{}_0D_t^{-\nu}\mathcal{N}(t),$ (8.51)

is provided as

$$\mathcal{N}(t) = \mathcal{N}_0 t^{\mu-1} \sum_{i=0}^{\infty} \left(-c^{\nu} t^{\nu}\right)^i \times {}^{\Gamma}\overline{I}_{u+1,v+1}^{r,s+1}\left[\mathfrak{h}t^{\eta} \left| \begin{matrix} (\mathfrak{g}_1, \alpha_1; \mathfrak{G}_1 : x), (1-\mu, \eta; 1), (\mathfrak{g}_j, \alpha_j; \mathfrak{G}_j)_{2,u} \\ (\mathfrak{h}_j, \beta_j; 1)_{1,r}, (\mathfrak{h}_j, \beta_j; \mathfrak{H}_j)_{r+1,v}, (1-\mu-\nu i, \eta; 1) \end{matrix} \right.\right]. \quad (8.52)$$

(iii) $\mathcal{N}(t) - \mathcal{N}_0 t^{\mu-1} \overline{\Gamma}_{u,v}^{r,s}\left[\mathfrak{h}\, t^{\eta}\right] = -c^{\nu}{}_0D_t^{-\nu}\mathcal{N}(t),$ (8.53)

is provided as

$$\mathcal{N}(t) = \mathcal{N}_0 t^{\mu-1} \sum_{i=0}^{\infty} \left(-c^{\nu} t^{\nu}\right)^i \times \overline{\Gamma}_{u+1,v+1}^{r,s+1}\left[\mathfrak{h}t^{\eta} \left| \begin{matrix} (\mathfrak{g}_1, \alpha_1; \mathfrak{G}_1 : x), (1-\mu, \eta; 1), (\mathfrak{g}_j, \alpha_j; \mathfrak{G}_j)_{2,s}, \\ (\mathfrak{h}_j, \beta_j)_{1,r}, (\mathfrak{h}_j, \beta_j; \mathfrak{H}_j)_{r+1,v}, (1-\mu-\nu i, \eta; 1) \end{matrix} \right.\right]. \quad (8.54)$$

(iv) $\mathcal{N}(t) - \mathcal{N}_0 t^{\mu-1} \Gamma_{u,v}^{r,s}\left[\mathfrak{h}t^{\eta}\right] = -c^{\nu}{}_0D_t^{-\nu}\mathcal{N}(t),$ (8.55)

is provided as

$$\mathcal{N}(t) = \mathcal{N}_0 t^{\mu-1} \sum_{i=0}^{\infty} \left(-c^{\nu} t^{\nu}\right)^i \Gamma_{u+1,v+1}^{r,s+1}\left[\mathfrak{h}t^{\eta} \left| \begin{matrix} (\mathfrak{g}_1, \mathfrak{G}_1, x), (1-\mu, \eta), (\mathfrak{g}_j, \mathfrak{G}_j)_{2,u} \\ (\mathfrak{h}_j, \mathfrak{H}_j)_{1,v}, (1-\mu-\nu i, \eta) \end{matrix} \right.\right]. \quad (8.56)$$

Solution. *If we set* $m = 1$, $\mathfrak{g} = 1$, $\zeta = 0$ *and* $A_{n,s} = \frac{s!}{(-n)_{ms}}$ *for* $s = 0$ *and* $A_{n,s} = 0$ *for* $s \neq 0$ *(i.e.,* $S_n^m\left[\mathfrak{g}t^{\zeta}\right] = 1$*) in* (8.24), (8.33), (8.39) *and* (8.45). *Thus, assertions of the example follow from Theorem 1, Corollary 1, Corollary 3, and Corollary 5.*

Remark 1

It is noteworthy that $x = 0$, $\mathfrak{G}_j = 1(j = 1,\cdots,u)$ *and* $\mathfrak{H}_j = 1\,(j = 1,\cdots,v)$, *and the kinetic equation and its solution given by* (8.49) *and* (8.50) *respectively yield the corresponding results given earlier by Choi and Kumar [2].*

Example 2

Show that the solution of:

$$\text{(i)}\quad \mathcal{N}(t) - \mathcal{N}_0 t^{\mu+\frac{n}{2}-1} H_n\left(\frac{1}{2\sqrt{t}}\right) {}^{\Gamma}I_{u,v}^{r,s}\left[\mathfrak{h}t^{\eta}\right] = -c^{\nu}{}_0D_t^{-\nu}\mathcal{N}(t), \tag{8.57}$$

is provided as

$$\mathcal{N}(t) = \mathcal{N}_0 t^{\mu-1}\sum_{i=0}^{\infty}\left(-c^{\nu}t^{\nu}\right)^i \sum_{k=0}^{[n/2]}\frac{(-1)^k t^k}{k!(n-2k)!} \times {}^{\Gamma}I_{u+1,v+1}^{r,s+1}\left[\mathfrak{h}t^{\eta}\middle|\begin{matrix}(\mathfrak{g}_1,\alpha_1;\mathfrak{G}_1:x),(1-\mu-k,\eta;1),(\mathfrak{g}_j,\alpha_j;\mathfrak{G}_j)_{2,u}\\ (\mathfrak{h}_j,\beta_j;\mathfrak{H}_j)_{1,v},(1-\mu-k-vi,\eta;1)\end{matrix}\right]. \tag{8.58}$$

$$\text{(ii)}\quad \mathcal{N}(t) - \mathcal{N}_0 t^{\mu+\frac{n}{2}-1} H_n\left(\frac{1}{2\sqrt{t}}\right) \bar{I}_{u,v}^{r,s}\left[\mathfrak{h}t^{\eta}\right] = -c^{\nu}{}_0D_t^{-\nu}\mathcal{N}(t), \tag{8.59}$$

is provided as

$$\mathcal{N}(t) = \mathcal{N}_0 t^{\mu-1}\sum_{i=0}^{\infty}\left(-c^{\nu}t^{\nu}\right)^i \sum_{k=0}^{[n/2]}\frac{(-1)^k t^k}{k!(n-2k)!} \times {}^{\Gamma}\bar{I}_{u+1,v+1}^{r,s+1}\left[\mathfrak{h}t^{\eta}\middle|\begin{matrix}(\mathfrak{g}_1,\alpha_1;\mathfrak{G}_1:x),(1-\mu-k,\eta;1),(\mathfrak{g}_j,\alpha_j;\mathfrak{G}_j)_{2,u}\\ (\mathfrak{h}_j,\beta_j;1)_{1,r},(\mathfrak{h}_j,\beta_j;\mathfrak{H}_j)_{r+1,v},(1-\mu-k-vi,\eta;1)\end{matrix}\right]. \tag{8.60}$$

$$\text{(iii)}\quad \mathcal{N}(t) - \mathcal{N}_0 t^{\mu+\frac{n}{2}-1} H_n\left(\frac{1}{2\sqrt{t}}\right) \bar{\Gamma}_{u,v}^{r,s}\left[\mathfrak{h}t^{\eta}\right] = -c^{\nu}{}_0D_t^{-\nu}\mathcal{N}(t), \tag{8.61}$$

is provided as

$$\mathcal{N}(\mathrm{t})=\mathcal{N}_0\mathrm{t}^{\mu-1}\sum_{i=0}^{\infty}\left(-c^{\nu}\mathrm{t}^{\nu}\right)^i\sum_{k=0}^{[n/2]}\frac{(-1)^k\mathrm{t}^k}{k!(n-2k)!}$$
$$\times\,\bar{\Gamma}^{r,s+1}_{u+1,v+1}\left[\mathfrak{h}\mathrm{t}^{\eta}\left|\begin{matrix}(\mathfrak{g}_1,\alpha\mathfrak{g}_1;A_1:x),(1-\mu-k,\eta;1),(\mathfrak{g}_j,\alpha_j;\mathfrak{G}_j)_{2,s},(\mathfrak{g}_j,\alpha_j)_{s+1,u}\\(\mathfrak{h}_j,\beta_j)_{1,r},(\mathfrak{h}_j,\beta_j;\mathfrak{H}_j)_{r+1,v},(1-\mu-k-\nu i,\eta;1)\end{matrix}\right.\right].\tag{8.62}$$

(iv) $\mathcal{N}(\mathrm{t})-\mathcal{N}_0\mathrm{t}^{\mu+\frac{n}{2}-1}\mathrm{H}_n\left(\frac{1}{2\sqrt{\mathrm{t}}}\right)\Gamma^{r,s}_{u,v}\left[\mathfrak{h}\mathrm{t}^{\eta}\right]=-c^{\nu}{}_0D_{\mathrm{t}}^{-\nu}\mathcal{N}(\mathrm{t}),$ (8.63)

is provided as

$$\mathcal{N}(\mathrm{t})=\mathcal{N}_0\mathrm{t}^{\mu-1}\sum_{i=0}^{\infty}\left(-c^{\nu}\mathrm{t}^{\nu}\right)^i\sum_{k=0}^{[n/2]}\frac{(-1)^k\mathrm{t}^k}{k!(n-2k)!}$$
$$\times\,\Gamma^{r,s+1}_{u+1,v+1}\left[\mathfrak{h}\mathrm{t}^{\eta}\left|\begin{matrix}(\mathfrak{g}_1,\mathfrak{G}_1,x),(1-\mu-k,\eta),(\mathfrak{g}_j,\mathfrak{G}_j)_{2,u}\\(\mathfrak{h}_j,\mathfrak{H}_j)_{1,v},(1-\mu-k-\nu i,\eta)\end{matrix}\right.\right].\tag{8.64}$$

Solution. *If we set* $m = 2$, $\mathfrak{g}=1$, $\zeta = 1$ *and* $A_{n,\,s} = (-1)^s$ *(i.e.,* $S_n^2[\mathrm{t}]=\mathrm{t}^{n/2}\mathrm{H}_n\left(1/2\sqrt{\mathrm{t}}\right)$, *where* $\mathrm{H}_n(\mathrm{t})$ *is a Hermite polynomial) in* (8.24), (8.33), (8.39), *and* (8.45). *Thus, assertions of the example follow from Theorem 1, Corollary 1, Corollary 3, and Corollary 5.*

Remark 2

From results (8.29), (8.35), (8.41), *and* (8.47), *we can derive a number of results.*

8.4 Conclusions

In this chapter, we investigated a new fractional generalization of the standard kinetic equation, which includes a family of polynomials and the incomplete *I*-functions, incomplete $\bar{I}$-functions, incomplete $\bar{H}$-functions, and incomplete *H*-functions, and presented their solutions using the well-known Laplace transform method. All of the derived results are of the natural type, yielding a wide range of fractional kinetic equations and solutions.

References

[1] R. Agarwal, Kritika and S.D. Purohit, Mathematical model pertaining to the effect of buffer over cytosolic calcium concentration distribution, *Chaos Solitons Fractals*, **143** (2021) 110610.

[2] J. Choi and D. Kumar, Solutions of generalized fractional kinetic equations involving Aleph functions, *Math. Communic.*, **20** (2015), 113–123.

[3] A. Chouhan, S.D. Purohit and S. Srivastava, An alternative method for solving generalized differential equations of fractional order, *Kragujevac J. Math.*, **37**(2) (2013), 299–306.

[4] H. Habenom, D.L. Suthar, D. Baleanu and S.D. Purohit, A numerical simulation on the effect of vaccination and treatments for the fractional hepatitis B model, *ASME. J. Comput. Nonlinear Dynam.* **16(1)** (2021), 011004.

[5] H.J. Haubold and A.M. Mathai, The fractional kinetic equation and thermonuclear functions, *Astrophys. Space Sci.*, **327** (2000), 53–63.

[6] K. Jangid, S. Bhatter, S. Meena, D. Baleanu, M.A. Qurashi and S.D. Purohit, Some fractional calculus findings associated with the incomplete *I*-functions, *Adv. Differ. Equ.*, **2020** (2020), 265.

[7] K. Jangid, S.D. Purohit, R. Agarwal and R.P. Agarwal, On the generalization of fractional kinetic equation comprising incomplete *H*-functions, *Kragujevac J. Math.*, **47**(5) (2023), 701–712.

[8] K. Jangid, S.D. Purohit, K.S. Nisar and T. Shefeeq, The internal blood pressure equation involving incomplete *I*-functions, *Inf. Sci. Lett.* **9(3)** (2020), 171–174.

[9] K. Jangid, S.D. Purohit and D.L. Suthar, Transformation formulas of incomplete hypergeometric functions via fractional calculus, *Bull. Transilv. Univ. Brasov, Ser. III, Math. Inform. Phys.* **13(62)** (2020), 571–580.

[10] N.K. Jangid, S. Joshi, K. Jangid, S. Araci and S.D. Purohit, Fractional calculus operators applied to the functions involving the product of Srivastava polynomials and incomplete *I*-functions, *Adv. Stud. Contemp. Math., Kyungshang.* **31(2)** (2021), 243–258.

[11] Kritika, R. Agarwal and S.D. Purohit, Mathematical model for anomalous subdiffusion using conformable operator, *Chaos Solitons Fractals*, **140** (2020) 110199.

[12] D. Kumar, S.D. Purohit, A. Secer and A. Atangana, On generalized fractional kinetic equations involving generalized Bessel function of the first kind, *Math. Probl. Engg.*, **2015** (2015), 7.

[13] A.M. Mathai and R.K. Saxena, *The H-function with Applications in Statistics and Other Disciplines*, Wiley Eastern Ltd. New Delhi and John Wiley and Sons, Inc. New York, 1978.

[14] A.M. Mathai, R.K. Saxena and H.J. Haubold, *The H-Functions: Theory and Applications*, Springer, New York, 2010.

[15] S. Meena, S. Bhatter, K. Jangid and S.D. Purohit, Certain integral transforms concerning the product of family of polynomials and generalized incomplete functions, *Moroccan J. Pure Appl. Anal.* (MJPAA), **6(2)** (2020), 243–254.

[16] K.S. Nisar, S.D. Purohit and S.R. Mondal, Generalized fractional kinetic equations involving generalized Struve function of the first kind, *J. King Saud Univ. Sci.*, **28**(2) (2016), 167–171.

[17] S.D. Purohit, A.M. Khan, D.L. Suthar and S. Dave, The impact on raise of environmental pollution and occurrence in biological populations pertaining to incomplete H-function, *Natl. Acad. Sci. Lett.*, **44(3)** (2021), 263–266.

[18] A.K. Rathie, A new generalization of generalized Hypergeometric functions, *Le Math.* **LII** (1997), 297–310.

[19] R.K. Saxena and S.L. Kalla, On the solutions of certain fractional kinetic equations, *Appl. Math. Comput.*, **199** (2008), 504–511.

[20] R.K. Saxena, A.M. Mathai and H.J. Haubold, On fractional kinetic equations, *Astrophys. Space Sci.*, **282** (2002), 281–287.

[21] R.K. Saxena, A.M. Mathai and H.J. Haubold, On generalized fractional kinetic equations, *Phys. A*, **344** (2004), 657–664.

[22] A. Shaikh, A. Tassaddiq, K.S. Nisar and D. Baleanu, Analysis of differential equations involving Caputo-Fabrizio fractional operator and its applications to reaction diffusion equations, *Adv. Difference Equ.*, **2019** (2019), 1–14.

[23] H. Singh, Analysis for fractional dynamics of Ebola virus model, *Chaos Solitons Fractals*, **138** (2020), 109992.

[24] H. Singh, Jacobi collocation method for the fractional advection-dispersion equation arising in porous media, *Numer. Methods Partial Differ. Equ.*, (2020), 1–18.

[25] H. Singh, Analysis of drug treatment of the fractional HIV infection model of CD4+ T-cells, *Chaos Solitons Fractals*, **146** (2021), 110868.

[26] H. Singh, Numerical Simulation for Fractional Delay Differential Equations, *Int. J. Dyn. Cont.*, **9**(2) (2021), 463–474.

[27] H. Singh, H.M. Srivastava and D. Kumar, A reliable algorithm for the approximate solution of the nonlinear Lane-Emden type equations arising in astrophysics, *Numer. Methods Partial Differ. Equ.*, **34**(5) (2018), 1524–1555.

[28] H. Singh, D. Kumar and D. Baleanu, *Methods of Mathematical Modelling: Fractional Differential Equations*, CRC Press Taylor and Francis, 2019.

[29] H. Singh, J. Singh, S.D. Purohit and D. Kumar, *Advanced Numerical Methods for Differential Equations: Applications in Science and Engineering*, CRC Press Taylor and Francis, 2021.

[30] H. Singh and H.M. Srivastava, Numerical investigation of the fractional-order Liénard and duffing equations arising in oscillating circuit theory, *Front. Phys.*, **8** (2020), 120.

[31] H. Singh and A.M. Wazwaz, Computational method for reaction diffusion-model arising in a spherical catalyst, *Int. J. Appl. Comput. Math.*, **7**(3) (2021), 65.

[32] H.M. Srivastava, A contour integral involving Fox's H-function. *Indian J. Math.*, **14** (1972), 1–6.

[33] H.M. Srivastava, M.A. Chaudhary and R.P. Agarwal, The incomplete Pochhammer symbols and their applications to hypergeometric and related functions, *Integral Transforms Spec. Funct.*, **23** (2012), 659–683.

[34] H.M. Srivastava, R.K. Saxena and R.K. Parmar, Some families of the incomplete H-functions and the incomplete $\overline{H}$-functions and associated integral transforms and operators of fractional calculus with applications. *Russ. J. Math. Phys.*, **25**(1) (2018), 116–138.

[35] D.L. Suthar, S.D. Purohit and S. Araci, Solution of fractional kinetic equations associated with the (p,q)-Mathieu-type series, *Discrete Dyn. Nat. Soc.*, **2020** (2020), 8645161, 7 pp.

9

Behavior of Slip Effects on Oscillating Flows of Fractional Second-Grade Fluid

Kashif Ali Abro
University of the Free State, South Africa

Ambreen Siyal
Mehran University of Engineering and Technology, Jamshoro, Pakistan

Abdon Atangana
University of the Free State, South Africa

CONTENTS

9.1 Introduction

The rheological impact of complicated fluids, such as blood, polymer solutions, and certain oils, are described by Newtonian constitutive equations that do not show any retardation or relaxation phenomena. Accordingly, distinct models have been suggested, and among these, differential types of model have received particular attention [1–3]. Owing to their complexity, several non-Newtonian fluids are recommended in the literature. One of the most

DOI: 10.1201/9781003263517-9

widely proposed non-Newtonian fluids is second-grade fluids [4–9]. Second-grade fluids represent the simplest occurrence of non-Newtonian fluids for acquiring exact solutions analytically. Recently, non-integer-order derivatives have encountered much success in characterizing complex dynamics. In particular, fractional calculus has been confirmed as a valuable tool for handling the properties of viscoelastic materials. Bagley [10], Friedrich [11], Junqi et al. [12], Guangyu et al. [13], Xu and Tan [14, 15], and Tan et al. [16–19] have all published fractional calculus approaches. Exact solutions to the flow of different fractionalized non-Newtonian fluids have recently been obtained [20–26]. In many applications the flow pattern corresponds to a slip flow, and the fluid shows loss of adhesion at the wetted wall, causing it to slide along the wall. In the study of fluid–solid surface interactions, the concept of slip of a fluid at a solid wall serves to describe macroscopic effects of certain molecular phenomena. The fluid exhibits non-continuum effects such as slip-flow, as demonstrated experimentally by Derek et al. [27]. Nearly 200 years ago, Navier [28] proposed boundary conditions that consider the possibility of fluid slip at a solid boundary. This condition states that the velocity of the fluid at the plane is linearly proportional to the shear stress at the plane. Beavers and Joseph [29] were the first to investigate fluid flow at the interface between a porous medium and fluid layer in an experimental study, proposing slip boundary conditions at the interface. Slipping flows have significant effects on the hysteresis phenomenon in non-Newtonian fluids. We therefore include recent studies of the slipping flow of non-Newtonian fluids [30–36], heat and mass transfer [42–47], control engineering theory [48–54], and others [55–61]. Additionally, in investigating governing partial differential equations in fractional form using the Caputo operator, a very distinct behavior is observed in fractionalized second-grade fluid, compared to ordinary second-grade fluid. As for checking, The velocity of the fluid is found to be very fast when the slip effects of fluids are examined. A novel approach to the solution of fractionalized equations has been introduced. Ndolane Sene [62] has contributed to the study of second-grade fluid by initiating a double integral method for approximating fractionalized energy, heat, and diffusion equations using the Caputo–Liouvile operator. Ndolane Sene [63] also suggests an approximate double integral method to the first Stokes solution, using the integral technique of heat balance to the generalized fractional derivative. Graphical impacts on the first Stokes equation in cubic or quadratic profile have been discussed. We conclude the literature review with some recent attempts [64–70]. The present chapter, seeks to provide exact analytical solutions for second-grade fluids with the fractional derivative approach, which is a more natural and appropriate tool for describing the complex behavior. The mathematical model is fractionalized and then tackled by means of the Laplace transform. The Wright generalized hypergeometric function is invoked to investigate optimal analytical solutions for velocity and shear stress. Newtonian and non-Newtonian solutions have been applied to show the effective influence of slipping and non-slipping oscillations of fractional and non-fractional second-grade fluids. The sub-solutions of both fractional and non-fractional second-grade fluids have been compared for slippage effects on the oscillations of fluid. Variable viscosity under the assumption of a slip condition on boundary conditions has produced sharp similarities and differences between non-integer and integer mathematical models on the basis of material and embedded parameters.

9.2 Mathematical Model of the Problem

The constitutive equations for an incompressible flow are [18]:

$$\nabla .V = 0, \nabla .T = \rho \frac{\partial V}{\partial t} + \rho (V.\nabla) V, \tag{9.1}$$

where ρ, V,t, ∇ and T are the functional parameters known as fluid density, velocity field, time, Nabla operator, and Cauchy stress respectively. Fluid motion is subject to an incompressible homogenous flow [4–9]

$$T = -p I + S, \; S = \mu A_1 + \alpha_1 A_2 + \alpha_2 A_1^2, \tag{9.2}$$

where $-p\ I$,S,μ, α_1,α_2 and A_1,A_2 are the rheological parameters; these parameters are termed indeterminate parts of the stress due to extra-stress tensor, dynamic viscosity, normal stress moduli, and tensors of kinematic respectively which are defined as:

$$A_1 = (\nabla V) + (\nabla V)^T, A_2 = \frac{dA_1}{dt} + A_1 (\nabla V) + (\nabla V)^T A_1, \tag{9.3}$$

For the problem under consideration we take V and S as velocity field and extra-stress tensor respectively, given as

$$V = V(y,t) = u(y,t) i, S = S(y,t), \tag{9.4}$$

where i is taken as the unit vector along the direction of the x coordinate. While the fluid is at rest because of $t = 0$, then

$$V = (y,0) = \mathbf{0}, S = (y,0) = \mathbf{0}, \tag{9.5}$$

Meaningful equations can be developed by tackling equations (9.1)–(9.5). We arrive at

$$\frac{\partial u(y,t)}{\partial t}=\left(\nu+\alpha\frac{\partial}{\partial t}\right)\frac{\partial^2 u(y,t)}{\partial y^2},\tau(y,t)=\left(\mu+\alpha_1\frac{\partial}{\partial t}\right)\frac{\partial u(y,t)}{\partial y}, \tag{9.6}$$

where $\tau(y,t)=S_{xy}(y,t)$ is the non-zero shear stress and $\nu=\mu/\rho$ is the kinematic viscosity and $\alpha=\alpha_1/\rho$ the viscoelastic parameter of the second-grade fluid. Due to the absence of a pressure gradient for fractionalized second-grade fluid, the governing equations are [18]

$$\frac{\partial u(y,t)}{\partial t}=\left(\nu+\alpha D_t^{\beta}\right)\frac{\partial^2 u(y,t)}{\partial y^2},\tau(y,t)=\left(\mu+\alpha_1 D_t^{\beta}\right)\frac{\partial u(y,t)}{\partial y}, \tag{9.7}$$

where the fractional parameter is defined as $0<\beta<1$. The fractional differential singular operator D_t^{β} is defined by [26, 27]

$$D_t^{\beta}f(t)=\begin{cases}\dfrac{1}{\Gamma(1-\beta)}\displaystyle\int_0^t\frac{f'(t)}{(t-\tau)^{\beta}}d\tau,\ 0\le\beta<1;\\ \dfrac{df(t)}{dt}, \qquad\qquad \beta=1\end{cases}. \tag{9.8}$$

By means of lengthy calculations for the contour integrals and residues, the solution of the system of fractional partial differential equations can be solved via Eq. (9.8) based on Laplace transform [16–24, 34]. In this context, the appropriate initial and boundary conditions are

$$u(y,0)=0,\ y>0, \tag{9.9}$$

$$u(0,t)=UH(t)\sin(\omega t)+\theta H(t)\left.\frac{\partial u(y,t)}{\partial y}\right|_{y=0} or\, UH(t)\cos(\omega t)+\theta H(t)\left.\frac{\partial u(y,t)}{\partial y}\right|_{y=0}, \tag{9.10}$$

where θ is the slip parameter and $H(t)$ is the Heaviside function [39]. Furthermore, the natural condition

$$u(y,t)\to 0 \text{ as } y\to\infty \text{ and } t>0. \tag{9.11}$$

9.3 Velocity Field via Singular Kernel

In order to obtain a second-order differential equation from equation $(9.7)_1$, we utilize Laplace transform [37–38, 41] with conditions (9.9) and $(9.10)_1$ to get

$$\left(\frac{\partial^2}{\partial y^2}-\frac{q}{\nu+\alpha q^{\beta}}\right)\bar{u}(y,q)=0, \tag{9.12}$$

subject to the conditions

$$\bar{u}(0,q)=\frac{U\omega}{q^2+\omega^2}+\theta\left.\frac{\partial\bar{u}(y,q)}{\partial y}\right|_{y=0},\ \text{and}\ \bar{u}(y,q)\to 0\ as\ y\to\infty, \tag{9.13}$$

where $\bar{u}(y,q)$ is the Laplace transform of $u(y,t)$ and q is a transform parameter. Solving Eq. (9.12), subject to boundary condition and natural conditions (9.13), we get

$$\overline{u_s}(y,q)=\frac{U\omega}{\left(q^2+\omega^2\right)\left\{1+\theta\left[\frac{q}{\nu+\alpha q^{\beta}}\right]^{\frac{1}{2}}\right\}}\exp\left\{-\left[\frac{q}{\nu+\alpha q^{\beta}}\right]^{\frac{1}{2}}y\right\}, \tag{9.14}$$

Before applying discrete inverse Laplace transform, first we rewrite Eq. (9.14) in series form and using the fact

$$(-1)^p\frac{\Gamma(\alpha+1)}{\Gamma(\alpha-p+1)}=\frac{\Gamma(p-\alpha)}{\Gamma(-\alpha)}, \tag{9.15}$$

we get

$$\overline{u_s}(y,q)=\frac{U\omega}{\left(q^2+\omega^2\right)}+U\omega\sum_{j=0}^{\infty}\left(-\omega^2\right)^j\sum_{k=1}^{\infty}\left(-\frac{\theta}{\sqrt{\alpha}}\right)^k$$

$$\sum_{n=1}^{\infty}\frac{\Gamma\left(n+\frac{k}{2}\right)\left(-\frac{\nu}{\alpha}\right)^n}{n!\Gamma\left(\frac{k}{2}\right)q^{(\beta-1)\frac{k}{2}+n\beta+2j+2}}+U\omega\sum_{j=0}^{\infty}\left(-\omega^2\right)^j\sum_{k=1}^{\infty}\left(-\frac{\theta}{\sqrt{\alpha}}\right)^k$$

$$\sum_{m=1}^{\infty}\left(-\frac{y}{\sqrt{\alpha}}\right)^m\sum_{n=0}^{\infty}\frac{\Gamma\left(n+\frac{k+m}{2}\right)\left(-\frac{\nu}{\alpha}\right)^n}{n!\Gamma\left(\frac{k+m}{2}\right)}\frac{1}{q^{(\beta-1)\frac{k+m}{2}+n\beta+2j+2}}, \tag{9.16}$$

Inverting equation (9.16) via Laplace transform, the resultant expression is:

$$u_s(y,t)=UH(t)\sin(\omega t)+U\omega H(t)\sum_{j=0}^{\infty}\left(-\omega^2\right)^j\sum_{k=0}^{\infty}\left(-\frac{\theta}{\sqrt{\alpha}}\right)^k$$

$$\times\sum_{n=0}^{\infty}\frac{\Gamma\left(n+\frac{k}{2}\right)\left(-\frac{\nu t^{\beta}}{\alpha}\right)^n t^{(\beta-1)\frac{k}{2}+2j+1}}{n!\Gamma\left(\frac{k}{2}\right)\Gamma\left((\beta-1)\frac{k}{2}+n\beta+2j+2\right)}+U\omega H(t)\sum_{j=0}^{\infty}\left(-\omega^2\right)^j\sum_{k=0}^{\infty}\left(-\frac{\theta}{\sqrt{\alpha}}\right)^k$$

$$\sum_{m=1}^{\infty}\frac{1}{m!}\left(-\frac{y}{\sqrt{\alpha}}\right)^m\times\sum_{n=0}^{\infty}\frac{\Gamma\left(n+\frac{k+m}{2}\right)\left(-\frac{\nu t^{\beta}}{\alpha}\right)^n t^{(\beta-1)\frac{k+m}{2}+2j+1}}{n!\Gamma\left(\frac{k+m}{2}\right)\Gamma\left((\beta-1)\frac{k}{2}+n\beta+2j+2\right)},\tag{9.17}$$

On working equation (9.17), velocity is expressed through the Wright generalized hypergeometric function [40] as:

$$u_s(y,t)=UH(t)\sin(\omega t)+U\omega H(t)\sum_{j=0}^{\infty}\left(-\omega^2\right)^j\sum_{k=0}^{\infty}\left(-\frac{\theta}{\sqrt{\alpha}}\right)^k t^{(\beta-1)\frac{k}{2}+2j+1}$$

$$\times\,{}_1\Psi_2\left[-\frac{\nu t^{\beta}}{\alpha}\middle|\begin{matrix}\left(\frac{k}{2},1\right)\\ \\ \left(\frac{k}{2},0\right),\left((\beta-1)\frac{k}{2}+2j+2,\beta\right)\end{matrix}\right]+U\omega H(t)\sum_{j=0}^{\infty}\left(-\omega^2\right)^j\sum_{k=0}^{\infty}\left(-\frac{\theta}{\sqrt{\alpha}}\right)^k$$

$$\sum_{m=0}^{\infty}\frac{1}{m!}\left(-\frac{y}{\sqrt{\alpha}}\right)^m\times t^{(\beta-1)\frac{k+m}{2}+2j+1}\;{}_1\Psi_2\left[-\frac{\nu t^{\beta}}{\alpha}\middle|\begin{matrix}\left(\frac{k+m}{2},1\right)\\ \\ \left(\frac{k+m}{2},0\right),\left((\beta-1)\frac{k+m}{2}+2j+2,\beta\right)\end{matrix}\right].\tag{9.18}$$

The special function is:

$$\sum_{n=0}^{\infty}\frac{(z)^n\prod_{j=1}^{p}\Gamma\left(a_j+A_jn\right)}{n!\prod_{j=1}^{q}\Gamma\left(b_j+B_jn\right)}={}_p\Psi_q\left[z\middle|\begin{matrix}(a_1,A_1),\ldots(a_p,A_p)\\ \\ (b_1,B_1),\ldots(b_q,B_q)\end{matrix}\right].\tag{9.19}$$

9.4 Shear Stress via Singular Kernel

Applying the Laplace transform to Eq. $(9.7)_2$, we find that

$$\bar{\tau}_s(y,q) = \left(\mu + \alpha_1 q^{\beta}\right)\frac{\partial \bar{u}(y,q)}{\partial y}, \tag{9.20}$$

Using Eq. (9.14) in (9.20), we find that

$$\bar{\tau}_s(y,q) = \frac{-\rho U\omega\sqrt{q}\left(\nu + \alpha q^{\beta}\right)^{\frac{1}{2}}}{\left(q^2+\omega^2\right)\left\{1+\theta\left[\frac{q}{\nu+\alpha q^{\beta}}\right]^{\frac{1}{2}}\right\}}\exp\left\{-\left[\frac{q}{\nu+\alpha q^{\beta}}\right]^{\frac{1}{2}} y\right\}, \tag{9.21}$$

Reworking on Eq. (9.21), we get a series format of equation (9.21) as:

$$\bar{\tau}_s(y,q) = -\rho U\omega\sqrt{\alpha}\sum_{j=0}^{\infty}\left(-\omega^2\right)^j\sum_{k=0}^{\infty}\left(-\frac{\theta}{\sqrt{\alpha}}\right)^k\sum_{m=0}^{\infty}\frac{1}{m!}\left(-\frac{y}{\sqrt{\alpha}}\right)^m \times\sum_{n=0}^{\infty}\frac{\Gamma\left(n+\frac{k+m-1}{2}\right)\left(-\frac{\nu}{\alpha}\right)^n}{n!\,\Gamma\left(\frac{k+m-1}{2}\right)q^{(\beta-1)\left(\frac{k+m-1}{2}\right)+n\beta+2j+1}}, \tag{9.22}$$

Inverting Eq. (9.22) by Laplace transforms, we have

$$\tau_s(y,t) = -\rho U\omega\sqrt{\alpha}H(t)\sum_{j=0}^{\infty}\left(-\omega^2\right)\sum_{k=0}^{\infty}\left(-\frac{\theta}{\sqrt{\alpha}}\right)^k\sum_{m=0}^{\infty}\frac{1}{m!}\left(-\frac{y}{\sqrt{\alpha}}\right)^m t^{(\beta-1)\left(\frac{m+k-1}{2}\right)+2j} \times {}_1\Psi_2\left[-\frac{\nu t^{\beta}}{\alpha}\left|\begin{matrix}\left(\frac{k+m-1}{2},1\right)\\ \left(\frac{k+m-1}{2},0\right),\left((\beta-1)\left(\frac{k+m-1}{2}\right)+2j+1,\beta\right)\end{matrix}\right.\right], \tag{9.23}$$

Similarly, shear stress corresponding to cosine oscillation can be obtained and it is

$$\tau_c(y,t) = -\rho U\sqrt{\alpha}H(t)\sum_{j=0}^{\infty}\left(-\omega^2\right)^j\sum_{k=0}^{\infty}\left(-\frac{\theta}{\sqrt{\alpha}}\right)^k\sum_{m=0}^{\infty}\frac{1}{m!}\left(-\frac{y}{\sqrt{\alpha}}\right)^m t^{(\beta-1)\left(\frac{m+k-1}{2}\right)+2j}$$
$$\times\, {}_1\Psi_2\left[-\frac{\nu t^\beta}{\alpha}\middle|\begin{matrix}\left(\frac{k+m-1}{2},1\right)\\ \\ \left(\frac{k+m-1}{2},0\right),\left((\beta-1)\left(\frac{k+m-1}{2}\right)+2j+1,\beta\right)\end{matrix}\right]. \tag{9.24}$$

9.5 Special Cases

9.5.1 Ordinary Second-Grade Fluid with $\beta \to 1$

Letting $\beta \to 1$ into Equations (9.18), (9.22), (9.23), and (9.24), we have solutions for classical second-grade fluid along the slip effect.

9.5.2 Fractionalized Second-Grade Fluid without $\theta \to 0$

Invoking $\theta \to 0$ into Equations (9.18), (9.22), (9.23), and (9.24), we have solutions

$$u_s(y,t) = UH(t)\sin(\omega t) + U\omega H(t)\sum_{j=0}^{\infty}\left(-\omega^2\right)^j\sum_{m=1}^{\infty}\frac{1}{m!}\left(-\frac{y}{\sqrt{\alpha}}\right)^m t^{(\beta-1)\frac{m}{2}+2j+1}$$
$$\times\, {}_1\Psi_2\left[-\frac{\nu t^\beta}{\alpha}\middle|\begin{matrix}\left(\frac{m}{2},1\right)\\ \\ \left(\frac{m}{2},0\right),\left((\beta-1)\frac{m}{2}+2j+2,\beta\right)\end{matrix}\right]. \tag{9.25}$$

$$u_c(y,t) = UH(t)\cos(\omega t) + UH(t)\sum_{j=0}^{\infty}\left(-\omega^2\right)^j \sum_{m=0}^{\infty}\frac{1}{m!}\left(-\frac{y}{\sqrt{\alpha}}\right)^m t^{(\beta-1)\frac{m}{2}+2j}$$

$$\times\ {}_1\Psi_2\left[-\frac{\nu t^{\beta}}{\alpha}\left|\begin{matrix}\left(\frac{m}{2},1\right)\\ \\ \left(\frac{m}{2},0\right),\left((\beta-1)\frac{m}{2}+2j+1,\beta\right)\end{matrix}\right.\right]. \quad (9.26)$$

$$\tau_s(y,t) = -\rho U\omega\sqrt{\alpha}H(t)\sum_{j=0}^{\infty}\left(-\omega^2\right)^j \sum_{m=0}^{\infty}\frac{1}{m!}\left(-\frac{y}{\sqrt{\alpha}}\right)^m t^{(\beta-1)\left(\frac{m-1}{2}\right)+2j}$$

$$\times\ {}_1\Psi_2\left[-\frac{\nu t^{\beta}}{\alpha}\left|\begin{matrix}\left(\frac{m-1}{2},1\right)\\ \\ \left(\frac{m-1}{2},0\right),\left((\beta-1)\left(\frac{m-1}{2}\right)+2j+1,\beta\right)\end{matrix}\right.\right]. \quad (9.27)$$

$$\tau_c(y,t) = -\rho U\sqrt{\alpha}H(t)\sum_{j=0}^{\infty}\left(-\omega^2\right)^j \sum_{m=0}^{\infty}\frac{1}{m!}\left(-\frac{y}{\sqrt{\alpha}}\right)^m t^{(\beta-1)\left(\frac{m-1}{2}\right)+2j}$$

$$\times\ {}_1\Psi_2\left[-\frac{\nu t^{\beta}}{\alpha}\left|\begin{matrix}\left(\frac{m-1}{2},1\right)\\ \\ \left(\frac{m-1}{2},0\right),\left((\beta-1)\left(\frac{m-1}{2}\right)+2j+1,\beta\right)\end{matrix}\right.\right]. \quad (9.28)$$

For Stokes' second problem of fractionalized second-grade fluid, it is also worth pointing out that substituting the frequency of oscillating plate $\omega = 0$, in Equations (9.27) and (9.28), we recovered the solutions [18] for Stokes' first problem of fractionalized second-grade fluid. Invoking $\alpha \to 0$ into Equations (9.14) and (9.21), the solutions for Newtonian fluid can be investigated.

9.6 Numerical Results and Discussion

Having obtained solutions for fractionalized second-grade fluid moving over the oscillating plane, our major objective has been achieved. In this section, we discuss the implementation and associated physical aspect of our obtained solutions, which are illustrated graphically. Attention has been focused on analyzing the difference between the velocity for sine oscillations in the presence and absence of slip effects, that is, $\theta = 5.0$ *or* 0.0 of fractionalized second-grade fluid for the flow generated by the oscillating plane. The results are interpreted with respect to the characteristics of pertinent parameters of interest. The diagrams for the field of velocity are carried out against y for the different values of time t, material constants (α,ν,θ), frequency ω, fractional parameter β and slipping values $\theta = 5.0$ or 0.0. In the interest of simplicity, graphs are traced by taking $U = 1$, $\nu = 0.295$, $\mu = 26$, $\alpha = 0.5$, $\beta = 0.2$ and using Mathcad software. It is clear from all figures that when the slip parameter θ at the plane increases, the velocity decreases. Figure 9.1 shows the influence of time on velocity profiles with and without slip effects. The difference in magnitude of velocity for slip parameter $\theta = 5.0$ *or* 0.0 is clear. As the plane oscillates, the amplitude of velocity sometimes increases and sometimes decreases (see Figure 9.1). The effect of material parameter α and kinematic viscosity ν are sketched in Figures 9.2 and 9.3 qualitatively. The influence of these parameters on fluid motion is identical, whether slip effects are present or not. The amplitude of velocity increases with increasing values of these parameters. It is important to mention the effect of fractional parameters on fluid motion. Whether slip effects are present or not, the amplitude of velocity field is a decreasing function as shown in Figures 9.4 and 9.5,

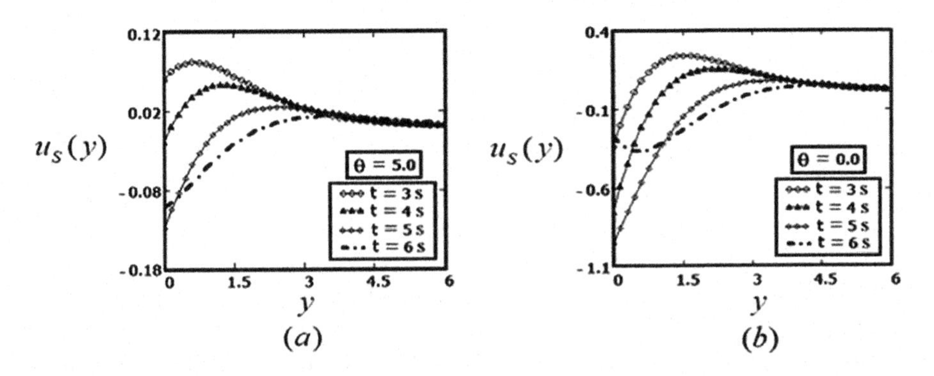

FIGURE 9.1
Profiles of the velocity field $u_s(y,t)$ for fractionalized second-grade fluid given by Eq. (9.18) for $U = 1$, $v = 0.295$, $\mu = 26$, $\alpha = 0.5$, $\beta = 0.2$, $\omega = 1$ and different values of t.

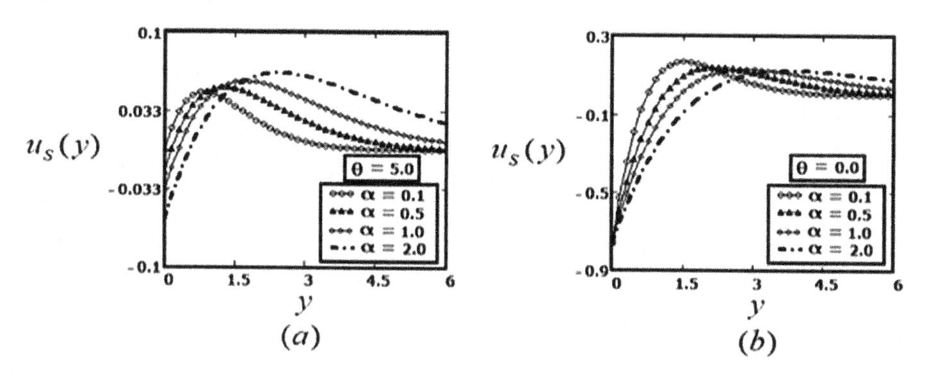

FIGURE 9.2
Profiles of the velocity field $u_s(y,t)$ for fractionalized second-grade fluid given by Eq. (9.18) for $U = 1$, $\rho = 88$, $\beta = 0.2$, $\omega = 1$, $t = 4$ s and different values of α.

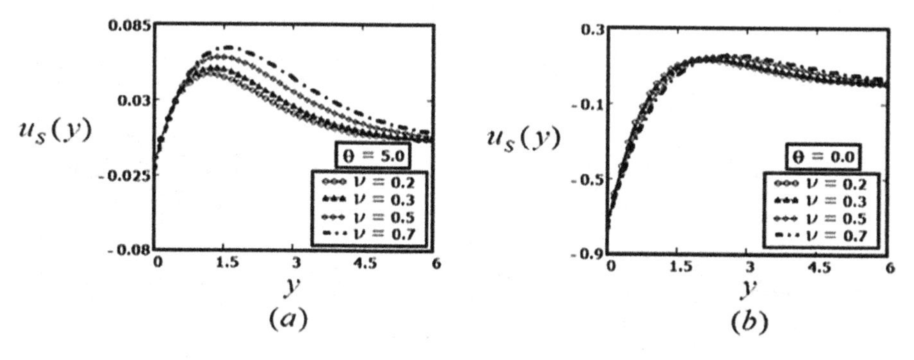

FIGURE 9.3
Profiles of the velocity field $u_s(y,t)$ for fractionalized second-grade fluid given by Eq. (9.18) for $U = 1$, $\rho = 88$, $\alpha = 0.5$, $\beta = 0.2$, $\omega = 1$, $t = 4$ s and different values of v.

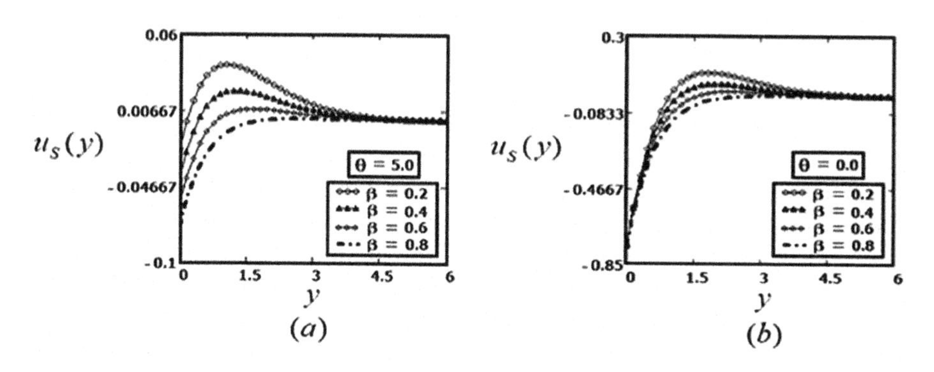

FIGURE 9.4
Profiles of the velocity field $u_s(y,t)$ for fractionalized second-grade fluid given by Eq. (9.18) for $U = 1$, $v = 0.295$, $\mu = 26$, $\alpha = 0.5$, $\omega = 1$, $t = 4$ s and different values of β.

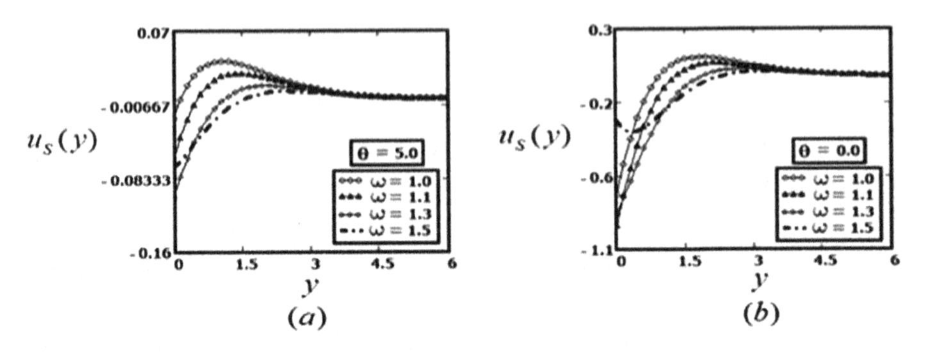

FIGURE 9.5
Profiles of the velocity field $u_s(y,t)$ for fractionalized second-grade fluid given by Eq. (9.18) for $U = 1$, $v = 0.295$, $\mu = 26$, $\alpha = 0.5$, $\beta = 0.2$, $t = 4$ s and different values of ω.

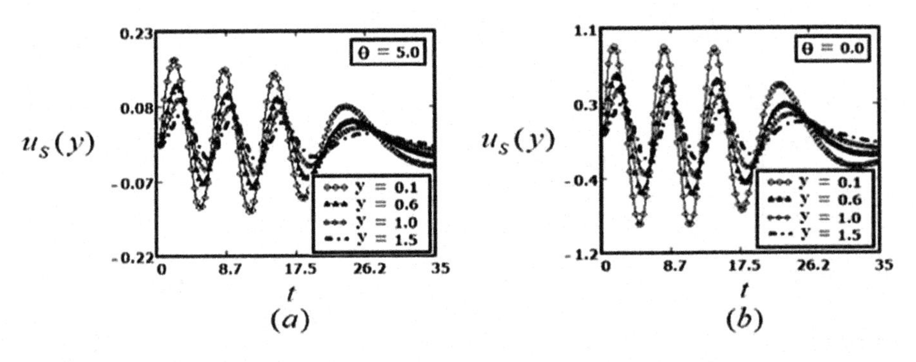

FIGURE 9.6
Profiles of the velocity field $u_s(y,t)$ for fractionalized second-grade fluid given by Eq. (9.18) for $U = 1$, $v = 0.295$, $\mu = 26$, $\alpha = 0.5$, $\beta = 0.2$, $\omega = 1$ and different values of y.

which illustrate the effect of the frequency ω of the oscillating plane on the fluid motion. These figures show the inverse relation between the amplitude of velocity field and the frequency of the oscillating plane. As expected, the velocity field approaches to zero away from the plane as shown in Figure 9.6. The velocity profiles equivalent to the four models (fractionalized second-grade for $\beta = 0.2, 0.5$ ordinary second-grade $\beta = 1$ Newtonian) are compared in Figure 9.7. It is clear from Figure 9.8 that the fractionalized second fluid is swiftest when slip effect is present and the Newtonian fluid is swiftest in the absence of slip effect. However ordinary second-grade fluid is the slowest, whether slip effect is present or not.

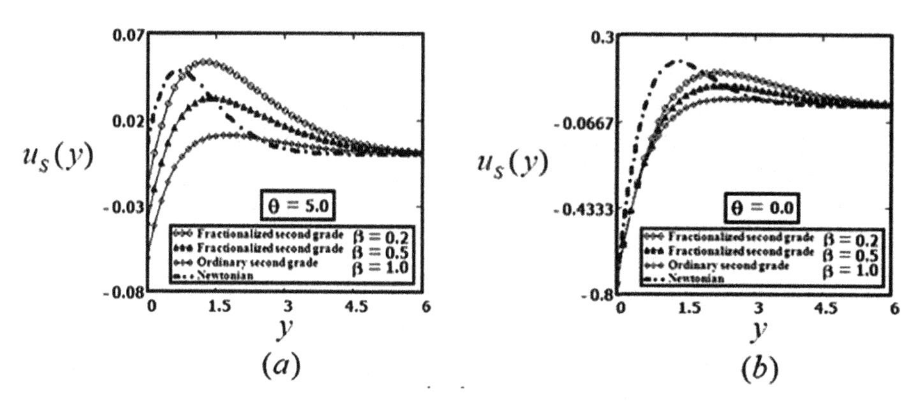

FIGURE 9.7
Profiles of the velocity field $u_s(y,t)$ for fractionalized second-grade, ordinary second-grade, and Newtonian fluids for $U = 1$, $v = 0.295$, $\mu = 26$, $\alpha = 0.5$, $\beta = 0.2$, 0.5 & 1, $\omega = 1$.

9.7 Concluding Remarks

The objective of the present chapter is to provide information on fractionalized and slipping second-grade fluid over an oscillating plane. The important conclusions from the present study are the following:

- The general solutions (9.18), (9.22), (9.23) and (9.24) demonstrate the sum of slippage and no-slippage contributions. These solutions can be specified to give similar solutions for ordinary second-grade fluid.
- The increasing values of slip parameter θ slow down the fluid motion.
- The amplitude of the velocity field may increase or decrease with respect to the time.
- The amplitude of the velocity field is an increasing function with respect to the material parameter α and kinematic viscosity ν, whether the slip effect is present or not.
- The fractional parameter β and frequency ω of the oscillating plane have similar effects on the fluid motion.
- The fractionalized second-grade and Newtonian fluids have greatest amplitude when slip effect is present and not present, respectively.

Future recommendations for the problem include employing fractal-fractional differential and integral operators.

Acknowledgment

All the authors are very grateful to Mehran University of Engineering and Technology, Jamshoro, Pakistan.

References

[1] Rajagopal K R, Longitudinal and torsional oscillations of a rod in a non-Newtonian fluid, *Acta Mech.*, 49 (1983) 281–285.
[2] Erdogan M E, Plane surface suddenly set in motion in a non-Newtonian fluid, *Acta Mech.*, 108 (1995) 179–187.
[3] Ghosh N C, Ghosh B C, Debnath L, The hydromagnetic flow of a dusty visco-elastic fluid between two infinite parallel plates, *Comput. Math. Appl.*, 39 (2000) 103–116.
[4] Fetecau C, Fetecau, Starting solutions for some unsteady unidirectional flows of a second grade fluid, *Int. J. Eng. Sci.*, 43 (2005) 781–789.
[5] Fetecau C, Corina F, Starting solutions for the motion of second grade fluid due to longitudinal and torsional oscillations of a circular cylinder, *Int. J. Eng. Sci.*, 44 (2006) 788–796.
[6] Fetecau C, Corina F, Zierep J, Decay of potential vortex and propagation of a heat wave in a second grade fluid, *Int. J. Non-linear Mech.*, 37 (2002) 1051–1056.
[7] Hayat T, Ellahi R, Asghar S, Siddiqui A M, Flow induced by a non-coaxial rotation of a porous disk executing non-torsional oscillations and a second grade fluid at infinity, *Appl. Math. Model.*, 28 (2004) 591–605.
[8] Hayat T, Khan M, Ayub M, Some analytical solutions for second grade fluid flows for cylindrical geometries, *Math. Comput. Modelling.*, 43 (2006) 16–29.
[9] Rajagopoal K R, On the creeping flow of the second order fluid, *J. Non-Newtonian Fluid Mech.*, 15 (1984) 239–246.
[10] Bagley R, A theoretical basis for the application of fractional calculus to visco-elasticity, *J. Rheol.*, 27 (1983) 201–210.
[11] Friedrich C H R, Relaxation and retardation function of the Maxwell model with fractional derivatives, *Rheol. Acta.*, 30 (1991) 151–158.
[12] Junki H, Guangyu H, Ciqun L, Analysis of general second order fluid flow in double cylinder rheometer, *Sci. China Series A*, 40 (1997) 183–190.
[13] Guangyu H, Junki H, Ciqun L, General second order fluid flow in a pipe, *Appl. Math. Mech.* (English Ed.) 16 (1995) 825–831.
[14] Xu M Y, Tan W, Theoretical analysis of the velocity field, stress field and vortex sheet of generalized second order fluid with fractional anomalous diffusion, *Sci. China Ser. A*, 44 (2001) 1387–1399.
[15] Xu M Y, Tan W C, Representation of the constitutive equation of viscoelastic materials by the generalized fracitonal element networks and its generalized solution, *Sci. China Ser. G.*, 46 (2003) 145–157.

[16] Tan W C, Pan W X, Xu M Y, A note on unsteady flows of a viscoelastic fluid with fractional Maxwell model between two parallel plates, *Int. J. Non-linear Mech.*, 38 (2003) 645–650.

[17] Tan W C, Xian F, Wei L, An exact solution of unsteady Couette flow of generalized second grade fluid, *Chin. Sci. Bullet.*, 47 (2002) 1783–1785.

[18] Tan W C, Xu M Y, The impulsive motion of flat plate in a generalized second grade fluid, *Mech. Res. Comm.*, 29 (2002) 3–9.

[19] Tan W C, Xu M Y, Unsteady flows of a generalized second grade fluid with the fractional derivative model between two parallel plates, *Acta Mech Sin.*, 20 (2004) 471–476.

[20] Mahmood A, Fetecau C, Khan N A, Jamil M, Some exact solutions of the oscillatory motion of a generalized second grade fluid in an annular region of two cylinders, *Acta Mech Sin.*, 26 (2010) 541–550.

[21] Fetecau C, Mahmood A, Jamil M, Exact solutions for the flow of a viscoelastic fluid induced by a circular cylinder subject to a time dependent shear stress, *Commun. Nonlinear Sci. Numer. Simulat.*, 15 (2010) 3931–3938.

[22] Liu Y, Zheng L, Zhang X, Unsteady MHD Couette flow of a generalized Oldroyd-B fluid with fractional derivative, *Comput. Math. Appl.*, 61 (2011) 443–450.

[23] Jamil M, Rauf A, Zafar A A, Khan N A, New exact analytical solutions for Stokes' first problem of Maxwell fluid with fractional derivative approach, *Comput. Math. Appl.*, 62 (2011) 1013–1023.

[24] Jamil M, Khan N A, Zafar A A, Translational flows of an Oldroyd-B fluid with fractional derivatives, *Comput. Math. Appl.*, 62 (2011) 1013–1023.

[25] Tripathi D, Peristaltic transport of a viscoelastic fluid in a channel, *Acta Astronautica.*, 68 (2011) 1379–1385.

[26] Tripathi D, Pandey S K, Das S, Peristaltic transport of a generalized Burgers fluid: Application to the movement of chime in small intestine, *Acta Astronautica.*, 69 (2011) 30–38.

[27] Derek C, Tretheway D C, Meinhart C D, A generating mechanism for apparent fluid slip in hydrophobic microchannels, *Phys. Fluid.*, 14 (2002) L9.

[28] Navier C L M H, Mémoire sur les Lois du Mouvement des Fluides, *Mem. Acad. Sci. Inst. France*, 1(1823) 414.

[29] Beavers G S, Joseph D D, Boundary conditions at a naturally permeable wall, *J. Fluid Mech.*, 30 (1967) 197–207.

[30] Khan M, Partial slip effects on the oscillatory flows of a fractional Jeffrey fluid in a porus medium, *J. Porus Media*, 10 (2007) 473–488.

[31] Ebaid A, Effects of magnetic field and wall slip conditions on the peristaltic transport of a Newtonian fluid in an asymmetric channel, *Phys. Lett. A*, 372 (2008) 4493–4499.

[32] Ellahi R, Hayat T, Mahomed F M, Generalized Couette flow of a third-grade fluid with slip: The exact solutions, *Z. Naturforsch.*, 65a (2010) 1071–1076.

[33] Hayat T, Najam S, Sajid M, Ayub M, Mesloub S, On exact solutions for oscillatory flows in a generalized burgers fluid with slip condition, *Z. Naturforsch.*, 65a (2010) 381–391.

[34] Zheng L, Liu Y, Zhang X, Slip effects on MHD flow of a generalized Oldroyd-B fluid with fractional derivative, *Nonlinear Anal. Real World Appl.*, 13 (2012) 513–523.

[35] Tripathi D, Gupta P K, Das S, Influence of slip condition on peristaltic transport of a viscoelastic fluid with fractional burgers model, *Ther. Sci.*, (2010) DOI: 10.2298/TSCI090801037G.
[36] Jamil M, Khan N A, Slip effects on fractional viscoelastic fluids, *Int. J. Differen. Equ.*, (2011) Article ID 193813.
[37] Podlubny I, *Fractional Differential Equations*, Academic Press, San Diego, (1999).
[38] Mainardi F, *Fractional Calculus and Waves in Linear Viscoelasticity: An Introduction to Mathematical Models*, Imperial College Press, London, (2010).
[39] Christov I C, Stokes' first problem for some non-Newtonian fluids: Results and mistakes, *Mech. Res. Comm.*, 37 (2010) 717–723.
[40] Mathai A M, Saxena R K, Haubold H J, *The H-Functions: Theory and Applications*, Springer, New York, (2010).
[41] Debnath L, Bhatta D, *Integral Transforms and their Applications* (2nd Ed.), Chapman & Hall/CRC, Boca Raton, (2007).
[42] Abro K A, Fractional characterization of fluid and synergistic effects of free convective flow in circular pipe through Hankel transform, *Phys. Fluid.*, 32, 123102 (2020). Doi: 10.1063/5.0029386
[43] Kashif A A, Imran Q, Ambreen S, Thermal transmittance and thermo-magnetization of unsteady free convection viscous fluid through non-singular differentiations, *Physica Scripta*, (2020). Doi: 10.1088/1402-4896/abc981.
[44] Abro K A, Abdon A, Dual fractional modeling of rate type fluid through non-local differentiation, *Numer. Methods Part. Different. Equ.*, 1–16 (2020). Doi: 10.1002/num.22633.
[45] Kashif A A, Abdon A, Numerical and mathematical analysis of induction motor by means of AB–fractal–fractional differentiation actuated by drilling system. Numer, *Methods Partial Different. Equ.*, 1–15 (2020). Doi: 10.1002/num.22618.
[46] Abro K A, Ambreen S, Basma S, Abdon A, Application of Statistical Method on Thermal Resistance and Conductance during Magnetization of Fractionalized *Free Convection Flow, Int. Commun. Heat Mass Trans.*, 119 (2020) 104971. Doi: 10.1016/j.icheatmasstransfer.2020.104971.
[47] Kashif A A, Mehwish S, Abdon A, Jose F G A, Thermophysical properties of Maxwell Nanoluids via fractional derivatives with regular kernel, *J Thermal Anal. Calor.*, (2020). DOI: 10.1007/s10973-020-10287-9
[48] Ali K A, Bhagwan D, A scientific report of non-singular techniques on microring resonators: An application to optical technology, *Optik-Int. J. Light Electron Opt.*, 224 (2020) 165696. https://doi.org/10.1016/j.ijleo.2020.165696.
[49] Qasim A, Samia R, Aziz U A, Kashif A A, Thermal investigation for electrified convection flow of Newtonian fluid subjected to damped thermal flux on a permeable medium, *Phys. Scr.*, (2020). https://doi.org/10.1088/1402-4896/abbc2e
[50] Abro K A, Role of fractal-fractional derivative on ferromagnetic fluid via fractal Laplace transform: A first problem via fractal–fractional differential operator, *Eur. J. Mech. B Fluids*, 85 (2021) 76–81. https://doi.org/10.1016/j.euromechflu.2020.09.002.
[51] Abro K A, Muzaffar H, Gomez-Aguilar J F, Application of Atangana-Baleanu fractional derivative to carbon nanotubes based non-Newtonian nanofluid: Applications in nanotechnology, *J. Appl. Comput. Mech.*, 6(SI) (2020) 1260–1269. https://doi.org/10.22055/JACM.2020.33461.2229

[52] Asıf Y, Hulya D, Abro K A, Dogan K, Role of Gilson–Pickering equation for the different types of soliton solutions: A nonlinear analysis, *Eur. Phys. J. Plus*, 135 (2020) 657. https://doi.org/10.1140/epjp/s13360-020-00646-8.
[53] Aziz U A, Muhammad T, Kashif A A, Multiple soliton solutions with chiral nonlinear Schrödinger's equation in (2+1)-dimensions, *Eur. J. Mech. B/Fluids*, 85, 68–75 (2020). https://doi.org/10.1016/j.euromechflu.2020.07.014.
[54] Kashif A A, Abdon A, Porous effects on the fractional modeling of magnetohydrodynamic pulsatile flow: an analytic study via strong kernels, *J. Ther. Anal. Calorimet.*, (2020). https://doi.org/10.1007/s10973-020-10027-z.
[55] Kashif A A, Atangana A, Numerical study and chaotic analysis of meminductor and memcapacitor through fractal-fractional differential operator, *Arab. J. Sci. Eng.*, 2020. https://doi.org/10.1007/s13369-020-04780-4.
[56] Ali A K, Atangana A, A comparative analysis of electromechanical model of piezoelectric actuator through Caputo–Fabrizio and Atangana–Baleanu fractional derivatives, *Math. Meth. Appl. Sci.*, (2020) 1–11. https://doi.org/10.1002/mma.6638.
[57] Ali A K, Jose F G A, Role of Fourier sine transform on the dynamical model of tensioned carbon nanotubes with fractional operator, *Math Meth. Appl. Sci.*, (2020) 1–11. https://doi.org/10.1002/mma.6655
[58] Abro K A, Atangana K, Mathematical analysis of memristor through fractal-fractional differential operators: A numerical study, *Math. Methods Appl. Sci.*, (2020) 1–18. https://doi.org/10.1002/mma.6378.
[59] Abro K A, Atangana A, A comparative study of convective fluid motion in rotating cavity via Atangana–Baleanu and Caputo–Fabrizio fractal–fractional differentiations, *Eur. Phys. J. Plus*, 135 (2020) 226–242. https://doi.org/10.1140/epjp/s13360-020-00136-x
[60] Kashif A A, Ambreen S, Abdon A, Thermal stratification of rotational second-grade fluid through fractional differential operators, *J. Ther. Anal. Calorimet.*, (2020). https://doi.org/10.1007/s10973-020-09312-8.
[61] Kashif A A, Abdon A, Role of Non-integer and Integer Order Differentiations on the Relaxation Phenomena of Viscoelastic Fluid, *Phys. Scr.*, 95 (2020) 035228. https://doi.org/10.1088/1402-4896/ab560c.
[62] Ndolane S, Second-grade fluid model with Caputo–Liouville generalized fractional derivative, *Chaos Solitons Fractal.*, 133 (2020) 109631.
[63] Ndolane S, Integral balance methods for Stokes' first equation described by the left generalized fractional derivative, *Physics*, 1(1) (2019) 154–166.
[64] Singh H, Wazwaz A M, Computational method for reaction diffusion-model arising in a spherical catalyst, *Int. J. Appl. Comput. Math.*, 7(3) (2021) 65.
[65] Singh H, Analysis for fractional dynamics of Ebola virus model, *Chaos Solitons Fractals*, 138 (2020) 109992.
[66] Singh H, Analysis of drug treatment of the fractional HIV infection model of CD4+ T-cells, *Chaos Solitons Fract.*, 146 (2021) 110868.
[67] Singh H, Jacobi collocation method for the fractional advection-dispersion equation arising in porous media, Numerical methods for partial differential equations, 2020.
[68] Singh H, Numerical simulation for fractional delay differential equations, *Int. J Dyn. Cont.*, (2020).

[69] Singh H, Srivastava H M, Devendra K, A reliable algorithm for the approximate solution of the nonlinear Lane-Emden type equations arising in astrophysics, *Numer. Methods Part. Different. Equ.*, 34(5) (2018) 1524–1555.
[70] Singh H, Srivastava H M, Numerical investigation of the fractional-order Liénard and duffing equations arising in oscillating circuit theory, *Front. Phys.*, 8 (2020) 120.

10

A Novel Fractional-Order System Described by the Caputo Derivative, Its Numerical Discretization, and Qualitative Properties

Ndolane Sene
Université Cheikh Anta Diop de Dakar, Dakar Senegal

CONTENTS

10.1 Introduction

Chaotic systems play an important role in modeling financial phenomena [1], physics [2], electrical circuits [3], biological diseases [4], and many others. Nowadays, fractional calculus has attracted many authors because of the various fractional derivatives that exist in this mathematical field. Fractional differential equations mean that in the classical differential equations described by integer derivatives, we replace the integer-order derivative in general with fractional operators, then examine their impact. With the introduction of chaos theory, many new chaotic and hyperchaotic systems were born. The importance of fractional operators in chaotic systems is that their influence can generate new types of attractors. The development of fractional calculus

DOI: 10.1201/9781003263517-10

with new fractional operators can probably generate attractors. By new types of operators, we mean the Caputo–Fabrizio [5] and the Atangana–Baleanu derivatives [6]. There have also been many investigations of non-singular and singular fractional operators; readers can refer to [7, 8]. For recent developments in the area of fractional calculus, see [9–17].

The literature related to fractional and classical versions of chaos systems is extensive. In [18], Diouf et al. proposed a numerical scheme to obtain the fractional financial chaotic system's phase portraits. Bifurcation analysis has been investigated to describe the impact of the parameters of the proposed model. In [19], Sene et al., investigating a fractional chaotic system, suggest the Lyapunov exponents in the context of fractional-order derivatives. They studied the impact of the fractional-order derivative on the phase portraits of Lu et al.'s chaotic system. In general, we find the order of the fractional derivative generates new types of attractors. In [20], in dimension four, Sene discusses the inadequacy of classical properties to characterize chaotic and hyperchaotic behavior using Lyapunov exponents. Sene has also provided phase portraits of a chaotic system in different orders using a numerical discretization of the Riemann–Liouville derivative; see also [21]. In [22], Li et al. propose a synchronization method for a fractional complex chaotic system. In [23], Lai et al. propose a new chaotic system, study its coexistence, and propose applications in electronics using the schematic circuit of the proposed model. In [24], Jiang et al. propose a new generalized combination complex synchronization between two fractional-order chaotic complex master systems and one fractional chaotic complex slave system. In [25], the authors propose the so-called local bifurcation of the Chen chaotic system. In [26], the authors analyze nonlinear dynamics and chaos in a delayed fractional-order chaotic system. In [27], the authors propose a new chaotic system with hyperbolic sinusoidal function and study all its properties in chaos theory. In [28], the authors propose a new fractional chaotic system and its control via Mittag–Leffler stability. For more investigations related to chaotic systems, see [22, 29–32].

In this work, we study the fractional novel four-scroll system described by the Caputo derivative. First, the phase portraits of the model are represented with the aid of a numerical discretization, including the discretization of the Riemann–Liouville integral. The fractional derivative's impact on the dynamics of the model is a particular focus, using the fractional version of the Lyapunov exponents. We will determine a threshold under which the stability of the equilibrium points of the model is obtained, and the famous Matignon criterion will be used. Classically, the initial condition has many impacts on the dynamics; in other words, new chaotic and hyperchaotic attractors are obtained with the aid of changes in the initial conditions. This influence will be analyzed and correlated with the impact of the fractional-order derivative. The Lyapunov exponents of the different attractors after the change of the initial conditions will be examined. Due to the variation of the model's parameters, the bifurcation maps are depicted, and the sectors of the obtention of hyperchaotic and chaotic behavior are determined. Note that the variation in the models' parameters can generate the loss of chaotic behavior; thus, bifurcation diagrams are indispensable. The phase portrait will support the bifurcation analysis where possible. In short, the impact of the fractional-order derivative will be focused in the paper by bifurcation maps, notably by the Lyapunov exponents in a fractional context. The coexisting attractors will be studied. For confirmation of the present work's applicability in real-world problems, the schematic circuit and its simulation using resistors and capacitors will be presented.

The chapter is structured as follows. In Section 10.2, we recall the fractional operators and properties necessary for our investigations. In Section 10.3, we present our novel fractional four-scroll system with nonlinearity described by the Caputo derivative. Section 10.4 briefly describes the numerical scheme used in this chapter and continues with the phase portraits at different values of the fractional-order derivative. In Section 10.5, we study the impact of the parameters of the model using the bifurcations maps. In Section 10.6, we study the nature of the chaos by using the Lyapunov exponents and calculating the Kaplan–Yorke dimension, where possible, for more clarity in the analysis. In Section 10.7, we characterize the influence of the initial condition on the proposed models' dynamics. In Section 10.8, we study the equilibrium points' local stability, and in Section 10.9, we suggest science and engineering models for our present model.

10.2 Preliminary Definitions in Fractional Calculus

This section recalls the derivatives with singular kernels used in fractional calculus. As derivatives in fractional calculus, they are the Caputo derivative, the Riemann–Liouville derivative, the Riemann–Liouville integral, the Atangana–Baleanu derivative, the Caputo–Fabrizio derivative, and many other derivatives. All the derivatives cited have their advantages and inconveniences; everything depends on the nature of the problem posed.

Definition 1

[33, 34] We describe the Riemann–Liouville integral operator of order $\alpha > 0$ for a continuous function x as the following form $y : [0, +\infty[\rightarrow R$

$$I^{\alpha}x(t) = \frac{1}{\Gamma(\alpha)}\int_0^t (t-s)^{\alpha-1} x(s)ds, \tag{10.1}$$

where the Euler function is represented by the form $\Gamma(\alpha) := \int_0^{\infty} e^{-u}u^{\alpha-1}du.$

After the fractional integral, we define the Caputo operator, with which we will continue in this investigation. We have the derivative in the following definition.

Definition 2

[33, 34] We describe the Caputo fractional derivative of order $\alpha \in (0.1)$ of the function $x : [0, +\infty[\rightarrow \mathrm{R}$ as the following form

$$D_c^{\alpha} x(t) = \frac{1}{\Gamma(1-\alpha)} \int_0^t (t-s)^{-\alpha} x'(s) ds, \tag{10.2}$$

where the Gamma function is defined as the following form $\Gamma(\alpha) := \int_0^{\infty} e^{-u} u^{\alpha-1}$.

There is also a fractional derivative directly associated with the Riemann–Liouville integral, called the Riemann–Liouville derivative. We recall it in the following definition.

Definition 3

[33, 34] We describe the Riemann–Liouville fractional derivative of order $\alpha \in (0.1)$ of the function $x : [0, +\infty[\rightarrow \mathbb{R}$ as the following form

$$D^{\alpha} x(t) = \frac{1}{\Gamma(1-\alpha)} \frac{d}{dt} \int_0^t (t-s)^{-\alpha} x(s) ds, \tag{10.3}$$

where the Gamma function is defined as the following form $\Gamma(\alpha) := \int_0^{\infty} e^{-u} u^{\alpha-1}$.

There are many properties, including these three operators, and readers are referred to the fractional calculus literature. The solution of the fractional differential equation defined by $D_t^{\alpha} x = f(x, t)$ with initial condition $x(0)$ is given by the following representation, which will play a crucial role in our numerical discretization

$$x(t) = x(0) + I^{\alpha} f(t). \tag{10.4}$$

The numerical scheme here will concern the numerical discretization of the fractional integral; the discretization procedure is not new and exists in the literature; see the investigation in [8].

10.3 Fractional Constructive Equations

For the rest of our investigation, we use the fractional version of the Caputo derivative for the model proposed in the literature by Sampath et al. [35]. The fractional differential system described by the Caputo derivative is given by the following form:

$$D_t^{\alpha} x = a\left(y - x\right) + byz, \tag{10.5}$$

$$D_t^{\alpha} y = -y + exz - dy^3, \tag{10.6}$$

$$D_t^{\alpha} z = cz - xy, \tag{10.7}$$

The initial conditions considered are described as the following, with values to be specified in the phase portrait section:

$$x\left(0\right) = x_0, \quad y\left(0\right) = y_0, \quad z\left(0\right) = z_0. \tag{10.8}$$

We set the parameters of model (10.5)–(10.7) as given by the following: $a = 3$, $b = 14$, $c = 3.9$, $d = 10$ and $e = 4$. Many others chaotic systems can be obtained with variations of these parameters and the variation of the Caputo derivative order. In our investigations these points will be discussed, for greater clarity, using bifurcation diagrams and fractional Lyapunov exponents. The fractional derivative is introduced in the modeling because we want to consider the memory effect in the modeling and obtain new chaotic systems versus the variation of the order of the fractional derivative.

10.4 Numerical Scheme Investigation and Applications

In this section, we propose the numerical procedure of the fractional differential system (10.5)–(10.7). The numerical scheme is already reported in the literature [36]. Here we use this numerical procedure to obtain the phase portraits of the fractional-order system. As will be noticed, this novel chaotic system also has interesting properties in a fractional context. The solutions of the fractional differential system (10.5)–(10.7) can be represented in the following form:

$$x\left(t\right) = x\left(0\right) + I^{\alpha}\phi\left(t, \tilde{x}\right), \tag{10.9}$$

$$y(t) = y(0) + I^{\alpha}\varpi(t, \tilde{x}), \tag{10.10}$$

$$z(t) = z(0) + I^{\alpha}\theta(t, \tilde{x}), \tag{10.11}$$

In the previous equation, the functions ϕ, φ and θ are represented in the following forms:

$$\phi(t, \tilde{x}) = a(y - x) + byz, \tag{10.12}$$

$$\varphi(t, \tilde{x}) = -dy^3 - y + exz, \tag{10.13}$$

$$\theta(t, \tilde{x}) = cz - xy. \tag{10.14}$$

We apply the solutions described in Eqs (10.9)–(10.11) at the point t_n. We obtain the following equations written with the aid of the Riemann–Liouville integral:

$$x(t_n) = x(0) + I^{\alpha}\phi(t_n, \tilde{x}), \tag{10.15}$$

$$y(t_n) = y(0) + I^{\alpha}\varphi(t_n, \tilde{x}), \tag{10.16}$$

$$z(t_n) = z(0) + I^{\alpha}\theta(t_n, \tilde{x}). \tag{10.17}$$

The second step consists of rewriting the time t_n as the following form $t_n = nh$, where h denotes the step size. Then we obtain the following integrals:

$$I^{\alpha}\phi(t_n, \tilde{x}) = h^{\alpha}\sum_{j=1}^{n} c_{n-j}\phi(t_j, \tilde{x}_j), \tag{10.18}$$

$$I^{\alpha}\varphi(t_n, \tilde{x}) = h^{\alpha}\sum_{j=1}^{n} c_{n-j}\varphi(t_j, \tilde{x}_j), \tag{10.19}$$

$$I^{\alpha}\theta(t_n, \tilde{x}) = h^{\alpha}\sum_{j=1}^{n} c_{n-j}\theta(t_j, \tilde{x}_j), \tag{10.20}$$

where the parameter is defined by the following:

$$c_{n-j} = \left((n-j+1)^{\alpha} - (n-j)^{\alpha} \right) / \frac{1}{\Gamma(1+\alpha)}.$$

To simplify the discretization in Eqs (10.18)–(10.20) we introduce the first-order interpolant polynomials of our functions $\phi(\tau, \tilde{x}(\tau))$, $\varphi(\tau, \tilde{x}(\tau))$ and $\theta(\tau, \tilde{x}(\tau))$. We have the following forms:

$$\phi(\tau, \tilde{x}(\tau)) = \varphi(t_{j+1}, \tilde{x}_{j+1}) + \frac{\tau - t_{j+1}}{h} \left[\phi(t_{j+1}, \tilde{x}_{j+1}) - \phi(t_j, \tilde{x}_j) \right], \tag{10.21}$$

$$\varphi(\tau, \tilde{x}(\tau)) = \varepsilon(t_{j+1}, \tilde{x}_{j+1}) + \frac{\tau - t_{j+1}}{h} \left[\varphi(t_{j+1}, \tilde{x}_{j+1}) - \varphi(t_j, \tilde{x}_j) \right], \tag{10.22}$$

$$\theta(\tau, \tilde{x}(\tau)) = \theta(t_{j+1}, \tilde{x}_{j+1}) + \frac{\tau - t_{j+1}}{h} \left[\theta(t_{j+1}, \tilde{x}_{j+1}) - \theta(t_j, \tilde{x}_j) \right]. \tag{10.23}$$

Using Eqs (10.21)–(10.23) into the equations described in Eqs (10.18)–(10.17), we have the following approximation of the fractional integrals:

$$I^{\alpha} \phi(t_n, \tilde{x}) = h^{\alpha} \left[\bar{c}_n^{(\alpha)} \varphi(0) + \sum_{j=1}^{n} c_{n-j}^{(\alpha)} \varphi(t_j, \tilde{x}_j) \right], \tag{10.24}$$

$$I^{\alpha} \varphi(t_n, \tilde{x}) = h^{\alpha} \left[\bar{c}_n^{(\alpha)} \varphi(0) + \sum_{j=1}^{n} c_{n-j}^{(\alpha)} \varphi(t_j, \tilde{x}_j) \right], \tag{10.25}$$

$$I^{\alpha} \theta(t_n, \tilde{x}) = h^{\alpha} \left[\bar{c}_n^{(\alpha)} \theta(0) + \sum_{j=1}^{n} c_{n-j}^{(\alpha)} \theta(t_j, \tilde{x}_j) \right], \tag{10.26}$$

In Eqs (10.24)–(10.26), the parameters of the discretization can be rewritten in the form

$$\bar{c}_n^{(\alpha)} = \frac{(n-1)^{\alpha} - n^{\alpha}(n-\alpha-1)}{\Gamma(2+\alpha)}, \tag{10.27}$$

and when the integer values describe $n = 1, 2, \ldots$, the parameters become the form

$$c_0^{(\alpha)} = \frac{1}{\Gamma(2+\alpha)} \text{ and } c_n^{(\alpha)} = \frac{(n-1)^{\alpha+1} - 2n^{\alpha+1} + (n+1)^{\alpha+1}}{\Gamma(2+\alpha)}. \tag{10.28}$$

Considering the previous reasoning and plugging Eqs. (10.24)–(10.26) into Eqs (10.15)–(10.17) we get the following approximations:

$$x(t_n) = x(0) + h^{\alpha}\left[\bar{c}_n^{(\alpha)}\varphi(0) + \sum_{j=1}^{n} c_{n-j}^{(\alpha)}\phi(t_j, \tilde{x}_j)\right], \tag{10.29}$$

$$y(t_n) = y(0) + h^{\alpha}\left[\bar{c}_n^{(\alpha)}\varphi(0) + \sum_{j=1}^{n} c_{n-j}^{(\alpha)}\varphi(t_j, \tilde{x}_j)\right], \tag{10.30}$$

$$z(t_n) = z(0) + h^{\alpha}\left[\bar{c}_n^{(\alpha)}\eta(0) + \sum_{j=1}^{n} c_{n-j}^{(\alpha)}\theta(t_j, \tilde{x}_j)\right]. \tag{10.31}$$

The numerical approximations of the functions φ, ϕ, and θ of our fractional system are represented as the following forms:

$$\phi(t_j, \tilde{x}_j) = a(y_j - x_j) + by_jz_j, \tag{10.32}$$

$$\varphi(t_j, \tilde{x}_j) = -dy_j^3 - y_j + ex_jz_j, \tag{10.33}$$

$$\theta(t_j, \tilde{x}_j) = cz_j - x_jy_j. \tag{10.34}$$

The convergence of the approximate solutions is not difficult to establish. We suppose that $x(t_n)$, $y(t_n)$ and $z(t_n)$ are the numerical approximations of the fractional system (10.5)–(10.7), and x_n, y_n and z_n are the exact solutions of the fractional-order system (10.5)–(10.7); then, in these context the error functions are expressed in the forms

$$\left|x(t_n)-x_n\right|=\mathcal{E}\left(h^{\min\{\alpha+1,2\}}\right), \tag{10.35}$$

$$\left|y(t_n)-y_n\right|=\mathcal{E}\left(h^{\min\{\alpha+1,2\}}\right), \tag{10.36}$$

$$\left|z(t_n)-z_n\right|=\mathcal{E}\left(h^{\min\{\alpha+1,2\}}\right). \tag{10.37}$$

Thus, the convergence of the approximated solution is obtained when the step size h converges to 0.

We continue with the phase portrait of the fractional-order four-scroll system by considering different orders of the Caputo derivative. The fractional-order derivative is described in this paper; it is important to see the novelty of this operator. In other words, what are the real advantages of considering the Caputo derivative instead of the integer-order derivative or the Riemann–Liouville derivative? We have not used the Riemann–Liouville derivative due to the initial conditions, which are not compatible with it. The second question will be answered after the phase portrait, but note that the fractional operators are to capture the memory effect. The orders considered in this section are $\alpha = 0.94$, $\alpha = 0.955$, $\alpha = 0.97$, and $\alpha = 0.985$. In our first phase portraits 1a, 1a, 2a, 2b will be illustrated at the order $\alpha = 0.94$ because before this value, chaotic behaviors are removed. This point will be discussed in detail in the next sections via the Lyapunov exponents (Figures 10.1 and 10.2).

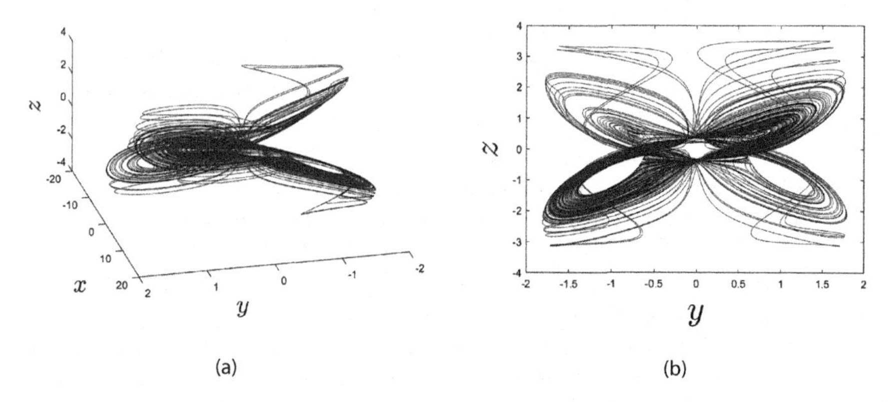

FIGURE 10.1
Dynamics of the solution of the fractional novel four-scroll system at $\alpha = 0.94$.

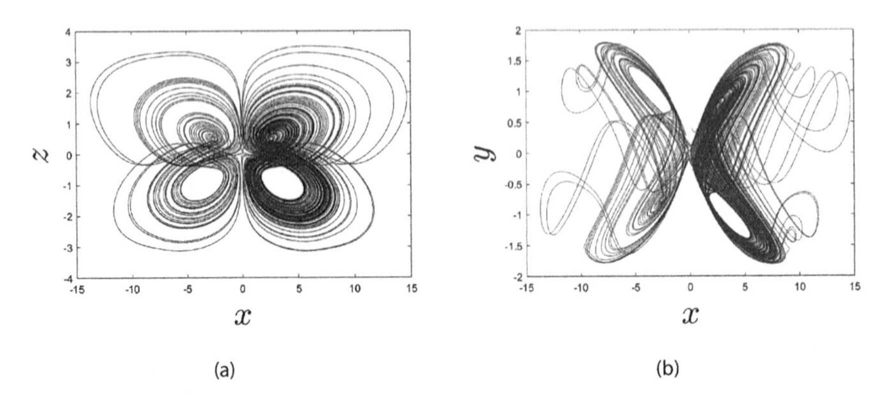

FIGURE 10.2
Dynamics of the solution of the fractional novel four-scroll system at $\alpha = 0.94$.

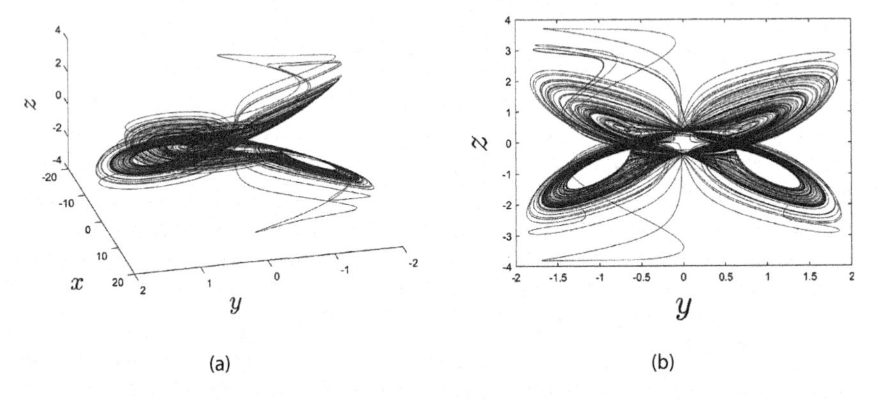

FIGURE 10.3
Dynamics of the solution of the fractional novel four-scroll system at $\alpha = 0.955$.

Let the order be $\alpha = 0.955$; we depict the phase portraits in Figures 10.3a, 10.3b, 10.4a, and 10.4b. As we shall see, the behaviors will be more complex.

We now take the order $\alpha = 0.97$and depict the phase portraits in Figures 10.5a, 10.5b, 10.6a, and 10.6b. The influence of the fractional order will be observed when we vary the values of the Caputo derivative's order.

We confirm the change in the dynamics with the order $\alpha = 0.985$. Figures 10.7a, 10.7b, 10.8a, and 10.8b represent the phase portraits for the fractional novel four-scroll model with Caputo derivative in different planes.

In regard to the phase portraits, we do not notice the same behaviors in the dynamics. When the fractional-order derivative varies (0.94, 1), we notice chaotic behavior in various contexts. In other words, we notice different behaviors in the dynamics depending on the choice of the value of the order of the Caputo derivative into (0.94, 1).

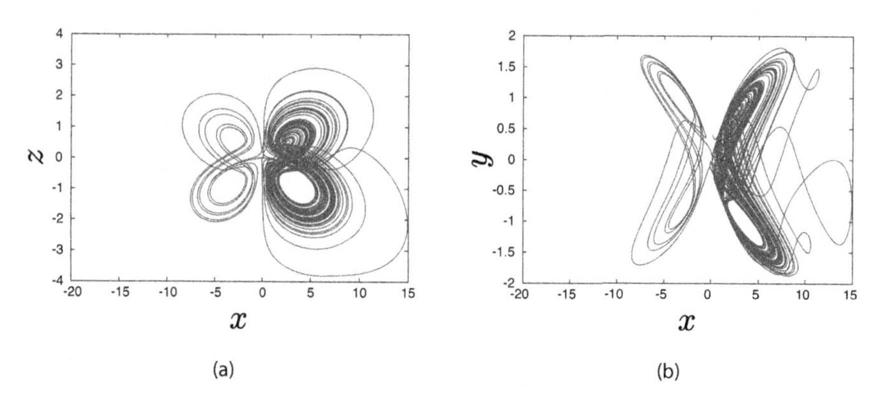

FIGURE 10.4
Dynamics of the solution of the fractional novel four-scroll system at $\alpha = 0.955$.

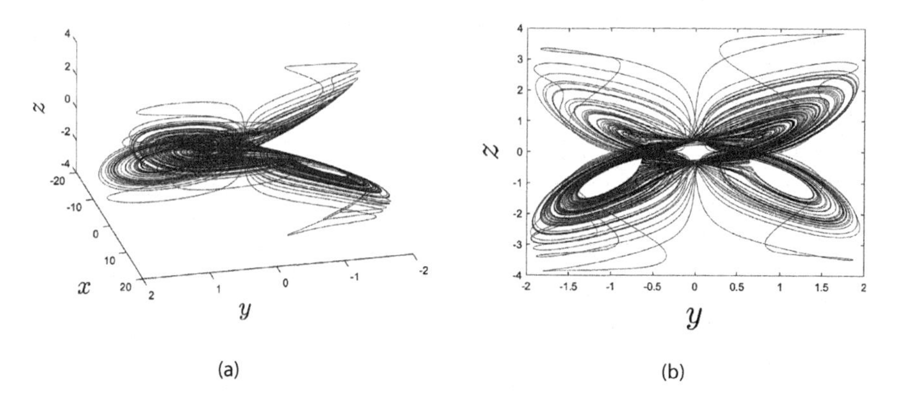

FIGURE 10.5
Dynamics of the solution of the fractional novel four-scroll system at $\alpha = 0.97$.

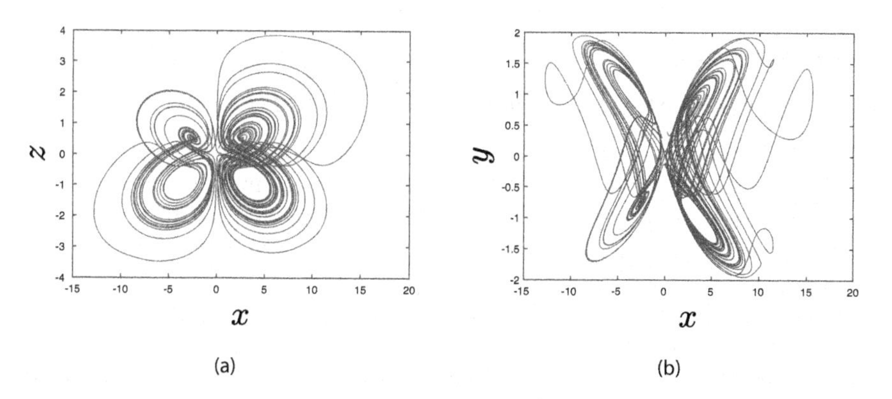

FIGURE 10.6
Dynamics of the solution of the fractional novel four-scroll system at $\alpha = 0.97$.

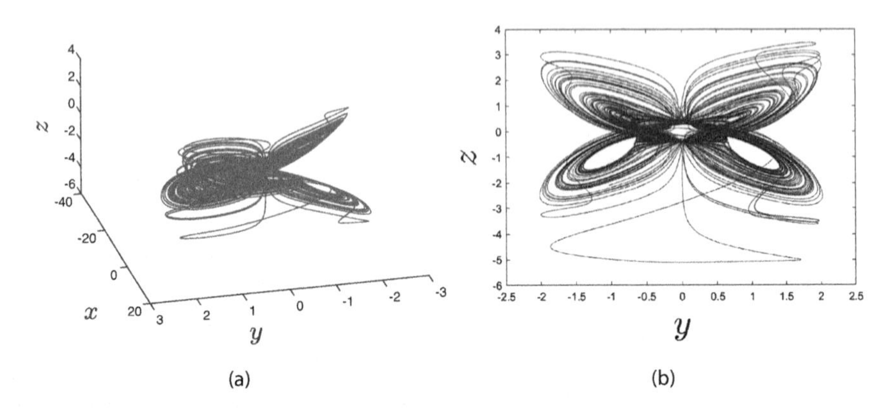

FIGURE 10.7
Dynamics of the solution of the fractional novel four-scroll system at $\alpha = 0.985$.

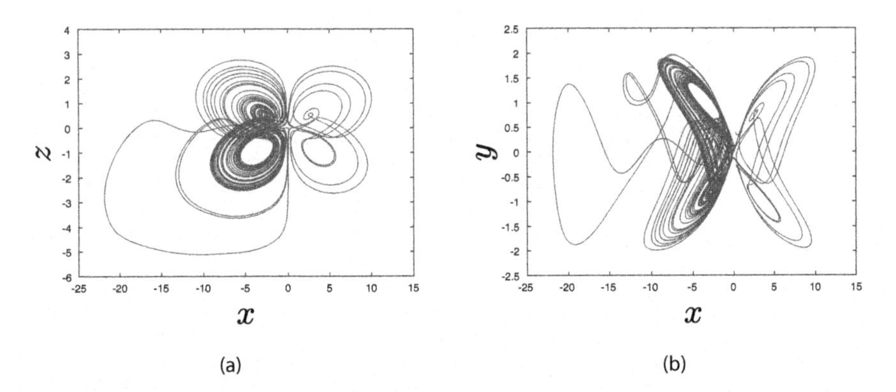

FIGURE 10.8
Dynamics of the solution of the fractional novel four-scroll system at $\alpha = 0.985$.

10.5 Influence of the Parameters of the Model

When the parameters vary in the chaotic system, the phase portraits change considerably and there are many options. The chaotic behavior can be preserved or hyperchaotic behavior can be obtained. Limit cycles can be obtained, as can periodic behaviors. Hyperchaotic or chaotic behaviors can also be lost. In our first analysis, we suppose the parameter a is influenced into $[0, 5]$; see the bifurcation diagrams in Figure 10.9a. For the rest of the section, we will use the order $\alpha = 0.98$.

We observe that the first chaotic region is the interval $[0.5, 1.5)$, where the system enters chaotic behavior with routine period doubling when the parameter a increases. We note a period-doubling route to chaos again when a is into $[1.5, 1.75)$. The fractional system's behaviors are chaotic into $[1.75, 5)$

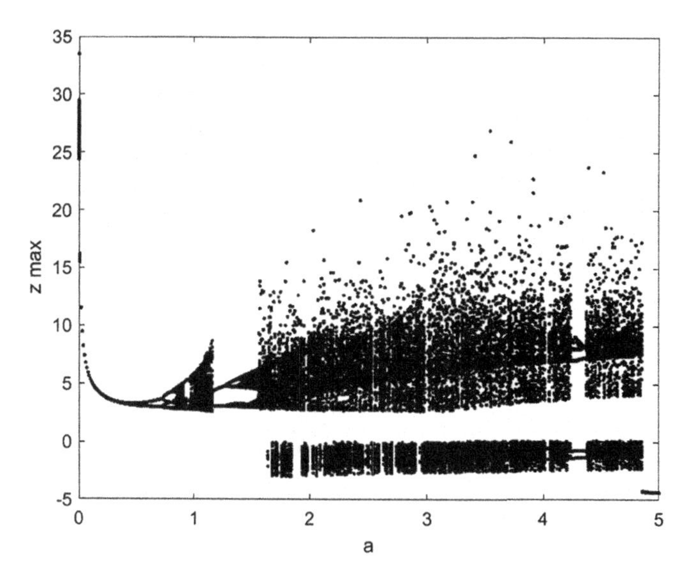

FIGURE 10.9
Bifurcation map versus the variation of the parameter a.

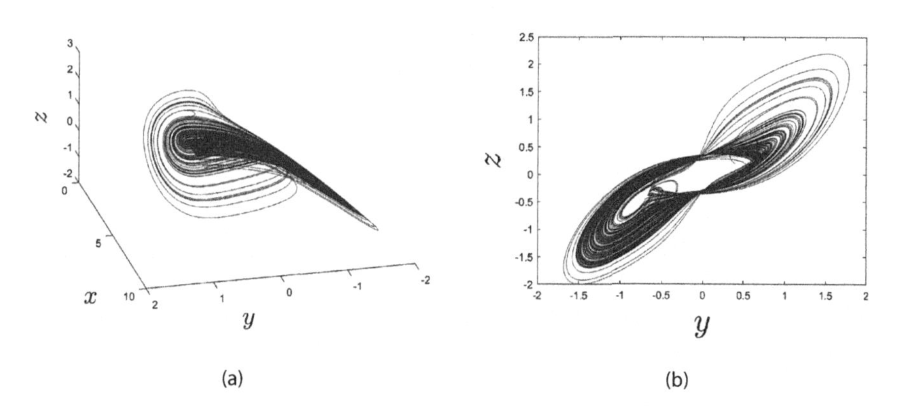

FIGURE 10.10
Dynamics of the fractional novel four-scroll system at $\alpha = 0.98$ and $a = 1.8$.

except the interval [4.1, 4.35), where the system loses its chaotic behavior. It is certain, however, that in this interval [4.1, 4.35), the system is neither chaotic nor hyperchaotic; the system generates limit cycles, as can be briefly observed later in the phase portraits in this interval. It is important to mention that in the neighborhood of 5, chaos is completely removed from the system. For clarity, we represent the phase portraits10a, 10b, 11a, and11b at $a = 1.8$; we can observe that we get chaotic behavior. However, a new type of chaos is generated compared to the behaviors observed in the previous section (Figures 10.10 and 10.11).

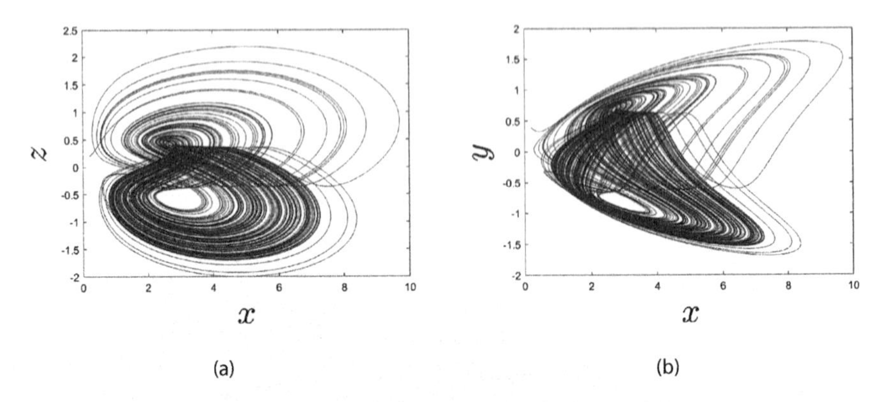

FIGURE 10.11
Dynamics of the fractional novel four-scroll system at $\alpha = 0.98$ and $a = 1.8$.

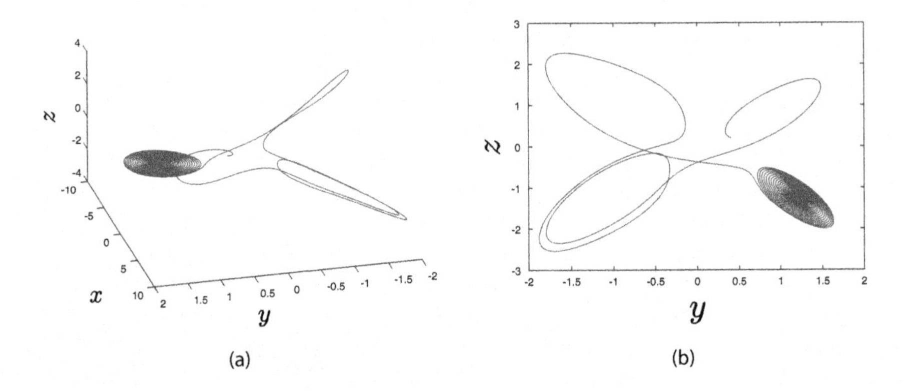

FIGURE 10.12
Dynamics of the fractional novel four-scroll system at $\alpha = 0.98$ and $a = 4.1$.

As previously mentioned, the chaotic behavior is removed at the $a = 4.1$; see Figures 10.12a and 10.12b for confirmation

We now consider the influence of the parameter b into [10, 15]. The bifurcation map is represented in Figure 10.13a, and the illustration in terms of phase portraits will be provided later. The value of the Caputo derivative's order does not change and is maintained at $\alpha = 0.98$.

Figure 10.13a shows that the system has chaotic behavior into the interval [10, 15]. The nature of chaos changes and the chaotic behavior is seriously reduced at the interval (10.2, 11.6). In the interval (11.6, 15), chaotic behaviors are observed later in the phase portraits. For confirmation of chaotic behavior

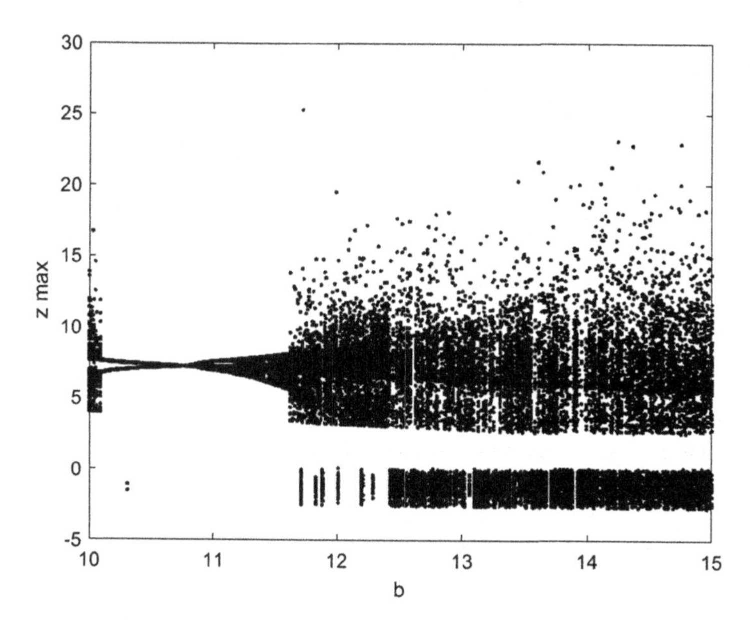

FIGURE 10.13
Bifurcation map versus the variation of the parameter b.

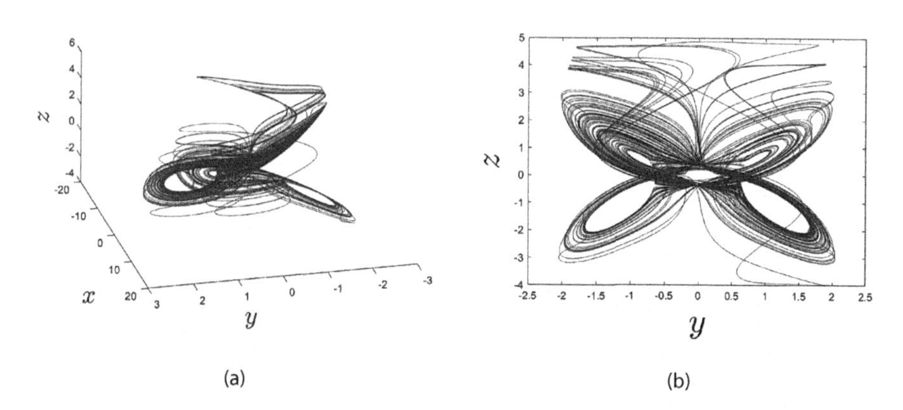

FIGURE 10.14
Dynamics of the fractional novel four-scroll system $\alpha = 0.98$ and $b = 10.8$.

in these intervals, we consider $b = 10.8$ and $b = 12$ (see Figures 10.14a, 10.14b, 10.15a, 10.15b, 10.16a, 10.16b, 10.17a, and 10.17b).

We finish with the study of the variation due to the parameter c in the interval [0, 5]. The bifurcation map is represented in Figure 10.18a, and the illustration in terms of phase portraits will be provided later. The value of the Caputo derivative's order does not change and is maintained at $\alpha = 0.98$.

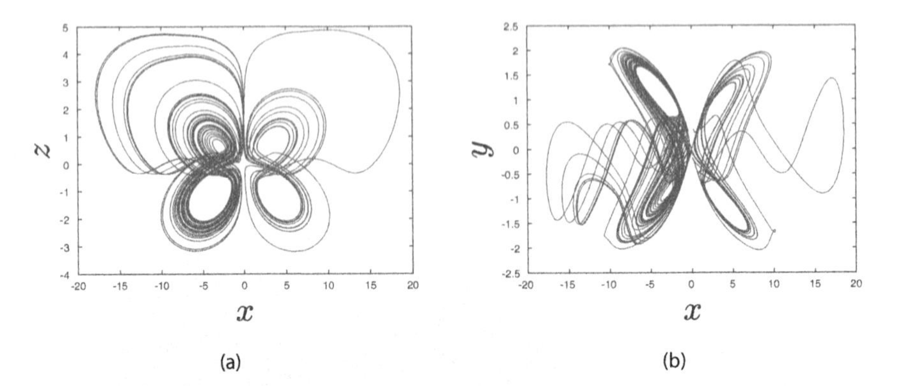

FIGURE 10.15
Dynamics of the fractional novel four-scroll system $\alpha = 0.98$ and $b = 10.8$.

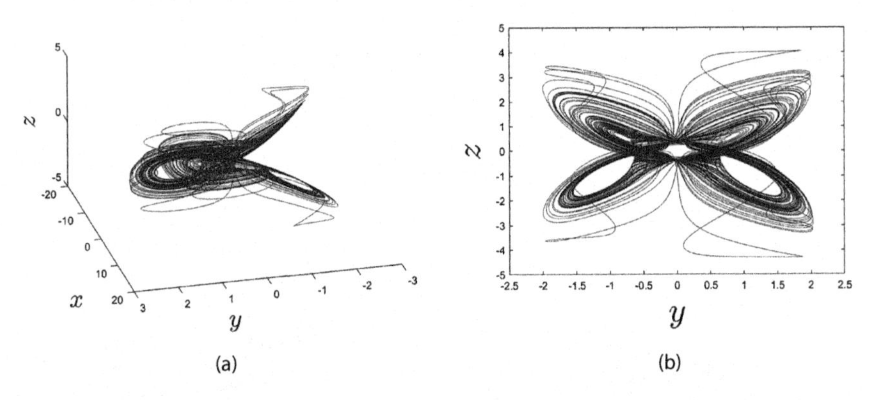

FIGURE 10.16
Dynamics of the fractional novel four-scroll system $\alpha = 0.98$ and $b = 12$.

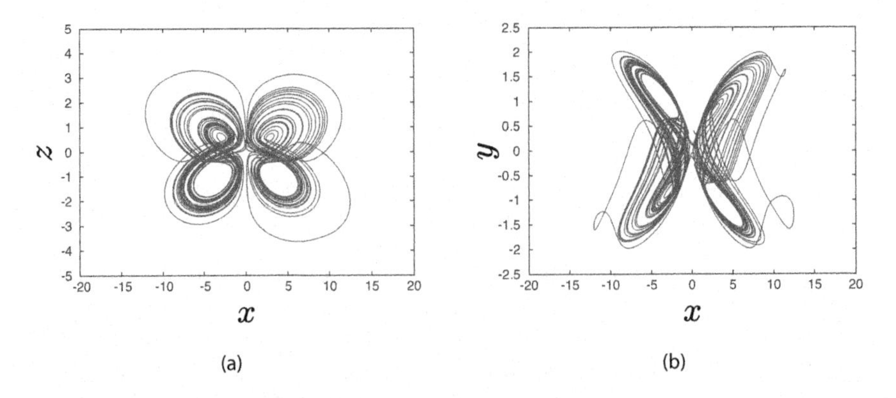

FIGURE 10.17
Dynamics of the fractional novel four-scroll system $\alpha = 0.98$ and $b = 12$.

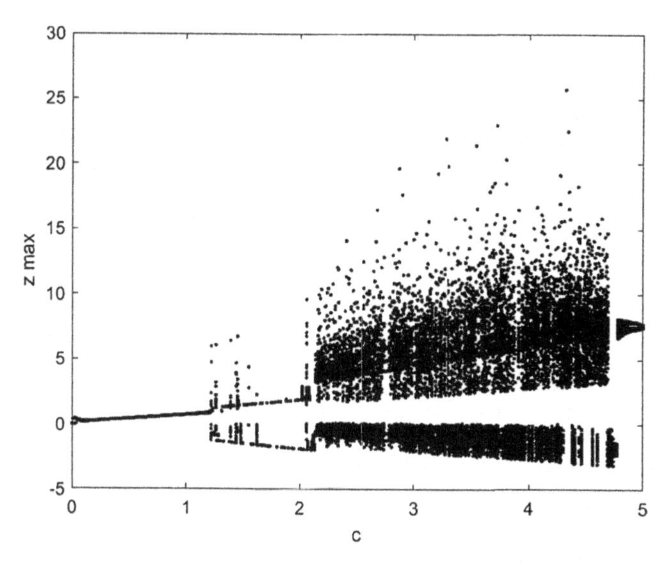

FIGURE 10.18
Bifurcation map versus the variation of the parameter c.

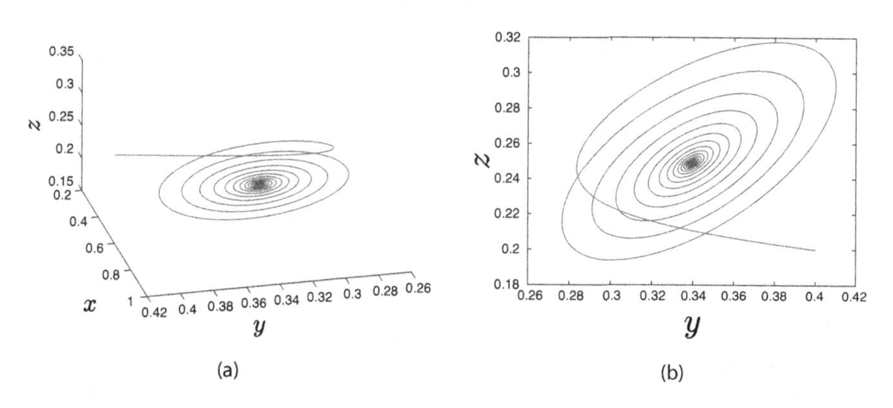

FIGURE 10.19
Dynamics of the fractional novel four-scroll system at $\alpha = 0.98$ and $c = 1$.

The bifurcation map informs us that in the interval [0, 2], the system generates limit cycles; that is, the chaotic behaviors are removed in the interval being considered. We notice that chaotic behaviors are obtained when the parameter c is into (2, 5). For confirmation of the previous analysis, we consider the phase portraits19a, 19b, 20a, 20b, 21a, 21b at the values $c = 1$ and $c = 4.5$ (Figures 10.19–10.21).

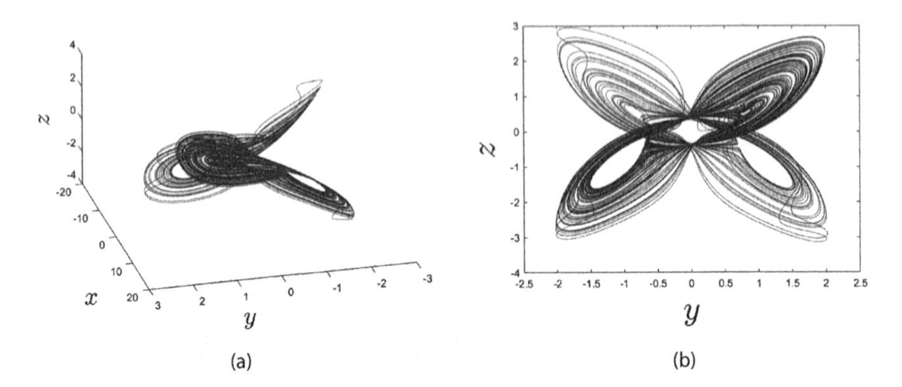

FIGURE 10.20
Dynamics of the fractional novel four-scroll system at $\alpha = 0.98$ and $c = 4.5$.

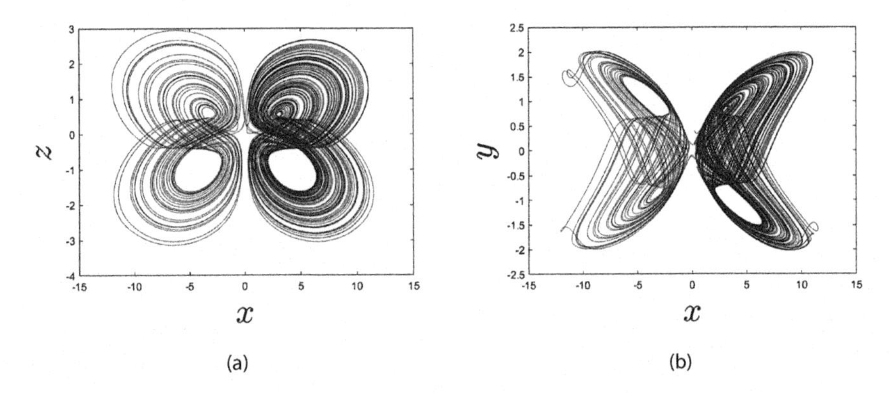

FIGURE 10.21
Dynamics of the fractional novel four-scroll system at $\alpha = 0.98$ and $c = 4.5$.

10.6 Detection of Chaos via Fractional Lyapunov Exponents

This section seeks to detect the types of chaos generated when we use the fractional-order derivative. For the purposes of the rest of our investigations, in this section, we need to recall the Jacobian matrix related to our model given by the following matrix:

$$J_{matrix} = \begin{pmatrix} -a & a+bz & by \\ ez & -1-3dy^2 & ex \\ -y & -x & c \end{pmatrix}. \tag{10.38}$$

The procedure used in this section to obtain the Lyapunov exponents is described in [30]. This procedure uses both the previous numerical method and the Jacobian matrix described in Eq. (10.38). In this section, we conserve the initial conditions fixed in the description of our model. In our first calculation, we consider the fractional-order derivative given by $\alpha = 0.94$. The values of the Lyapunov exponents with initial condition (0.2, 0.4, 0.2) are represented as follows:

$$Lya1 = 0.8934, \qquad Lya2 = -0.3001, \qquad Lya3 = -29.2920. \tag{10.39}$$

We can observe that the sum of the first and second Lyapunov exponents is positive, and the sum of all Lyapunov exponents is negative, which specifically means our system is dissipative, and the dimension of the Lyapunov exponents can be calculated, respectively; the following gives the Kaplan–Yorke value:

$$\dim(Lya) = 2 + \frac{Lya1 + Lya2}{|Lya3|} = 2.02025. \tag{10.40}$$

Form the positivity of one Lyapunov exponent $Lya1 = 0.8934$ and the value of the Kaplan–Yorke dimension, which is fractional, we conclude that the fractional novel four-scroll system is chaotic at the order $\alpha = 0.94$. This conclusion is important in the fractional version because the Lyapunov exponents' theory in the fractional context has many surprising results; see, for example, in Danca [30].

We consider the fractional-order derivative given by $\alpha = 0.955$. Using the same procedure as in the previous order, the values of the Lyapunov exponents with initial condition (0.2, 0.4, 0.2) are represented as follows:

$$Lya1 = 1.3875, \qquad Lya2 = -0.4874, \qquad Lya3 = -28.5137. \tag{10.41}$$

We can observe that the sum of the first and second Lyapunov exponents is positive, and the sum of all Lyapunov exponents is negative, which particularly means that our system is dissipative and the dimension of the Lyapunov exponent can be calculated, respectively. The Kaplan–Yorke value is given by:

$$\dim(Lya) = 2 + \frac{Lya1 + Lya2}{|Lya3|} = 2.03156. \tag{10.42}$$

With the positivity of one Lyapunov exponent $Lya1 = 1.3875$ and the value of the Kaplan–Yorke dimension, which is fractional, we conclude that the fractional novel four-scroll system is chaotic at the order $\alpha = 0.955$. Compared with the previous phase portraits, we can observe that the chaotic behavior is higher at the order $\alpha = 0.955$ because the positive Lyapunov exponent is larger. This difference in the chaotic behavior can also be observed in the dimension of the Lyapunov exponents.

We continue with the fractional-order derivative given by $\alpha = 0.97$. Using the same procedure as with the previous orders, the values of the Lyapunov exponents with initial condition (0.2, 0.4, 0.2) are represented as follows:

$$Lya1 = 1.0546, \qquad Lya2 = -0.6252, \qquad Lya3 = -23.7268. \quad (10.43)$$

We can observe that the sum of the first and second Lyapunov exponent is positive, and the sum of all Lyapunov exponents is negative, which particularly means that our system is dissipative and the dimension of the Lyapunov exponent can be calculated. The following gives the Kaplan–Yorke value:

$$\dim(Lya) = 2 + \frac{Lya1 + Lya2}{|Lya3|} = 2.01809. \quad (10.44)$$

With the positivity of one Lyapunov exponent $Lya1 = 1.0546$ and the value of the Kaplan– Yorke dimension, which is fractional, we conclude that the fractional novel four-scroll system is chaotic at the order $\alpha = 0.97$. The conclusion is the same as the previous conclusion. No significant difference in chaotic behaviors exists with the order $\alpha = 0.955$.

The last order to be characterized is the fractional order given by $\alpha = 0.98$. Using the same procedure as with the previous orders, the values of the Lyapunov exponents with initial condition (0.2, 0.4, 0.2) are represented as follows:

$$Lya1 = 0.8371, \qquad Lya2 = -0.6964, \qquad Lya3 = -23.8880. \quad (10.45)$$

We can observe that the sum of the first and the second Lyapunov exponent is positive, but the sum of all Lyapunov exponents is negative, which particularly means that our system is dissipative, and the dimension of the Lyapunov exponent can be calculated. The following gives the Kaplan–Yorke value:

$$\dim(Lya) = 2 + \frac{Lya1 + Lya2}{|Lya3|} = 2.0058. \quad (10.46)$$

With the positivity of one Lyapunov exponent $Lya1 = 0.8371$ and the value of the Kaplan–Yorke dimension, which is fractional, we conclude that the fractional novel four-scroll system is chaotic at the order $\alpha = 0.98$. We notice a change in chaotic behaviors by observing the associated phase portraits. This difference from the other order can be observed in the Kaplan–Yorke number, which is fractional.

10.7 Initial Condition Influences and Coexistence Attractors

In this section, we focus on the impacts of the initial conditions in the model's dynamics (10.5)–10.7). We influence the initial conditions one by one; we focus the change by calculating the Lyapunov exponents at the same time, because chaotic systems are very sensitive to the initial conditions. We particularly consider the impacts generated by the simultaneous influence of two initial conditions on the behaviors of the solutions.

Let the order $\alpha = 0.98$ and we first influence the condition x_0. The influence is represented in Figure 10.22a. In Figure 10.22a, we observe that the initial condition's influence is not seen at the beginning of the process. However, after a certain time, as can be observed in the figure, the initial condition's impact is visible. The red line represents the influenced initial condition (0.20001, 0.4, 0.2) and the curve in the blue line represents the primary initial condition (0.2, 0.4, 0.2).

To see the initial conditions' influence, we are interested in the Lyapunov exponents and its Kaplan–Yorke dimension. Calculating the Lyapunov exponents at the new initial condition (0.20001, 0.4, 0.2), we have the following results:

$$Lya1 = 0.8372, \qquad Lya2 = -0.6964, \qquad Lya3 = -23.8875. \qquad (10.47)$$

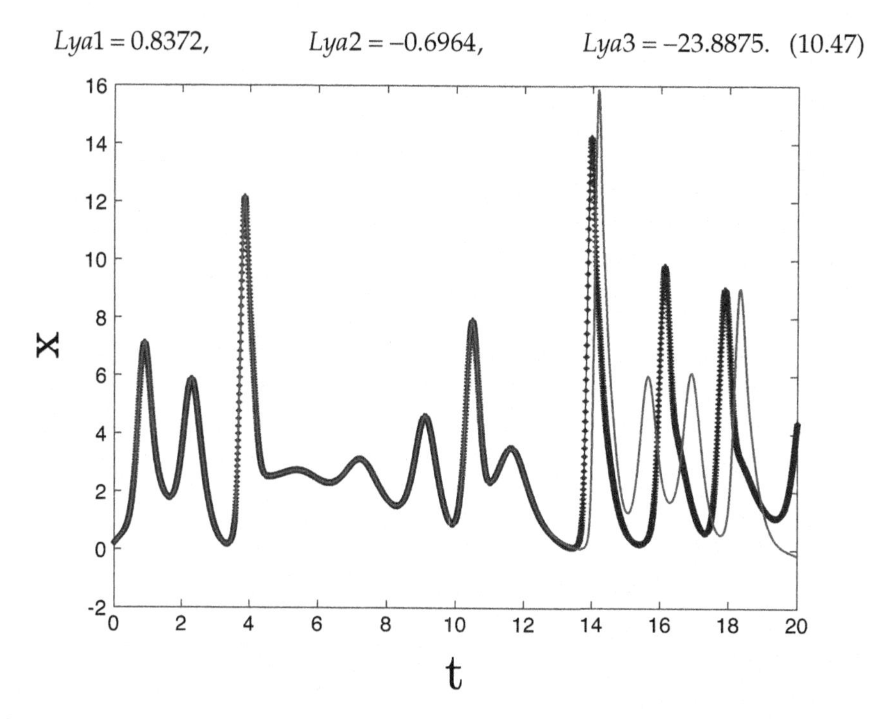

FIGURE 10.22
Sensitivity to the variable x_0 with the order $\alpha = 0.985$.

Its Kaplan–Yorke dimension is given by

$$\dim(Lya) = 2 + \frac{Lya1 + Lya2}{|Lya3|} = 2.0058. \tag{10.48}$$

The perturbation's influence due to the first initial condition can be observed in the first Lyapunov exponent and the last Lyapunov exponent, which have changed in comparison with the Lyapunov exponents at the order $\alpha = 0.985$ and with the initial condition (0.2, 0.4, 0.2).

We now perturb the second initial condition y_0 but significantly, that is, we consider the initial condition (0.2, 0.41, 0.2). In the figure, the red curve represents the influenced initial condition (0.2, 0.41, 0.2) and the blue curve represents the primary initial condition (0.2, 0.4, 0.2)23a.

Let us observe the impact of the initial condition by calculating the Lyapunov exponents and the Kaplan–Yorke dimension, giving us the following values:

$$Lya1 = 0.9578, \qquad Lya2 = -0.6946, \qquad Lya3 = -22.6425. \tag{10.49}$$

Its Kaplan–Yorke dimension is given by

$$\dim(Lya) = 2 + \frac{Lya1 + Lya2}{|Lya3|} = 2.01162. \tag{10.50}$$

We observe at (0.2, 0.41, 0.2) the Lyapunov exponents change considerably comparing with the Lyapunov exponents at the initial condition (0.2, 0.4, 0.2). This influence is clearly noticed in Figure 10.23a. The red curve represents the influenced initial condition (0.2, 0.41, 0.2) and the blue curve represents the primary initial condition (0.2, 0.4, 0.2).

We perturb two initial conditions, the second y_0 and the last z_0. We consider the initial condition given by (0.2, 0.41, 0.21); the behaviors are depicted in Figure 10.24a.

Let us observe the impact of the initial condition by calculating the Lyapunov exponents and the Kaplan–Yorke dimension; we have the following values:

$$Lya1 = 0.9740, \qquad Lya2 = -0.7040, \qquad Lya3 = -22.8413. \tag{10.51}$$

Its Kaplan–Yorke dimension is given by:

$$\dim(Lya) = 2 + \frac{Lya1 + Lya2}{|Lya3|} = 2.01182. \tag{10.52}$$

We observe at (0.2, 0.41, 0.21) that the Lyapunov exponents change considerably compared with the Lyapunov exponents at the initial condition (0.2, 0.4, 0.2). This influence can be clearly seen in Figure 10.24a. The red curve represents the influenced initial condition (0.2, 0.41, 0.21) and the curved blue line represents the primary initial condition (0.2, 0.4, 0.2).

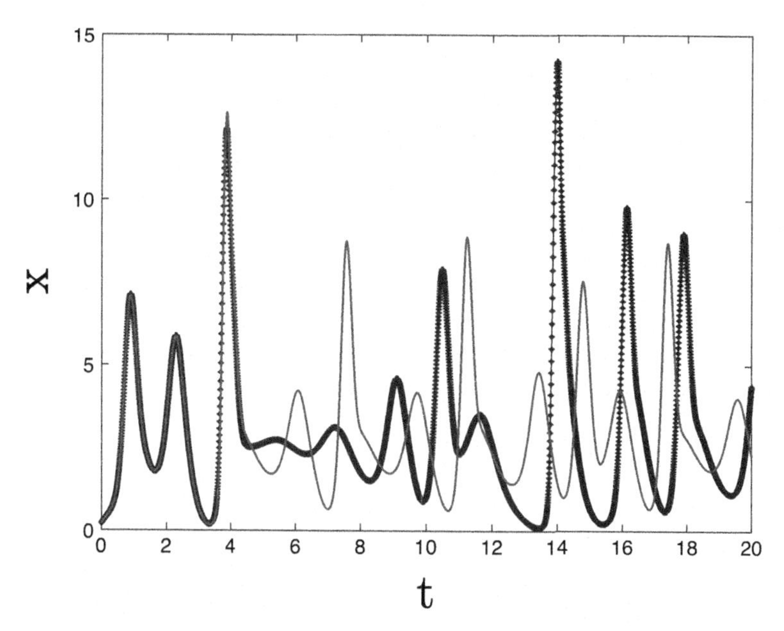

FIGURE 10.23
Sensitivity to the variable y_0 with the order $\alpha = 0.98$.

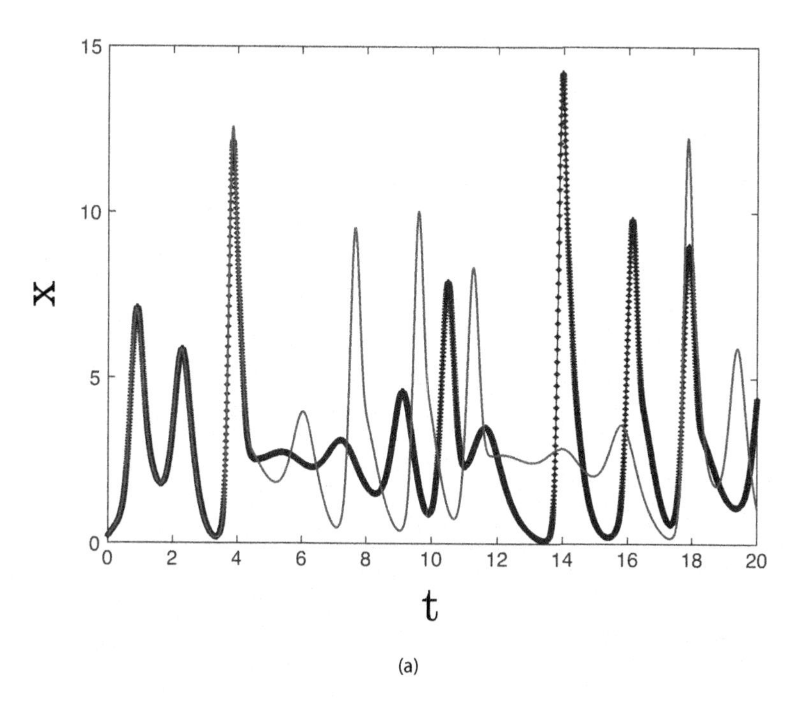

FIGURE 10.24
Sensitivity of the variable y_0 with the order $\alpha = 0.98$.

For more clarity in the initial conditions' influence, we depict the phase portraits at the different initial conditions previously considered (Figures 10.25a, 10.25b, 10.26a, 10.26b, 10.27a, 10.27b, 10.28a, 10.28b).

First, we fix the parameters of the model as the form $a = 1.8$, $b = 13$ and $c = 3.9$, and study the existence of a pair of attractors at the initial conditions (0.2, 0.4, 0.2) and (0.2, −0.4, 0.2) (Figures 10.29a, 10.29b, 10.30a, 10.30b). Second, we fix the parameters of the model as the form $a = 3$, $b = 13$ and $c = 3.9$, and study the existence of a pair of attractors at the initial conditions (0.2, 0.4, 0.2) and (0.2, −0.4, 0.2) (Figures 10.31a, 10.31b, 10.32a, 10.32b). In chaos this principle is called the coexistence attractors; it will prove the influence of the initial conditions. Due to the present model's symmetry, the attractors' coexistence is visible and symmetric for our model.

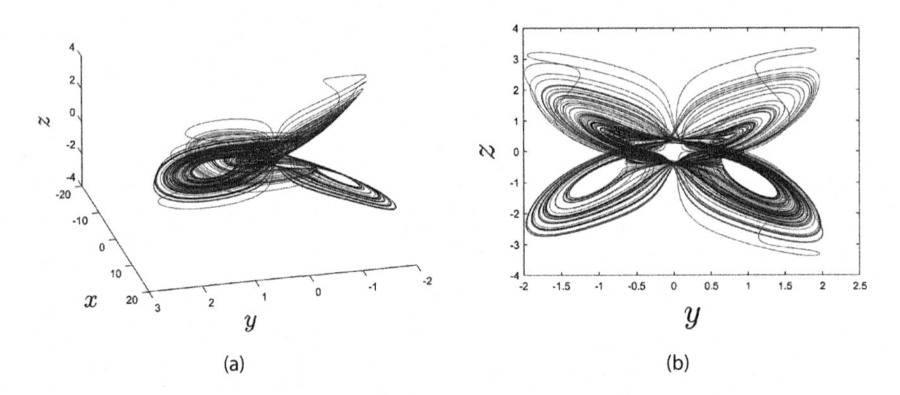

FIGURE 10.25
Dynamics of the fractional novel four-scroll system at $\alpha = 0.98$ and (0.2, 0.41, 0.2).

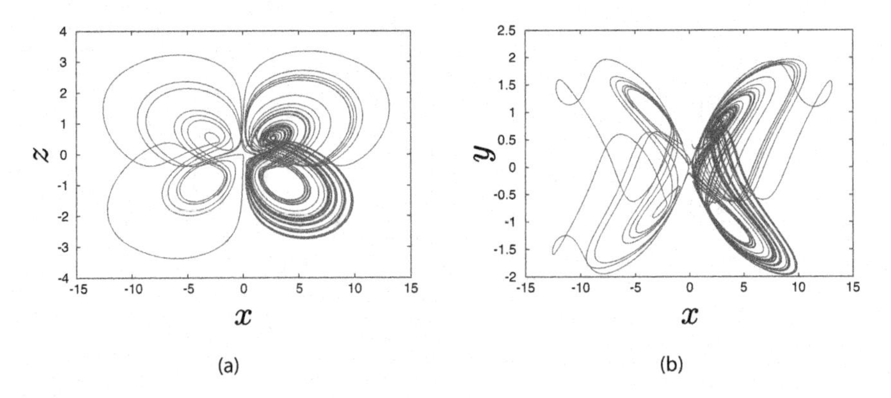

FIGURE 10.26
Dynamics of the fractional novel four-scroll system at $\alpha = 0.98$ and (0.2, 0.41, 0.21).

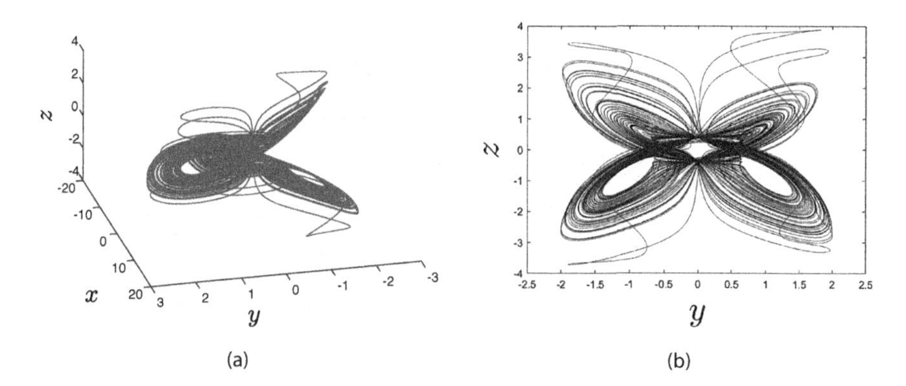

FIGURE 10.27
Dynamics of the fractional novel four-scroll system $\alpha = 0.98$ and (0.2, 0.41, 0.21).

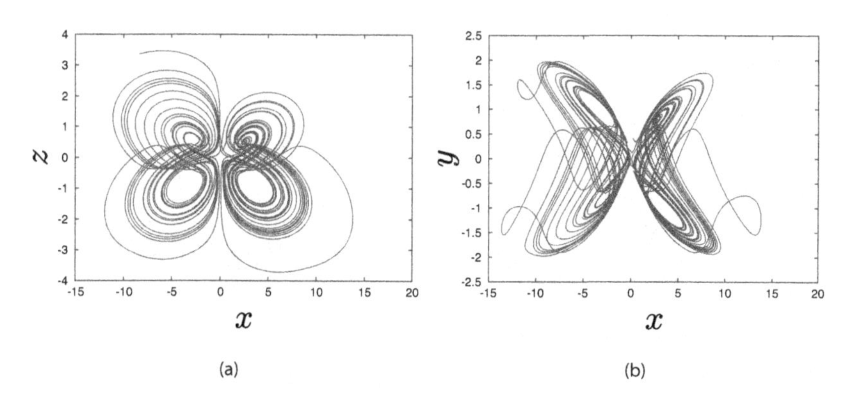

FIGURE 10.28
Dynamics of the fractional novel four-scroll system $\alpha = 0.98$ and (0.2, 0.41, 0.2).

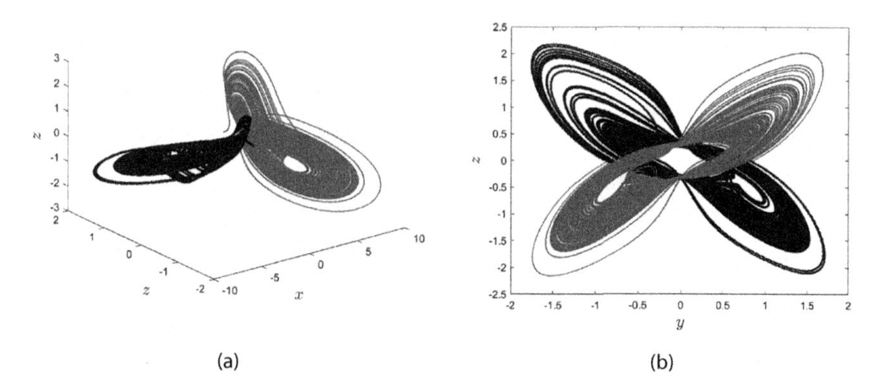

FIGURE 10.29
Coexistence in (x-y) plane and (x-z) plane for $\alpha = 0.98$ at (0.2, 0.4, 0.2) (red) and (0.2, −0.4, 0.2) (blue), respectively.

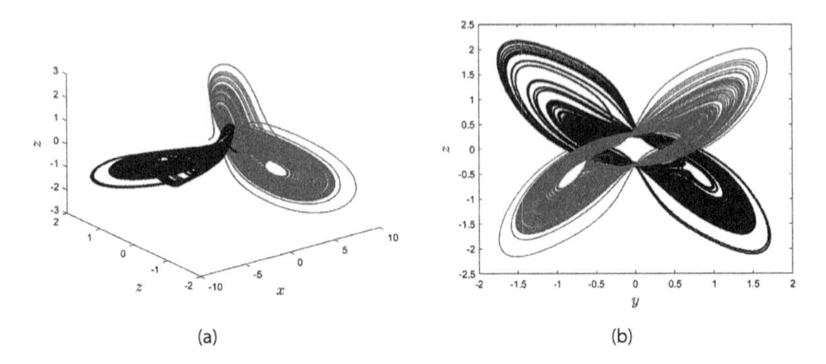

FIGURE 10.30
Coexistence in (x-y-z) plane and (y-z) plane for $\alpha = 0.98$ at (0.2, 0.4, 0.2) (red) and (0.2, −0.4, 0.2) (blue), respectively.

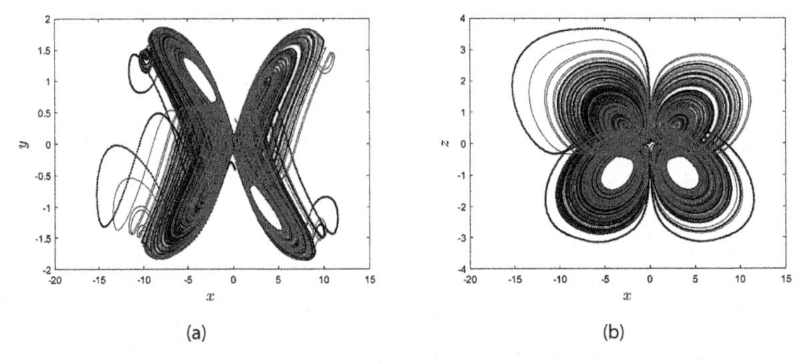

FIGURE 10.31
Coexistence in $(x - y)$ plane and $(x - z)$ plane for $\alpha = 0.98$ at (0.2, 0.4, 0.2) (red) and (0.2, −0.4, 0.2) (blue), respectively.

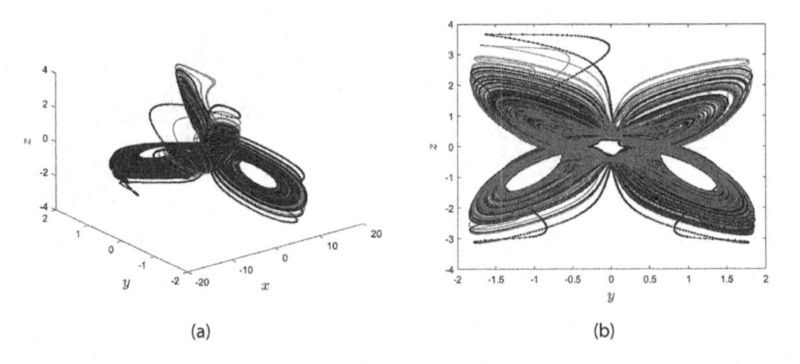

FIGURE 10.32
Coexistence in $(x - y - z)$ plane and $(y - z)$ plane for $\alpha = 0.98$ at (0.2, 0.4, 0.2) (red) and (0.2, −0.4, 0.2) (blue), respectively.

10.8 Fractional Local Stability with Matignon Criterion

In this section, we investigate the local stability of the proposed model's equilibrium points via the Jacobian matrix. The first equilibrium points generated by our fractional differential system (10.5)–(10.7) are given by the following points: $A_0 = (0, 0, 0)$, $A_1 = (0.26, 0.77, 0.51)$, $A_2 = (-0.26, -0.77, 0.51)$, $A_3 = (3.41, -1.04, -0.91)$ and $A_4 = (-3.41, 1.04, -0.91)$.

We take the Jacobian matrix represented in Eq. (10.38) at the first equilibrium point $A_0 = (0, 0, 0)$, and after determining the eigenvalues via Matlab, we obtain

$$J_{matrix} = \begin{pmatrix} -3 & 3 & 0 \\ 0 & -1 & 0 \\ 0 & 0 & 3.9 \end{pmatrix}. \quad (10.53)$$

The following values give the eigenvalues of the above matrix $\lambda_1 = -3$ $\lambda_2 = -1$ and $\lambda_3 = 3.9$. We can observe that the first and the second eigenvalues have negative real parts. Thus, they immediately satisfy the Matignon criterion utilized to study stability in fractional contexts [37–43]. The last eigenvalue satisfies the condition $|\arg(\lambda_3)| = 0 < \alpha\pi/2$. That is, the last eigenvalue does not respect the Matignon criterion.

We repeat the same reasoning as previously made with the first point. We take the Jacobian matrix represented in Eq. (10.38) at the second equilibrium point $A_1 = (0.26, 0.77, 0.51)$, and after determining the eigenvalues via Matlab, we have

$$J_{matrix} = \begin{pmatrix} -3 & 10.14 & 10.78 \\ 0.51 & -2.7787 & 0.26 \\ -0.77 & -0.26 & 3.9 \end{pmatrix}. \quad (10.54)$$

The following values give the eigenvalues of the above matrix $\lambda_1 = -4.7974$ $\lambda_2 = 1.4594 + 0.9320i$ and $\lambda_3 = 1.4594 - 0.9320i$. We can observe that the first eigenvalue has a negative real part; thus, it immediately satisfies the Matignon criterion utilized to study the fractional context's stability. The second and the last are conjugated and satisfy the condition $|\arg(\lambda_{2,3})| = 33/180\pi < \alpha\pi/2$ for all α in (0.4, 1). Thus, the second equilibrium point is not stable.

We take the Jacobian matrix represented in Eq. (10.38) at the second equilibrium point $A_2 = (-0.26, -0.77, 0.51)$, and after determining the eigenvalues via Matlab, we have

$$J_{matrix} = \begin{pmatrix} -3 & 10.14 & -10.78 \\ 0.51 & -2.7787 & -0.26 \\ 0.77 & 0.26 & 3.9 \end{pmatrix}. \quad (10.55)$$

The following values give the eigenvalues of the matrix $\lambda_1 = -4.7974$ $\lambda_2 = 1.4594 + 0.9320i$ and $\lambda_3 = 1.4594 - 0.9320i$. We have the same eigenvalues as the previous point; utilizing the same procedure, we conclude that the third equilibrium point is not stable.

We take the Jacobian matrix represented in Eq. (10.38) at the second equilibrium point $A_3 = (3.41, -1.04, -0.91)$ and after determining the eigenvalue via Matlab, we have

$$J_{matrix} = \begin{pmatrix} -3 & -9.74 & -14.56 \\ -0.91 & -4.2448 & 3.41 \\ 1.04 & -3.41 & 3.9 \end{pmatrix}. \tag{10.56}$$

The eigenvalue of the matrix is given by the following values $\lambda_1 = -6.6693$ $\lambda_2 = 1.6622 + 4.6686i$ and $\lambda_3 = 1.6622 + 4.6686i$. The first eigenvalue satisfies trivially the Matignon criterion with the condition (10.38). The second and the last eigenvalues are conjugated and satisfy the condition $|\arg(\lambda_{2,3})| = 35/90\pi < \alpha\pi/2$ for all α in $(0.7, 1)$. Thus, the equilibrium point $A_3 = (3.41, -1.04, -0.91)$ is not stable; the same conclusion is obtained with the last point $A_4 = (-3.41, 1.04, -0.91)$.

10.9 Applications in Science and Engineering

This section gives the schematic circuit of the model considered by looking at the fractional-order derivative. We begin with the order $\alpha = 1$. This section will tell us what happens in real applications and confirm the results obtained with Matlab software. The main idea will be to use the resistors and capacitors to simulate the model we are considering. We describe the circuit's schematic in the following forms: we use 12 resistors, 3 capacitors, 5 multipliers, and 5 analog multipliers. The schematic of our present model with integer-order derivative is described in Figure 10.33.

Figure 10.33 is associated with the following scaled model, including the capacitors and resistors:

$$\begin{aligned} x' &= \frac{1}{R_1C_1}y - \frac{1}{R_2C_1}x + \frac{1}{R_3C_1}yz, \\ y' &= -\frac{1}{R_4C_2}y + \frac{1}{R_5C_2}xz - \frac{1}{R_6C_2}y^3, \\ z' &= \frac{1}{R_7C_3}z - \frac{1}{R_8C_3}xy. \end{aligned} \tag{10.57}$$

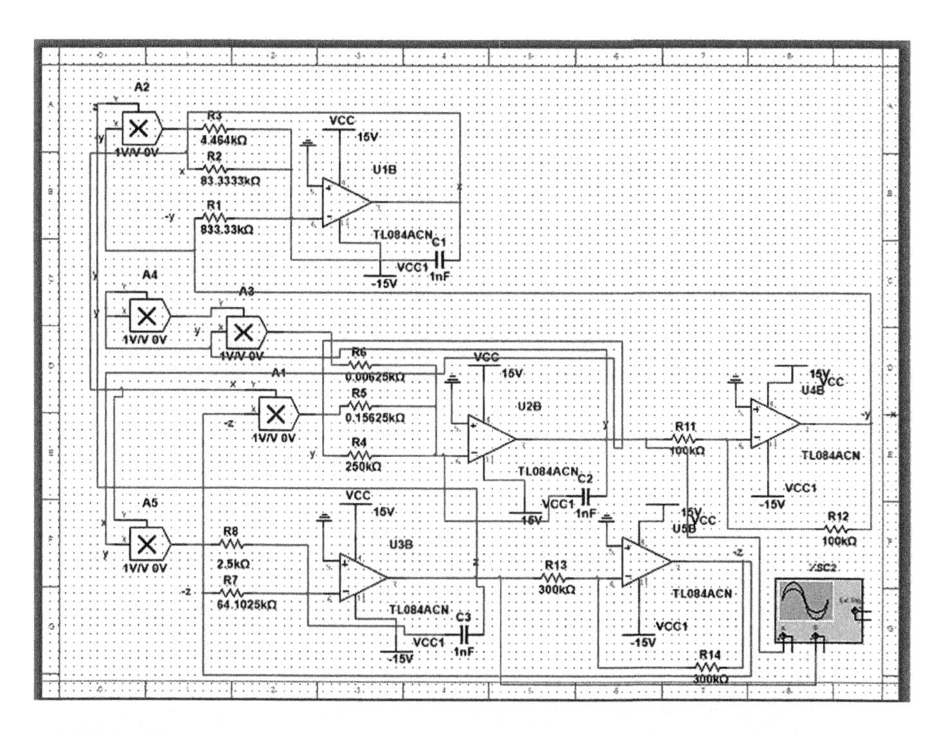

FIGURE 10.33
Circuit schematic of the novel chaotic system with integer-order derivative.

We set the following values for the tools of the model (10.5)–(10.7): $R_1 = 833.33k$, $R_2 = 83.33k$, $R_3 = 4.464k$, $R_4 = 250k$, $R_5 = 0.15625k$, $R_6 = 0.00625k$, $R_7 = 64.1025k$, $R_8 = 2.5k$ and $C_1 = 1nF$, $C_2 = 1nF$, $C_3 = 1nF$. The oscilloscope results are presented in Figures 10.34–10.36.

The errors in the resistors' values, which are not exact values, generate small modifications in the dynamics, but in general, the phase portraits in the oscilloscope and MatLab are the same. In conclusion, we can observe from Figures 10.34a, 10.35a, and 10.36a that the theoretical results with MatLab are in good agreement with the real application results obtained in the oscilloscope.

In the second part, we give the schematic circuit when the considered fractional-order chaotic system is given by

$$D_t^{\alpha} x = a(y - x) + byz, \tag{10.58}$$

$$y' = -y + exz - dy^3, \tag{10.59}$$

$$z' = cz - xy, \tag{10.60}$$

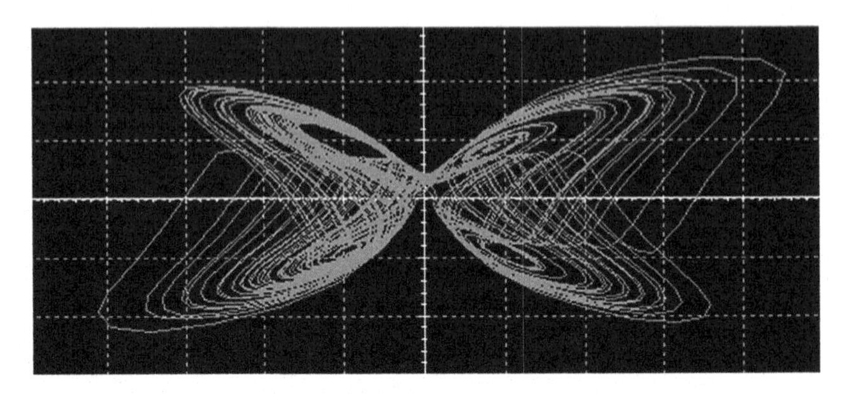

FIGURE 10.34
Dynamics in the oscilloscope trough (x, y) plane.

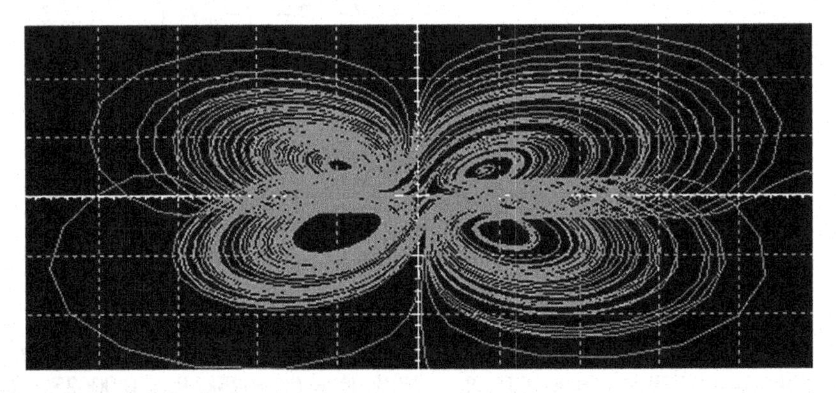

FIGURE 10.35
Dynamics in the oscilloscope trough (x, z) plane.

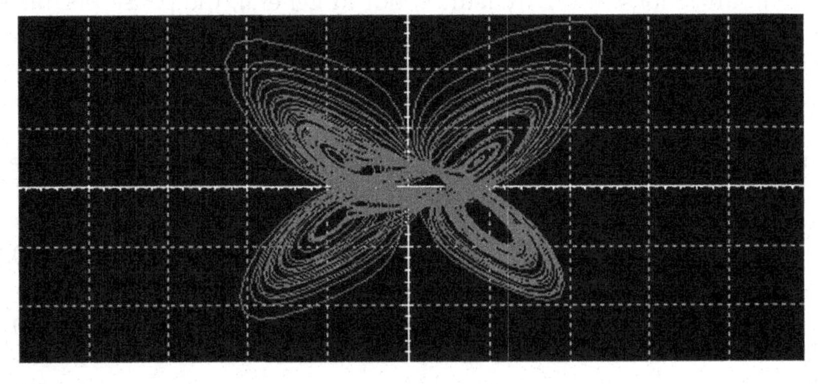

FIGURE 10.36
Dynamics in the oscilloscope trough (y, z) plane.

To draw the circuit, we need the fractional integrator. For simplification, we consider the order $\alpha = 0.9$, then the approximation of the fractional-order integrator is given by the following expression:

$$\frac{1}{s^{0.98}} \approx \frac{1.2234s^2 + 1463.28s + 4893.2}{(s+0.0106)(s+3.7716)(s+1341.4)}. \tag{10.61}$$

Equation (10.61) can be rewritten using the resistors and capacitors, which means we will use the transfer function associated with the fractional-order integrator; we have the following representation:

$$Z(s) = \frac{\frac{1}{C_1}}{s+\frac{1}{R_1C_1}} + \frac{\frac{1}{C_2}}{s+\frac{1}{R_2C_2}} + \frac{\frac{1}{C_3}}{s+\frac{1}{R_3C_3}}. \tag{10.62}$$

Now, combining Eq. (10.61) and Eq. (10.62), we recover the values of the capacitors and resistors given by the following forms $C_4 = 1.0656nF$, $C_5 = 8.5245nF$, $C_6 = 7.596nF$ and $R_9 = 91.17M$, $R_{14} = 32.046k$, $R_{15} = 101.12$. The circuit associated with the fractional differential equation Eqs. (10.58)–(10.60) is then represented in Figure 10.37.

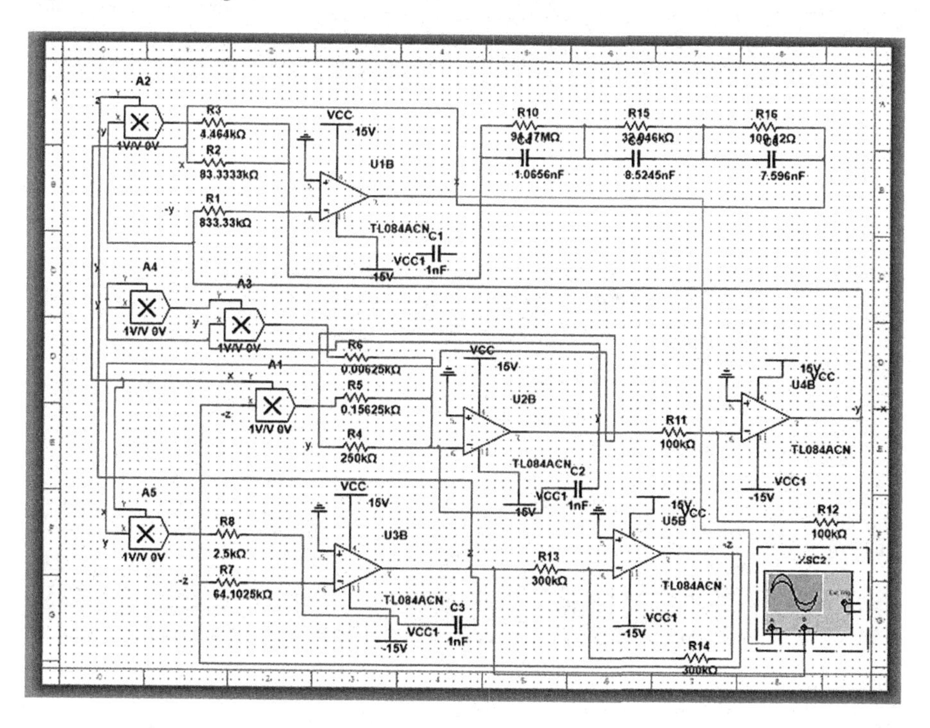

FIGURE 10.37
Circuit schematic of the novel chaotic system with fractional-order derivative.

The results of the simulations are assigned in Figures 10.38–10.40. We have the phase portraits38a, 39a, 40a, observed in the oscilloscopes.

For comparison, we depict the phase portraits of Eqs. (10.58)–(10.60) using the numerical scheme presented in this chapter. The figures are represented as follows: 10.41a, 10.41b, 10.42a, 10.42b.

We can clearly observe that the results obtained with the numerical scheme 40, 40, 41, 41 and the phase portraits in the oscilloscopes are in good agreement.

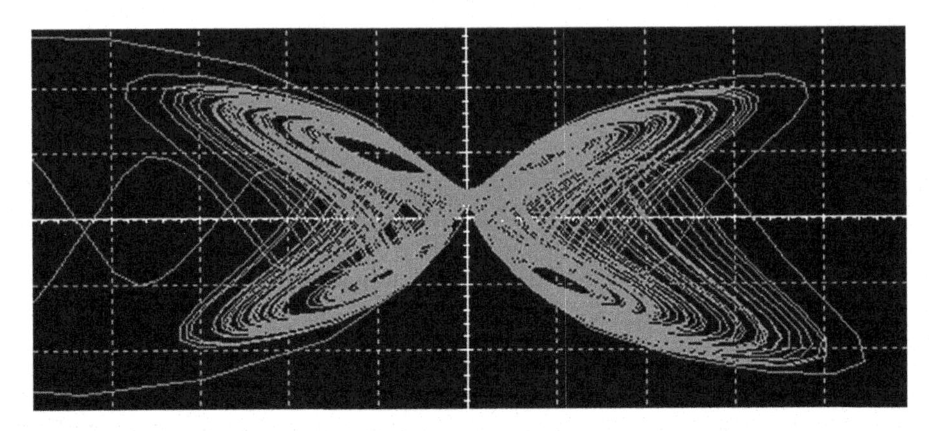

FIGURE 10.38
Dynamics in the oscilloscope trough (x, y) plane.

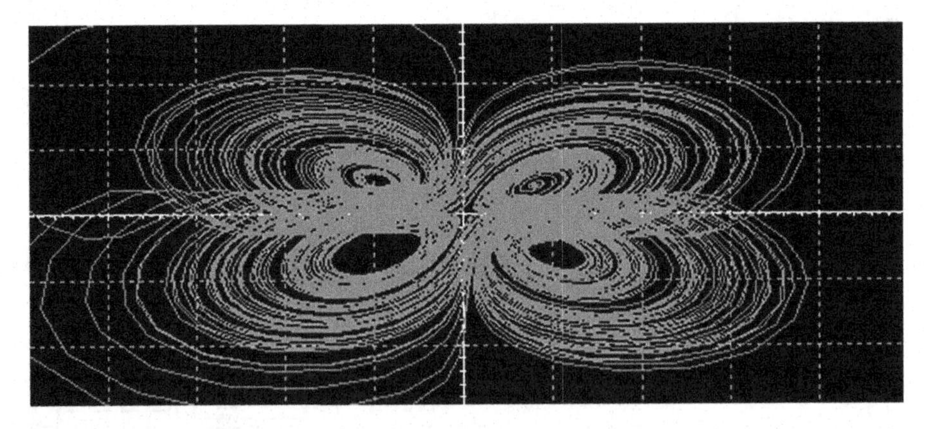

FIGURE 10.39
Dynamics in the oscilloscope trough (x, z) plane.

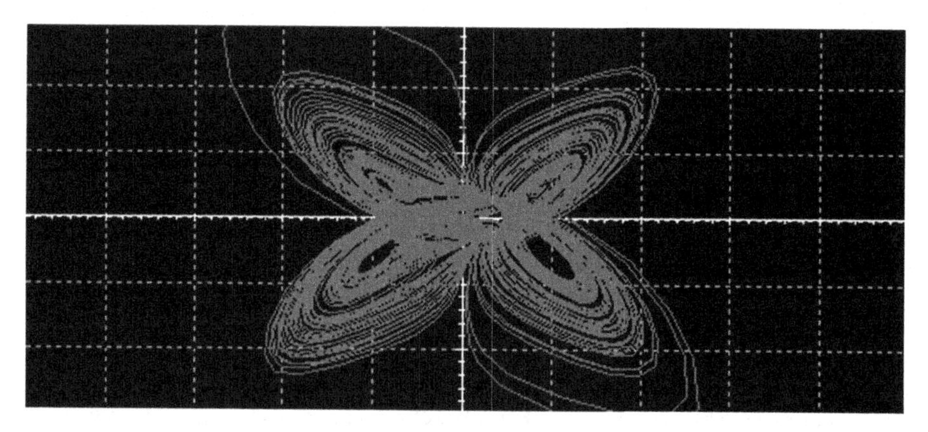

FIGURE 10.40
Dynamics in the oscilloscope trough (y, z) plane.

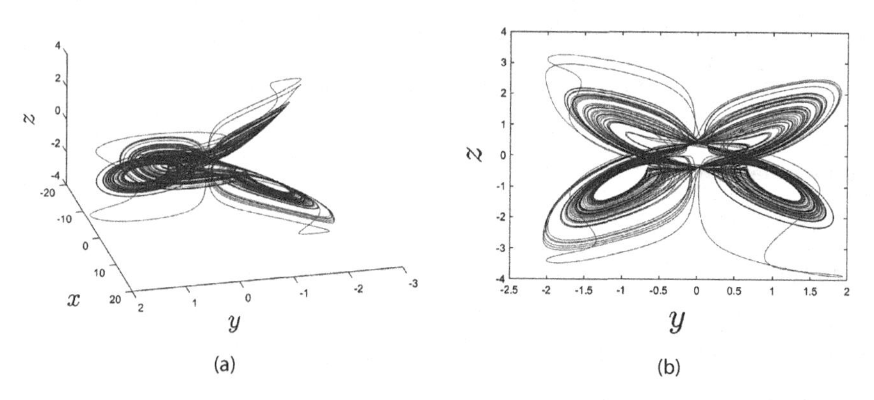

FIGURE 10.41
Dynamics of the fractional novel chaotic system at $\alpha = 0.98$ for Eqs. (10.58)–(10.60).

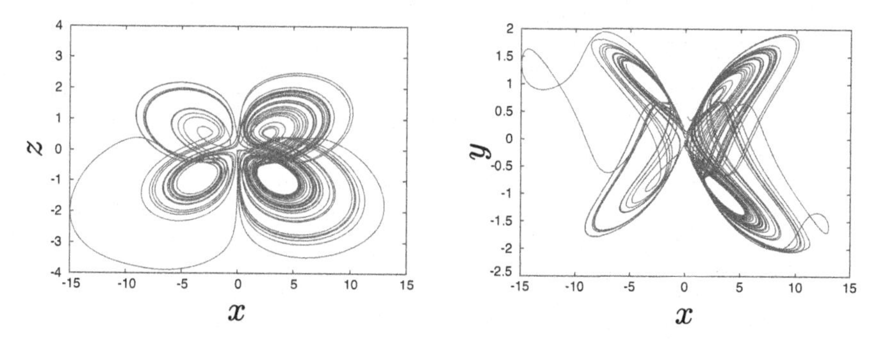

FIGURE 10.42
Dynamics of the fractional novel chaotic system $\alpha = 0.98$ for Eqs. (10.58)–(10.60).

10.10 Conclusion

This chapter investigated the numerical discretization of fractional differential equations which was applied to obtain the phase portraits, the Lyapunov exponents, the bifurcation diagrams, the coexistence, and the impact of the initial conditions of the novel four-scroll chaotic system described by the Caputo derivative. The effect of the order of the fractional-order derivative impacts the fractional chaotic model and creates new types of attractors when the order is in the chaotic region. The stability analysis for the equilibrium points was presented via the Matignon criterion used in fractional calculus for the local stability. The circuit implementations and oscilloscopes results confirm the work. The future direction of research will be to consider the incommensurate fractional order and to investigate the bifurcation maps and the Lyapunov exponents in the fractional context. Determination of the chaotic region is an open problem for the present fractional-order chaotic system with incommensurate order. The numerical scheme and the phase portraits for the incommensurate fractional-order chaotic system are the major research directions motivated by the present chapter.

Conflict of Interest

The authors declare that they have no conflict of interest.

References

[1] D. Kumar, and S. Kumar, Construction of four dimensional chaotic finance model and its applications, *International Journal of Pure and Applied Mathematics* (2018), **118(22)**, 1171–1187.

[2] E. Ahmed, A. M. A. El-Sayed, and H. A. A. El-Saka, On some Routh–Hurwitz conditions for fractional order differential equations and their applications in Lorenz, Rossler, Chua and Chen systems, *Physics Letters A* (2006), **358**, 1–4.

[3] S. Vaidyanathan, C. Volos, V. T. Pham, and K. Madhavan, Analysis, adaptive control and synchronization of a novel 4-D hyperchaotic hyperjerk system and its SPICE implementation, *Archives of Control Sciences* (2015), **25(1)**, 135–158.

[4] J. F. Gomez Aguilar et al., Chaos in a cancer model via fractional derivatives with exponential decay and Mittag-Leffler Law, *Entropy* (2017), **19**, 681.

[5] M. Caputo, and M. Fabrizio, A new definition of fractional derivative without singular kernel, *Progress in Fractional Differentiation and Applications* (2015), **1(2)**, 1–15.

[6] A. Atangana, and D. Baleanu, New fractional derivatives with nonlocal and non-singular kernel: Theory and application to heat transfer model, *Thermal Science* (2016), **20(2)**, 763–769.

[7] N. Sene, and A. N. Fall, Homotopy perturbation ρ-Laplace transform method and its application to the fractional diffusion equation and the fractional diffusion-reaction equation, *Fractal Fractional* (2019), **3**, 14.
[8] M. Yavuz, and N. Sene, Approximate solutions of the model describing fluid flow using generalized ρ Laplace transform method and heat balance integral method, *Axioms* (2020), **9(4)**, 123.
[9] H. Singh, Analysis for fractional dynamics of Ebola virus model, *Chaos, Solitons & Fractals* (2020), **168**, 109992.
[10] H. Singh, Analysis of drug treatment of the fractional HIV infection model of CD4+ T-cells, *Chaos, Solitons & Fractals* (2020), **146**, 110868.
[11] H. Singh, Jacobi collocation method for the fractional advection-dispersion equation arising in porous media, *Numerical Methods for Partial Differential Equations* (2020). https://doi.org/10.1002/num.22674.
[12] H. Singh, Numerical simulation for fractional delay differential equations, *International Journal of Dynamics and Control* (2021), **9**, 463–474.
[13] H. Singh, and A. M. Wazwaz, Computational method for reaction diffusion-model arising in a spherical catalyst, *International Journal of Applied and Computational Mathematics* (2021), **7(65)**. https://doi.org/10.1007/s40819-021-00993-9.
[14] H. Singh, and H. M. Srivastava, Numerical investigation of the fractional-order Liénard and Duffing equations arising in oscillating circuit theory, *Frontiers in Physics* (2020), **8(120)**. doi: 10.3389/fphy.2020.00120
[15] H. Singh, H.M. Srivastava, and D. Kumar, A reliable algorithm for the approximate solution of the nonlinear Lane-Emden type equations arising in astrophysics, *Numerical Methods for Partial Differential Equations* (2017), **34(5)**, 1524–1555.
[16] H. Singh, D. Kumar, and D. Baleanu, *Methods of Mathematical Modelling: Fractional Differential Equations*, CRC Press Taylor and Francis (2019).
[17] H. Singh, J. Singh, S. D. Purohit, and D. Kumar, *Advanced Numerical Methods for Differential Equations: Applications in Science and Engineering*, CRC Press Taylor and Francis (2021).
[18] M. Diouf, and N. Sene, Analysis of the financial chaotic model with the fractional derivative operator, *Complexity* (2020), **2020**, 14. https://doi.org/10.1155/2020/9845031.
[19] N. Sene, and A. Ndiaye, On class of fractional-order chaotic or hyperchaotic systems in the context of the Caputo fractional-order derivative, *Journal of Mathematics* (2020), **2020**, 15. https://doi.org/10.1155/2020/8815377.
[20] N. Sene, Analysis of a four-dimensional hyperchaotic system described by the Caputo–Liouville fractional derivative, *Complexity* (2020), **2020**, 20. https://doi.org/10.1155/2020/8889831.
[21] N. Sene, Analysis of a fractional-order chaotic system in the context of the Caputo fractional derivative via bifurcation and Lyapunov exponents, *Journal of King Saud University – Science* (2020), 33 (1), 101275.
[22] Z. Li, T. Xia, and C. Jiang, Synchronization of fractional-order complex chaotic systems based on observers, *Entropy* (2019), **21**, 481.
[23] Q. Lai, A. Akgul, C. Li, G. Xu, and U. Çavusoglu, A new chaotic system with multiple attractors: Dynamic analysis, Circuit Realization and S-Box Design, *Entropy* (2018), **20**, 12.
[24] C. Jiang, S. Liu, and D. Wang, Generalized combination complex synchronization for fractional-order chaotic complex systems, *Entropy* (2015), **17**, 5199–5217.

[25] J. Lu, T. Zhou, G. Chen, and S. Zhang, Local bifurcations of the Chen system, *International Journal of Bifurcation and Chaos* (2002), **12(10)**, 2257–2270.
[26] Z. Wang, H. Xia, and S. Guodong, Analysis of nonlinear dynamics and chaos in a fractional order financial system with time delay, *Computer and Mathematics with Applications* (2011), **62**, 1531–1539.
[27] S. Mobayen, C. K. Volos, U. Çavusoglu, and S. Kaçar, A simple chaotic flow with hyperbolic sinusoidal function and its application to voice encryption, *Symmetry* (2020), **12**, 2047.
[28] Z. Hammouch, and T. Mekkaoui, Control of a new chaotic fractional-order system using Mittag-Leffler stability, *Nonlinear Studies* (2015), **22(4)**, 1–13.
[29] A. K. Alomari, A novel solution for fractional chaotic Chen system, *Journal of Nonlinear Science and Applications* (2015), **8**, 478–488.
[30] M. F. Danca, and N. Kuznetsov, Matlab code for Lyapunov exponents of fractional-order systems, *International Journal of Bifurcation and Chaos* (2018), **28(5)**, 1850067.
[31] I. Petras, A note on the fractional-order Chua's system, *Chaos, Solitons & Fractals* (2008), **38**, 140–147.
[32] X. Yang, H. Liu, and S. Li, Synchronization of fractional-order and integer-order chaotic (hyper-chaotic) systems with different dimensions, *Advances in Difference Equations* (2017), **2017(344)**.
[33] A. A. Kilbas, H. M. Srivastava, and J. J. Trujillo, *Theory and Applications of Fractional Differential Equations*, North-Holland Mathematics Studies, Elsevier, Amsterdam, The Netherlands (2006), 204.
[34] I. Podlubny, *Fractional Differential Equations*, Mathematics in Science and Engineering, Academic Press, New York, NY (1999), 198.
[35] S. Sampath, S. Vaidyanathan, Ch. K. Volos, and V.-T. Pham, An eight-term novel four-scroll chaotic system with cubic nonlinearity and its circuit simulation, *Journal of Engineering Science and Technology Review* (2015), **8(2)**, 1–6.
[36] R. Garrappa, Numerical solution of fractional differential equations: A survey and a software tutorial 2018, *Mathematics* (2018), **6**, 16.
[37] J. Fahd, T. Abdeljawad, and D. Baleanu, On the generalized fractional derivatives and their Caputo modification, *Journal of Nonlinear Science and Applications* (2017), **10**, 2607–2619.
[38] C. Jiang, S. Liu, and C. Luo, A new fractional-order chaotic complex system and its antisynchronization, *Abstract and Applied Analysis* (2014), **2014(326354)**, 12.
[39] N. Sene, Stability analysis of the generalized fractional differential equations with and without exogenous inputs, *Journal of Nonlinear Science and Applications* (2019), **12**, 562–572.
[40] N. Sene, Global asymptotic stability of the fractional differential equations, *Journal of Nonlinear Science and Applications* (2020), **13**, 171–175.
[41] N. Sene, SIR epidemic model with Mittag-Leffler fractional derivative, *Chaos, Solitons & Fractals* (2020), **137**, 109833.
[42] N. Sene, and G. Srivastava, Generalized Mittag-Leffler input stability of the fractional differential equations, *Symmetry* (2019), **11**, 608.
[43] M. Yavuz, Characterization of two different fractional operators without singular kernel, *Fractional Order Mathematical Models in Physical Sciences* (2019) **14(3)**, 302.

11

Extraction of Deeper Properties of the Conformable Gross–Pitaevskii Equation via Two Powerful Approaches

Haci Mehmet Baskonus
Harran University, Sanliurfa, Turkey

Gulnur Yel
Final International University, Kyrenia Mersin, Turkey

Hasan Bulut
Firat University, Elazig, Turkey

Fayık Değirmenci
Harran University, Sanliurfa, Turkey

CONTENTS

11.1 Introduction

Recently, fractional analysis has played an important role in nonlinear dynamics in a wide variety of fields from mathematics to medicine. In fact, the study of fractional calculus started with L'Hôpital's question 'what does

DOI: 10.1201/9781003263517-11

$d^n f/dx^n$ mean if $n = 1/2$?' in 1695. Ever since then, fractional calculus has attracted the attention of both mathematicians and applied scientists. There are many applications of fractional models to real-world phenomena in science and engineering. Examples include: power law attenuation in the biomedical and underwater sediment fields; acoustic wave propagation in porous media; diffusion equations in biology, engineering, and cosmology; thermoelasticity heat conduction models; control systems in bioinformatics in flexible structures and in software; and image processing in artificial intelligence [1]. The literature describes many operators of the fractional derivative such as Riemann–Liouville, Caputo, Grünwald–Letnikov, and Jumarie. These definitions are based on the integral form. At least some of the properties of classic calculus, such as the derivative of the quotient of two functions, the chain rule, and the product of two functions, are not included in these definitions or are complicated by them. Khalila et al. [2], therefore, introduced a new definition of the fractional derivative called conformable fractional derivative, using the basic limit definition of the derivative. This new definition is more suitable for understanding behaviors in nature. In addition, finding general solutions and evaluations of differential equations are so easy with a conformable derivative. For these reasons, we have used the conformable derivative definitions to determine the general solutions of the models mentioned here.

11.2 Some Preliminary Remarks on the Conformable

Definition

Let h: $[0, \infty) \to \mathbb{R}$ be a given function, the conformable derivative of h of order α is defined as

$$L_\alpha(h)(t) = \lim_{\varepsilon \to 0} \frac{h(t + \varepsilon t^{1-\alpha}) - h(t)}{\varepsilon},$$

for all $t > 0$, $\alpha \in (0, 1)$ [2].

Theorem

Let L_α be the derivative operator with order α and $\alpha \in (0,1]$ and h, k be α-differentiable at a point $t > 0$. Then [2, 3], we have the following

i. $L_\alpha(ah + bk) = aL_\alpha(h) + bL_\alpha(k), \forall\, a, b \in \mathbb{R}$.
ii. $L_\alpha(t^p) = pt^{p-}\alpha, \forall\, p \in \mathbb{R}$.

iii. $L_\alpha(hk) = hL_\alpha(g) + kL_\alpha(f)$.

iv. $L_\alpha\left(\frac{h}{k}\right) = \frac{kL_\alpha(h) - hL_\alpha(k)}{k^2}$.

v. $L_\alpha(\lambda) = 0$, for all constant functions $h(t) = \lambda$.

vi. If h is differentiable then $L_\alpha(h)(t) = t^{1-\alpha}\frac{dh}{dt}(t)$.

Conformable Derivatives for Some Special Functions

1. $L_\alpha(1) = 0$.
2. $L_\alpha(e^{cx}) = cx^{1-\alpha}e^{cx}$, $c \in \mathbb{R}$.
3. $L_\alpha(\sin bx) = bx^{1-\alpha}\cos bx$, $b \in \mathbb{R}$.
4. $L_\alpha(\cos bx) = -bx^{1-\alpha}\sin bx$, $b \in \mathbb{R}$.
5. $L_\alpha\left(\frac{t^\alpha}{\alpha}\right) = 1$.

11.3 General Properties of the Approaches

11.3.1 Fundamental Properties of SGEM

In this section, we describe the sine-Gordon expansion method. Before giving the general features of the sine-Gordon expansion method, we need to explain how we get two significant equations. First, let us suppose that the sine-Gordon equation is given as the following [4–6]:

$$u_{xx} - u_{tt} = m^2 \sin(u), \tag{11.1}$$

where $u = u(x,t)$, m is a real constant. Applying the wave transform $u = u(x,t) = U(\xi)$, $\xi = \mu(x - ct)$to Eq. (11.1),

$$\begin{aligned} u_x &= \frac{dU}{d\xi}.\frac{d\xi}{dx} = \mu.U', u_{xx} = \frac{d(u_x)}{d\xi}.\frac{d\xi}{dx} = \mu^2 U'', \\ u_t &= \frac{dU}{d\xi}.\frac{d\xi}{dt} = -\mu.c.U', u_{tt} = \frac{d(u_t)}{d\xi}.\frac{d\xi}{dt} = c^2\mu^2 U'', \end{aligned} \tag{11.2}$$

we get the following nonlinear ordinary differential equation:

$$U'' = \frac{m^2}{\mu^2(1-c^2)}\sin(U), \tag{11.3}$$

where $U = U(\xi)$, μ is the amplitude of the traveling wave and c is the velocity of the traveling wave. We can fully simplify Eq. (11.3) and it can be written as follows:

$$\left[\left(\frac{U}{2}\right)'\right]^2 = \frac{m^2}{\mu^2\left(1-c^2\right)}\sin^2\left(\frac{U}{2}\right)+K, \tag{11.4}$$

where K is the constant of integration. Substituting $K = 0$, $w(\xi) = U/2$ and $a^2 = m^2/\mu^2(1 - c^2)$ in Eq. (11.4) gives

$$w' = a\sin\left(w\right), \tag{11.5}$$

setting $a = 1$ in Eq. (11.5) gives:

$$w' = \sin\left(w\right). \tag{11.6}$$

Solving Eq. (11.6) by separable variables gives the following two significant properties:

$$\sin\left(w\right) = \sin\left(w\left(\xi\right)\right) = \frac{2pe^{\xi}}{p^2e^{2\xi}+1}\bigg|_{p=1} = \operatorname{sec}h\left(\xi\right), \tag{11.7}$$

$$\cos\left(w\right) = \cos\left(w\left(\xi\right)\right) = \frac{p^2e^{2\xi}-1}{p^2e^{2\xi}+1}\bigg|_{p=1} = \tan h\left(\xi\right), \tag{11.8}$$

where p is the integral constant and non-zero. After these two important properties, when it comes to the description of the sine-Gordon expansion method, to obtain the solution of the nonlinear partial differential equation of the following form:

$$P\left(u, u_x, u_t, u_{xx}, u_{tt}, u^2, \cdots\right) = 0, \tag{11.9}$$

we consider

$$U\left(\xi\right) = \sum_{i=1}^{n}\tanh^{i-1}\left(\xi\right)\left[B_i\operatorname{sec}h\left(\xi\right)+A_i\tanh\left(\xi\right)\right]+A_0. \tag{11.10}$$

Equation (11.10) can be rearranged according to Eqs. (11.7) and (11.8) as follows:

$$U\left(w\right) = \sum_{i=1}^{n}\cos^{i-1}\left(w\right)\left[B_i\sin\left(w\right)+A_i\cos\left(w\right)\right]+A_0. \tag{11.11}$$

We apply the balance principle to determine the value of n under the highest power nonlinear term and highest derivative in the ordinary differential equation. We suppose that the summation of coefficients of $\sin^i(w)\cos^j(w)$ with the same power is zero; this yields a system of equations. With the aid of software, we solve the system of equations to obtain the values of A_i, B_i, μ and c. At the end of the method, substituting the values of A_i, B_i, μ and c into Eq. (11.10), we get the new traveling wave solutions to Eq. (11.9).

11.3.2 Fundamental Properties of MEFM

This section examines the general structure of the modified $\exp(-\Omega(\zeta))$-expansion function method. This method [7, 8] is an improved form of the $\exp(-\Omega(\zeta))$-expansion function method.

Let us consider nonlinear partial differential equations to apply this method as follows:

$$P\left(u, u_x, u_t^{\gamma}, u_{xx}, u_{tt}^{2\gamma}, u_{tx}^{\gamma}, \cdots\right) = 0, \tag{11.12}$$

where $u = u(x,t)$ is an unknown function, P is a polynomial that has $u(x,t)$ function and its partial derivatives in respect to x and t, and $\gamma \in (0,1]$ is the order of the conformable derivative.

Step 1: Let suppose the traveling wave transformation as

$$u(x,t) = U(\zeta), \zeta = x - \frac{lt^{\gamma}}{\gamma}, \tag{11.13}$$

where l is a non-zero constant that can be determined later. Using partial derivatives of Eq. (11.13) into Eq. (11.12), Eq. (11.12) is converted to a nonlinear ordinary differential equation defined as:

$$N\left(U, U', U'', U''', \cdots\right) = 0, \tag{11.14}$$

where N is a polynomial dependent on U.

Step 2: We suppose the traveling wave solution of Eq. (11.14) can be expressed as follows:

$$U(\zeta) = \frac{\sum_{i=0}^{N} A_i \left[\exp(-\Omega(\zeta))\right]^i}{\sum_{j=0}^{M} B_j \left[\exp(-\Omega(\zeta))\right]^j} = \frac{A_0 + A_1\exp(-\Omega) + \cdots + A_N \exp\left(N(-\Omega)\right)}{B_0 + B_1\exp(-\Omega) + \cdots + B_M \exp\left(M(-\Omega)\right)}, \tag{11.15}$$

where A_i, B_j, $(0 \leq i \leq N, 0 \leq j \leq M)$ are constants that can be determined later. $A_N \neq 0$, $B_M \neq 0$, and $\Omega = \Omega(\zeta)$ solves the following ordinary differential equation:

$$\Omega'(\zeta) = \exp(-\Omega(\zeta)) + \mu \exp(\Omega(\zeta)) + \lambda. \tag{11.16}$$

When we solve Eq. (11.16), we reach the following five solution families [7, 8]:

Family 1: When $\mu \neq 0$, $\lambda^2 - 4\mu > 0$,

$$\Omega(\zeta) = \ln\left(\frac{-\sqrt{\lambda^2 - 4\mu}}{2\mu} \tanh\left(\frac{\sqrt{\lambda^2 - 4\mu}}{2}(\zeta + E)\right) - \frac{\lambda}{2\mu}\right). \tag{11.17}$$

Family 2: When $\mu \neq 0$, $\lambda^2 - 4\mu < 0$,

$$\Omega(\zeta) = \ln\left(\frac{\sqrt{-\lambda^2 + 4\mu}}{2\mu} \tan\left(\frac{\sqrt{-\lambda^2 + 4\mu}}{2}(\zeta + E)\right) - \frac{\lambda}{2\mu}\right). \tag{11.18}$$

Family 3: When $\mu = 0$, $\lambda \neq 0$, and $\lambda^2 - 4\mu > 0$,

$$\Omega(\zeta) = -\ln\left(\frac{\lambda}{\exp(\lambda(\zeta + E)) - 1}\right). \tag{11.19}$$

Family 4: When $\mu \neq 0$, $\lambda \neq 0$, and $\lambda^2 - 4\mu = 0$,

$$\Omega(\zeta) = \ln\left(-\frac{2\lambda(\zeta + E) + 4}{\lambda^2(\zeta + E)}\right). \tag{11.20}$$

Family 5: When $\mu = 0$, $\lambda = 0$, and $\lambda^2 - 4\mu = 0$,

$$\Omega(\zeta) = \ln(\zeta + E). \tag{11.21}$$

where $A_0, A_1, \ldots, A_N, B_0, B_1, \ldots, B_M, E, \lambda, \mu$ are constants and can be determined later. Using the homogeneous balance principle between the highest nonlinear terms with the highest-order derivatives of Uin Eq. (11.14), a relationship can be found between N and M.

Step 3: Substituting Eq. (11.146) along with solution families into Eq. (11.14),we have a polynomial of $e^{\Omega(\zeta)}$. After all coefficients of the similar power of $e^{\Omega(\zeta)}$ are equated to zero, it yields an algebraic equation system in terms of $A_0, A_1, \ldots, A_N, B_0, B_1, \ldots, B_M, E, \lambda, \mu$. At the end of this procedure, the values of coefficients substituting into Eq. (11.14) that are obtained provide the traveling wave solutions to the governing model.

11.4 Applications of the Approaches

11.4.1 SGEM for the Conformable Gross–Pitaevskii Equation

This section considers the conformable Gross–Pitaevskii equation, which is given as the following:

$$i\frac{\partial}{\partial t}\Psi(x,t) = -\frac{\partial^2}{\partial x^2}\Psi(x,t) - \frac{1}{2}\lambda^2 x^2 \Psi(x,t) + 2N\frac{a_s}{l_\perp}\left|\Psi(x,t)\right|^2 \Psi(x,t), \quad (11.22)$$

where $\lambda \equiv 2|\omega_0|/\omega_\perp \ll 1$ and the time t and coordinate x have been measured in units $2/\omega_\perp$ and $l_\perp$, respectively [9]. ω_0, $\omega_\perp$, and a_s are the axial and transverse harmonic oscillator frequencies, and the s-wave scattering length, respectively. N is a constant value. Eq. (11.22) describes the rogue wave in Bose–Einstein condensates (BECs) in an expulsive potential [9, 10, 11]. The Gross–Pitaevskii equation describes the evolution of the macroscopic wave function of BECs at the mean-field level. Eq. (11.22) is converted to following standard form:

$$\Psi(x,t) = q(X,T)e^{\lambda t/2 - i\lambda x^2/4} \quad (11.23)$$

here $X = xe^{\lambda t}$ and $T = 2\int_0^t e^{2\lambda\tau}d\tau$.

$$i\frac{\partial q}{\partial T} + \frac{1}{2}\frac{\partial^2 q}{\partial X^2} - \left(\frac{Na_0}{l_\perp}\right)|q|^2 q = 0. \quad (11.24)$$

Equation (11.24) shows that in BECs the non-autonomous solitons formed by magnetically tuning the interatomic interaction can be obtained from the autonomous solitons [9]. We suppose Eq. (11.24) with conformable sense obtains more general soliton solutions which explains the behaviors of interatomic interactions. Its standard form is given below:

$$i\frac{\partial q^\theta}{\partial T} + \frac{1}{2}\frac{\partial^2 q}{\partial X^2} - \Upsilon|q|^2 q = 0, \quad (11.25)$$

where $\Upsilon = Na_0/l_\perp$, θ is the conformable derivative order in $0 < \theta \leq 1$. First, we consider the following traveling wave transformation for converting the nonlinear partial differential equation Eq. (11.25) to a linear ordinary differential equation.

$$q(X,T) = U(\zeta)e^{i\varphi},\ \zeta = \alpha X - \frac{\beta T^{\theta}}{\theta}, \varphi = pX - \kappa\frac{T^{\theta}}{\theta}, \tag{11.26}$$

where α, β, p, κ are non-zero constants. The following correspond to the real part and imaginary part, respectively.

$$\left(\kappa - \frac{p^2}{2}\right)U + \frac{\alpha^2}{2}U'' - \Upsilon U^3 = 0, \tag{11.27}$$

$$\beta = \alpha p. \tag{11.28}$$

Using the homogeneous balance principle between U'' and U^3, we get $n = 1$. Putting$n = 1$ into Eq. (11.11) gives

$$U(\zeta) = B_1 \sin(w) + A_1 \cos(w) + A_0. \tag{11.29}$$

Substituting Eq. (11.29) and its second-order derivative into Eq. (11.27), we obtain a trigonometric function with different degrees. Equating to zero all sums of coefficients of the same power of the trigonometric functions, we get an algebraic equation system. The solution of this algebraic equation system gives the coefficients. After this we have the following cases:

Case 1

When we choose these values of coefficients $A_0 = 0, B_1 = iA_1, \alpha = 2\sqrt{\Upsilon}A_1, p = -\sqrt{2\left(\kappa - \Upsilon A_1^2\right)}$, the solution takes the form

$$\Psi_1(x,t) = e^{i\left(px - \kappa\frac{t^{\theta}}{\theta} - \lambda x^2/4\right) + \lambda t/2} A_1\left(i\,\mathrm{sec}\,h\left[2\sqrt{\Upsilon}A_1 x + \varpi t^{\theta}\right] + \tanh\left[2\sqrt{\Upsilon}A_1 x + \varpi t^{\theta}\right]\right), \tag{11.30}$$

where $\varpi = \dfrac{2\sqrt{2\Upsilon\left(\kappa - \Upsilon A_1^2\right)}A_1}{\theta}$. The solution can be plotted using suitable parameter values as shown in Figures 11.1–11.14.

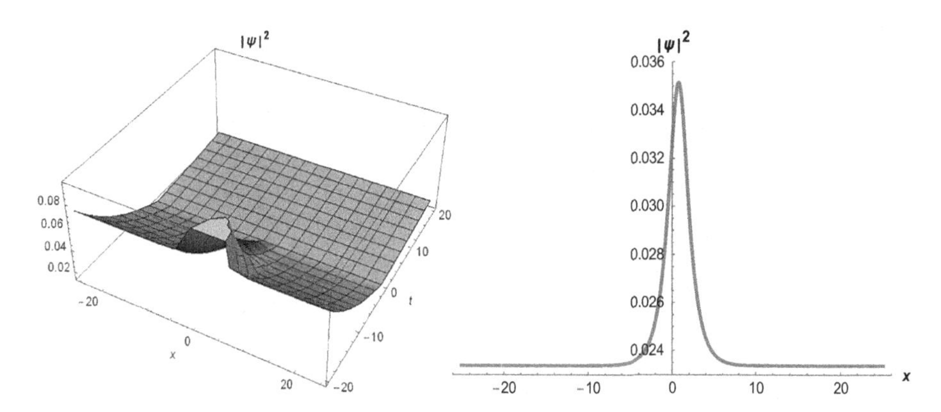

FIGURE 11.1
2D and 3D figures of Eq. (11.30).

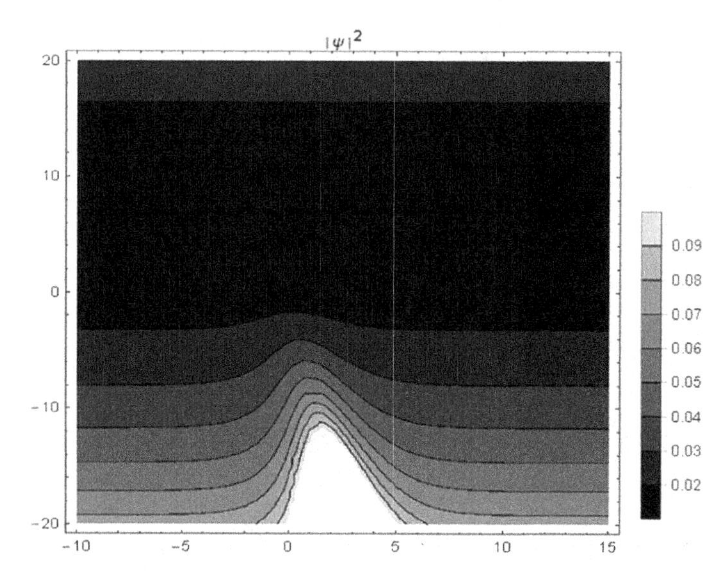

FIGURE 11.2
Contour figures of Eq. (11.30).

Case 2

When $A_0 = 0, B_1 = 0, \alpha = \sqrt{\Upsilon} A_1, p = -\sqrt{2\left(\kappa - \Upsilon A_1^2\right)}$, it produces

$$\Psi_2\left(x,t\right) = e^{i\left(-\sqrt{2\left(\kappa - \Upsilon A_1^2\right)}x - \kappa\frac{t^\theta}{\theta} - \lambda x^2/4\right) + \lambda t/2} A_1 \text{Tanh}\left[A_1\sqrt{\Upsilon}x + \frac{\sqrt{2\Upsilon\left(\kappa - \Upsilon A_1^2\right)}A_1 t^\theta}{\theta}\right]. \tag{11.31}$$

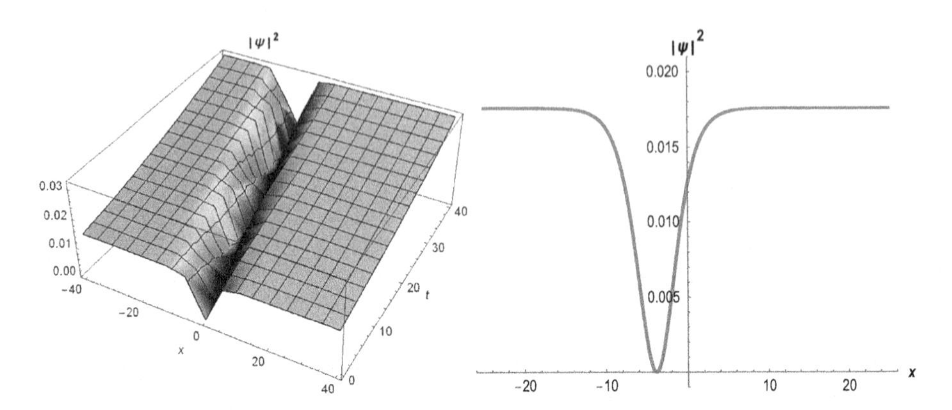

FIGURE 11.3
2D and 3D figures of Eq. (11.31).

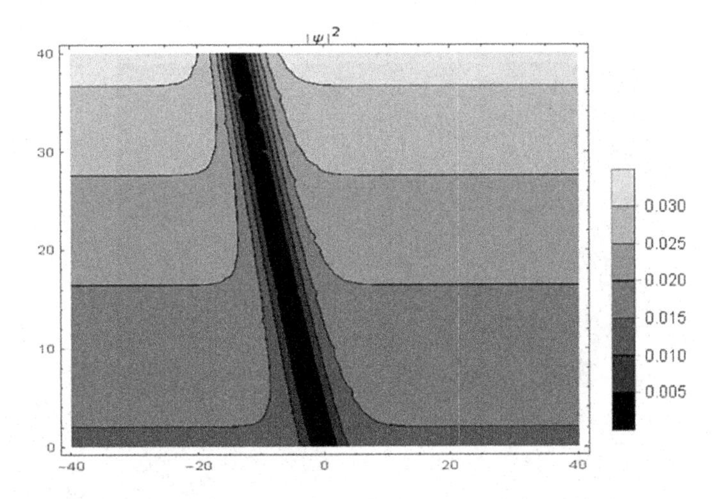

FIGURE 11.4
Contour figures of Eq. (11.31).

Case 3

If $A_0 = 0, A_1 = -\dfrac{\alpha}{2\sqrt{\Upsilon}}, B_1 = -\dfrac{i\alpha}{2\sqrt{\Upsilon}}, \kappa = \dfrac{\alpha^2 + 2p^2}{4}$, it gives

$$\Psi_3(x,t) = -e^{i\left(px - \frac{\alpha^2+2p^2}{4}\frac{t^\theta}{\theta} - \lambda x^2/4\right) + \lambda t/2} \frac{\alpha}{2\sqrt{\Upsilon}}\left(i\,\mathrm{Sech}\left[\alpha x - \frac{\alpha p t^\theta}{\theta}\right] + \mathrm{Tanh}\left[\alpha x - \frac{\alpha p t^\theta}{\theta}\right]\right) \tag{11.32}$$

We can plot various simulations of Eq. (11.32) by using a computational program as shown in Figures 11.5 and 11.6.

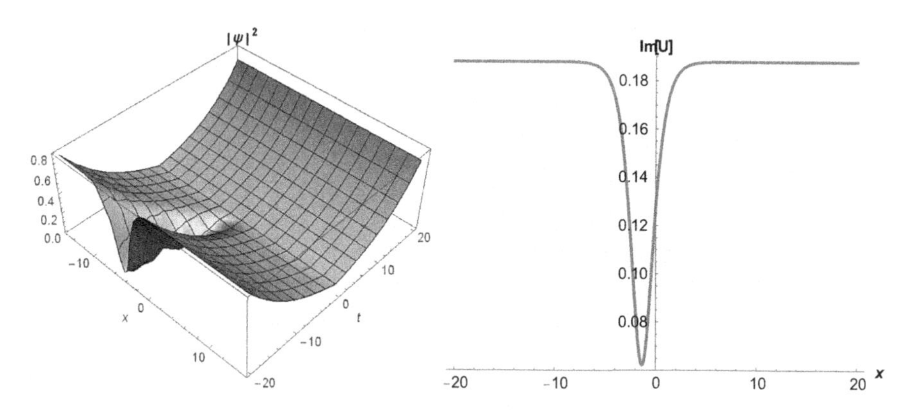

FIGURE 11.5
2D and 3D figures of Eq. (11.32).

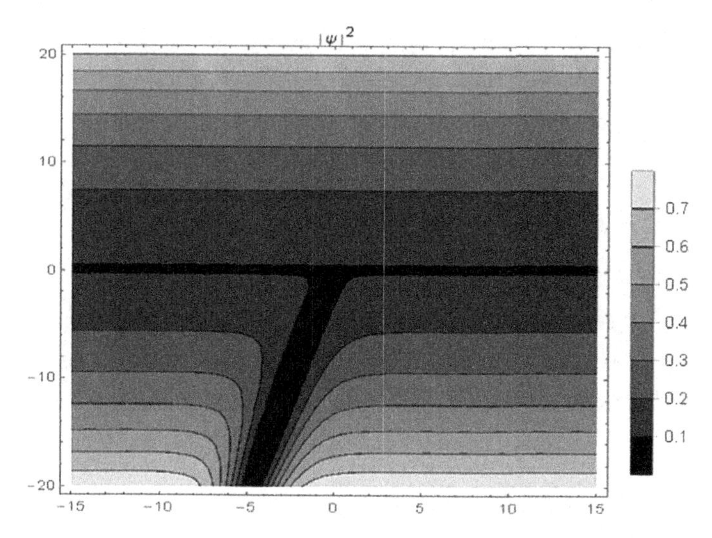

FIGURE 11.6
Contour figures of Eq. (11.32).

Case 4

Taking as $A_0 = 0, A_1 = 0, \alpha = -iB_1\sqrt{\Upsilon}, \kappa = \dfrac{\Upsilon B_1{}^2 + p^2}{2}$, results in

$$\Psi_4(x,t) = e^{i\left(px - \frac{\left(\Upsilon B_1{}^2 + p^2\right)t^\theta}{2\theta} - \lambda x^2/4\right) + \lambda t/2} \sec\left[B_1\sqrt{\Upsilon}x - \frac{pB_1\sqrt{\Upsilon}t^\theta}{\theta}\right]B_1. \tag{11.33}$$

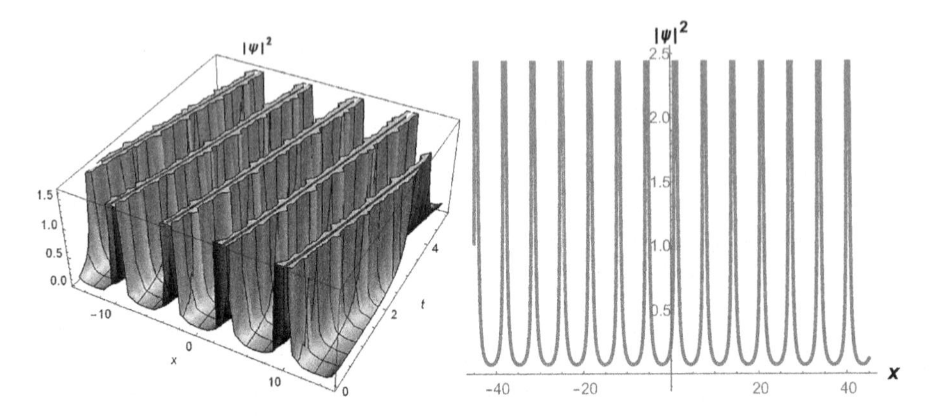

FIGURE 11.7
2D and 3D figures of Eq. (11.33).

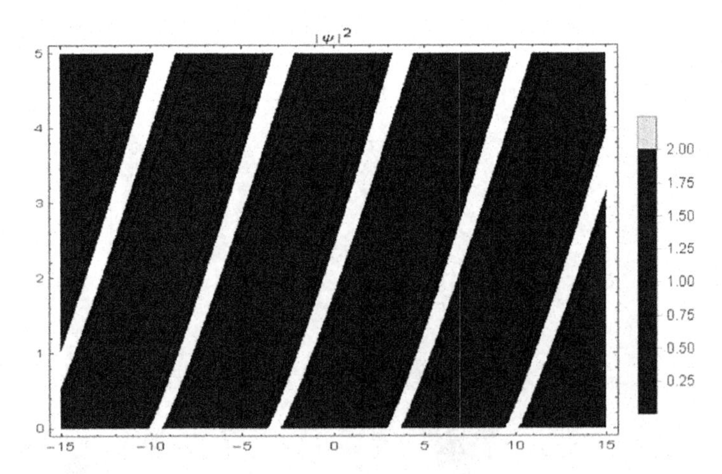

FIGURE 11.8
Contour figures of Eq. (11.33).

Case 5

Considering $A_0 = 0, A_1 = -iB_1, p = -\sqrt{2\left(\kappa + \Upsilon B_1^{\,2}\right)}, \alpha = 2iB_1\sqrt{\Upsilon}$, we extract

$$\Psi_5\left(x,t\right) = e^{i\left(-\sqrt{2\left(\kappa+\Upsilon B_1^{\,2}\right)}x-\kappa\frac{t^\theta}{\theta}-\lambda x^2/4\right)+\lambda t/2} B_1\left(\sec\left[2\sqrt{\Psi}B_1x+\varpi t^\theta\right]+\tan\left[2\sqrt{\Psi}B_1x+\varpi t^\theta\right]\right), \tag{11.34}$$

where $\varpi = 2/\theta\sqrt{2\Upsilon\left(\kappa+\Upsilon B_1^{\,2}\right)}A_1$. With the help of computational programs, we can plot Eq. (11.34) as shown in Figures 11.9 and 11.10.

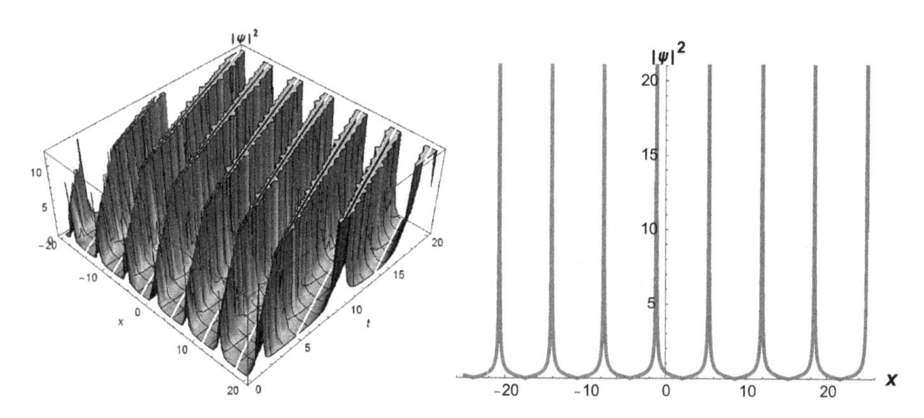

FIGURE 11.9
2D and 3D figures of Eq. (11.34).

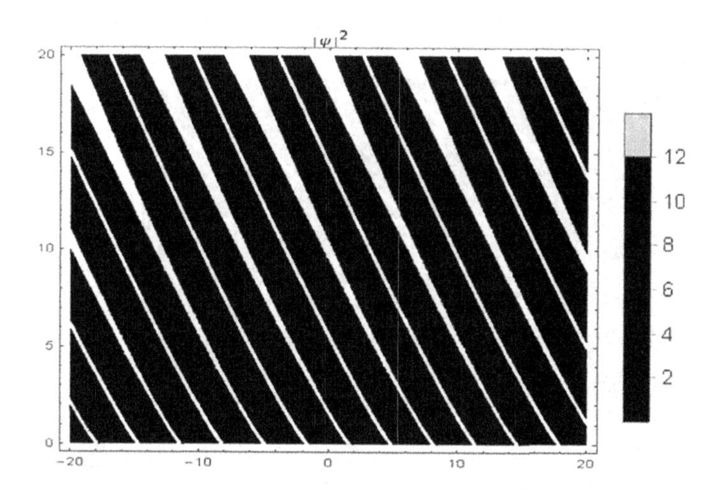

FIGURE 11.10
Contour figures of Eq. (11.34).

Case 6

Using an alternative coefficient $A_0 = 0, A_1 = iB_1, \alpha = 2iB_1\sqrt{\Upsilon}, k = p^2 - 2\Upsilon B_1^2 / 2$, we find the analytical solution

$$\Psi_6 = e^{i\left(px - \frac{\left(p^2 - 2\Upsilon B_1^2\right)t^\theta}{2\theta} - \lambda x^2/4\right) + \lambda t/2} B_1\left(\sec\left[2\sqrt{\Psi}B_1 x - \frac{2p\sqrt{\Upsilon}B_1 t^\theta}{\theta}\right] - \tan\left[2\sqrt{\Psi}B_1 x - \frac{2p\sqrt{\Upsilon}B_1 t^\theta}{\theta}\right]\right). \quad (11.35)$$

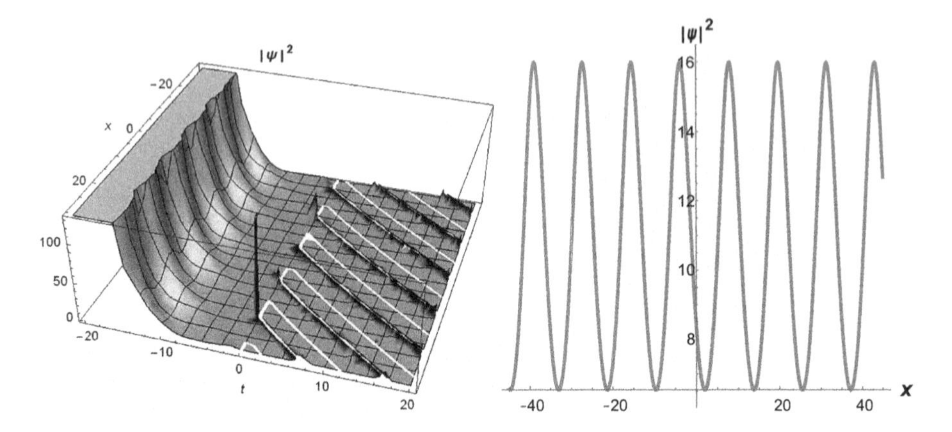

FIGURE 11.11
2D and 3D figures of Eq. (11.35).

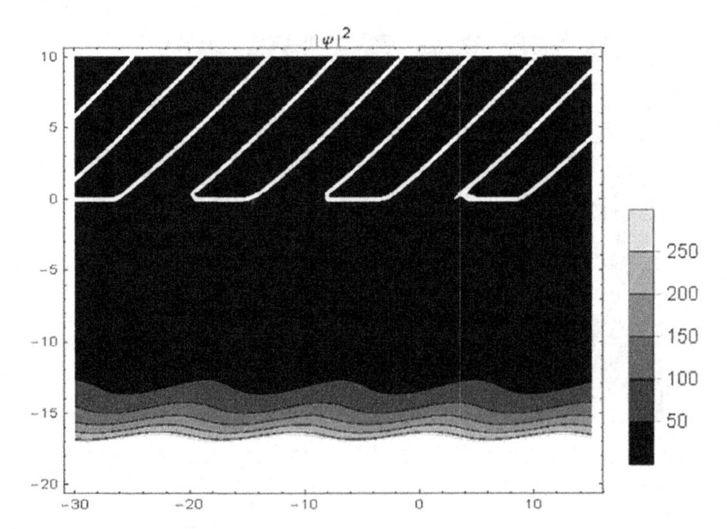

FIGURE 11.12
Contour figures of Eq. (11.35).

Case 7

With coefficients such as $A_0 = 0, A_1 = -\frac{\alpha}{2\sqrt{\Upsilon}}, B_1 = \frac{i\alpha}{2\sqrt{\Upsilon}}, p = -\sqrt{-\frac{\alpha^2}{2} + 2\kappa}$, we reach

$$\Psi_7(x,t) = ie^{i\left(-\tau x - \frac{\kappa t^\theta}{\theta} - \lambda x^2/4\right) + \lambda t/2} \frac{\alpha}{2\sqrt{\Upsilon}}\left(\text{Sech}\left[\alpha x + \sigma t^\theta\right] + i\,\text{Tanh}\left[\alpha x + \sigma t^\theta\right]\right), \tag{11.36}$$

in which $\sigma = \frac{\alpha\sqrt{-\alpha^2 + 4\kappa}}{\theta\sqrt{2}}, \tau = \sqrt{-\frac{\alpha^2}{2} + 2\kappa}$.

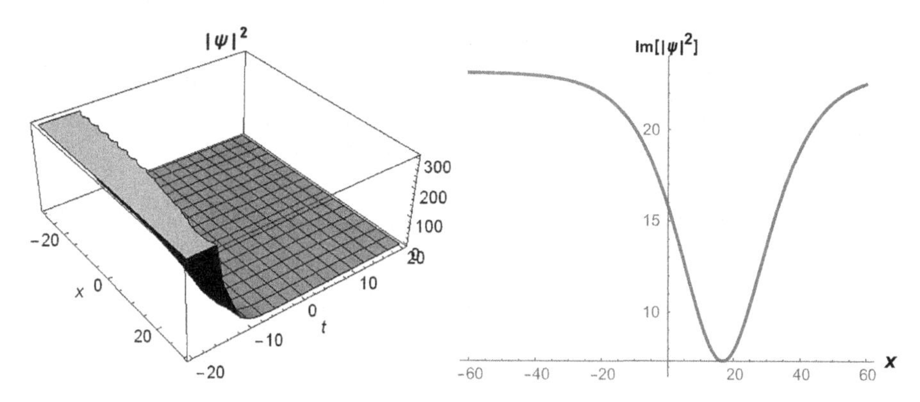

FIGURE 11.13
2D and 3D figures of Eq. (11.36).

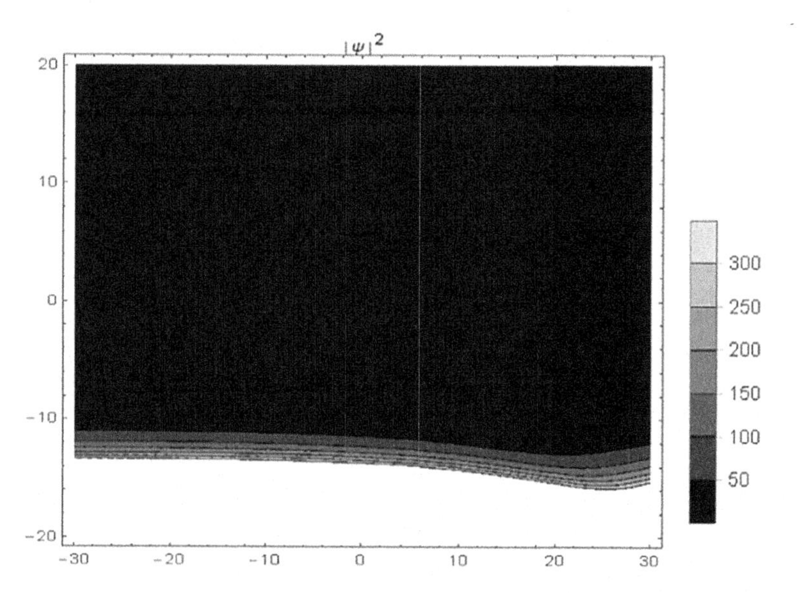

FIGURE 11.14
Contour figures of Eq. (11.36).

11.4.2 MEFM for the Conformable Gross–Pitaevskii Equation

In this section, we investigate the soliton solutions Eq. (11.22) by the modified $\exp(-\Omega(\zeta))$-expansion function method. We use the traveling wave transform, which is given as

$$q(X,T)=U(\zeta)e^{i\varphi},\ \zeta=\alpha x-\frac{\beta t^{\theta}}{\theta},\varphi=px-\kappa\frac{t^{\theta}}{\theta}. \tag{11.37}$$

The following nonlinear ordinary differential equation corresponds to the real and imaginary parts, respectively.

$$\left(\kappa-\frac{p^2}{2}\right)U+\frac{\alpha^2}{2}U''-\Upsilon U^3=0, \tag{11.38}$$

$$\beta=\alpha p. \tag{11.39}$$

Using the homogeneous balance principle between U'' and U^3, we get a relationship between M and N of

$$M+1=N.$$

For suitable integer values of M and N, one can achieve different cases. When the values $M=1$ and $N=2$ are chosen, the solution form yields

$$U(\zeta)=\frac{A_0+A_1e^{-\Omega(\zeta)}+A_2e^{-2\Omega(\zeta)}}{B_0+B_1e^{-\Omega(\zeta)}}, \tag{11.40}$$

When Eq. (11.40) and its second-order derivative are substituted into Eq. (11.39), the soliton solutions shown below are extracted.

Case 1

When we select for the coefficients of Eq. (11.40)

$$A_0=-\frac{2\alpha\mu B_0}{\lambda\sqrt{\Upsilon}},A_1=-\frac{2\alpha B_0}{\sqrt{\Upsilon}},A_2=-\frac{2\alpha B_0}{\lambda\sqrt{\Upsilon}},B_1=\frac{2B_0}{\lambda},\kappa=\frac{1}{2}\left(p^2-\alpha^2\left(\lambda^2-4\mu\right)\right),$$

we obtain the following solution families.

Family 1: When we consider Family 1 conditions, we find the following solution (Figures 11.15–11.36)

$$\Psi_{1,1}(x,t)=\frac{2e^{i\left(px-\varepsilon t^{\theta}-\lambda x^{2}/4\right)+\lambda t/2}\alpha\left(\lambda^{2}-4\mu\right)\mu\,\mathrm{Sech}^{2}\left[f(\zeta)\right]}{\sqrt{\Upsilon}\left(\lambda+\sqrt{\lambda^{2}-4\mu}\,\mathrm{Tanh}\left[f(\zeta)\right]\right)\left(\lambda^{2}-4\mu+\lambda\sqrt{\lambda^{2}-4\mu}\,\mathrm{Tanh}\left[f(\zeta)\right]\right)}, \tag{11.41}$$

where $\varepsilon=\frac{1}{2\theta}\left(p^{2}-\alpha^{2}\left(\lambda^{2}-4\mu\right)\right), f(\zeta)=\frac{1}{2}(EE+\zeta)\sqrt{\lambda^{2}-4\mu}, \lambda^{2}-4\mu>0.$

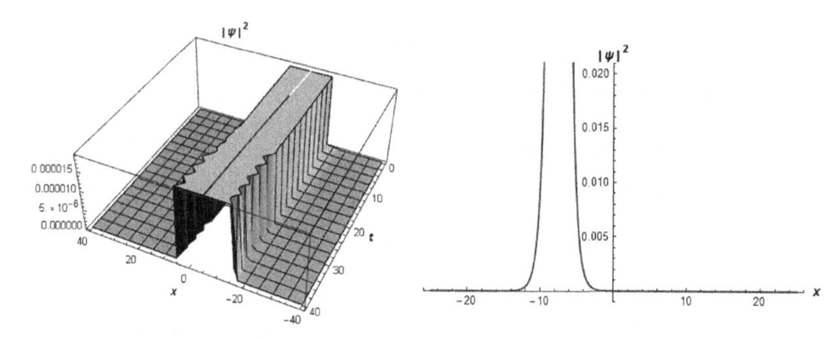

FIGURE 11.15
2D and 3D figures of Eq. (11.41).

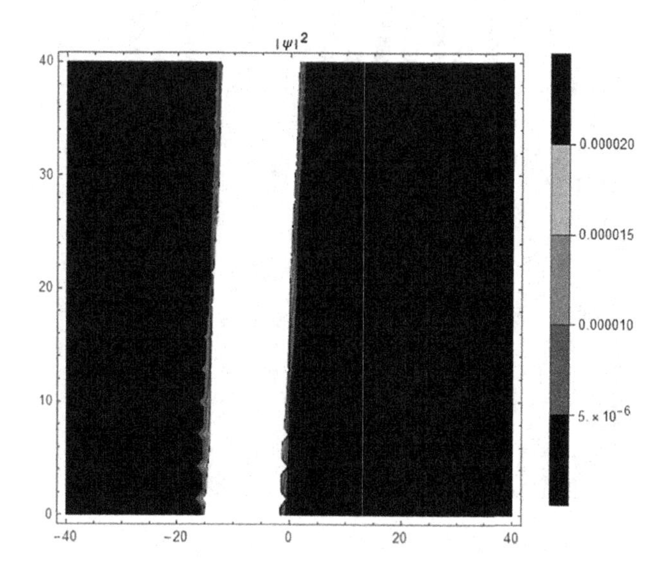

FIGURE 11.16
Contour figures of Eq. (11.41).

Family 2: With the Family 2 conditions, we obtain

$$\Psi_{1,2}(x,t)=\frac{2\alpha e^{i\left(px-\gamma t^{\theta}-\lambda x^{2}/4\right)+\lambda t/2}\left(\lambda^{2}-4\mu\right)\mu\sec^{2}\left[f(\zeta)\right]}{\sqrt{\Upsilon}\left(\lambda-\sqrt{-\lambda^{2}+4\mu}\tan\left[f(\zeta)\right]\right)\left(\lambda^{2}-4\mu-\lambda\sqrt{-\lambda^{2}+4\mu}\tan\left[f(\zeta)\right]\right)}, \tag{11.42}$$

in which $f(\zeta)=\frac{1}{2}(EE+\zeta)\sqrt{-\lambda^{2}+4\mu}, \gamma=\frac{1}{2\theta}\left(p^{2}-\alpha^{2}\left(\lambda^{2}-4\mu\right)\right), \lambda^{2}-4\mu<0.$

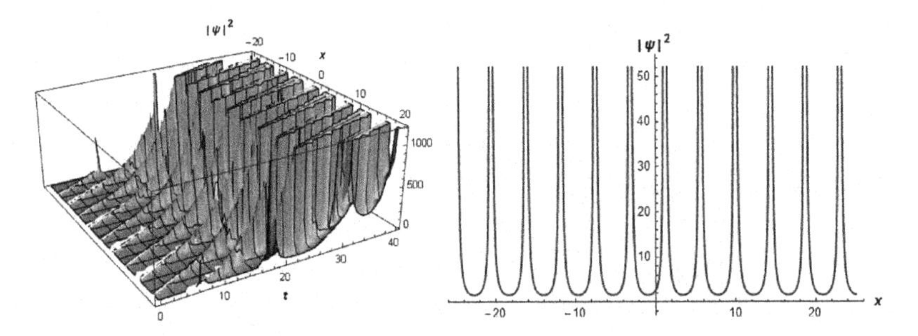

FIGURE 11.17
2D and 3D figures of Eq. (11.42).

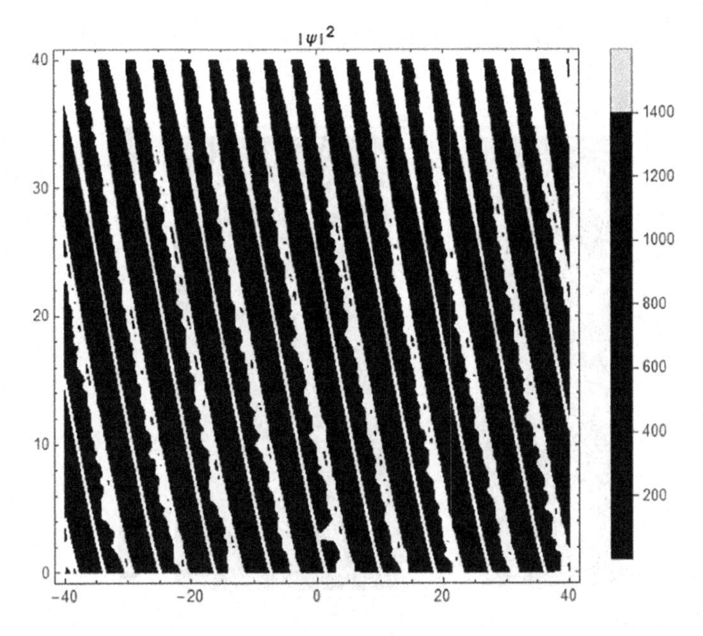

FIGURE 11.18
Contour figures of Eq. (11.42).

Family 4: Via Family 4 conditions, we gain

$$\Psi_{1,3}(x,t)=\frac{e^{i\left(px-\frac{p^2t^\theta}{2\theta}-\sqrt{\mu}x^2/2\right)+t\sqrt{\mu}}\alpha\theta\sqrt{\mu}}{\sqrt{\Upsilon}\left(\theta\left(-1+\sqrt{\mu}\left(EE+\zeta\right)\right)-\alpha p\sqrt{\mu}t^\theta\right)}, \tag{11.43}$$

when $\mu \neq 0$, $\lambda \neq 0$, and $\lambda^2 - 4\mu = 0$.

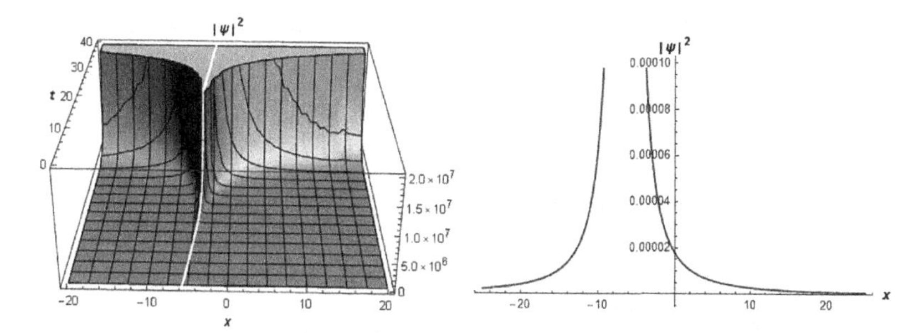

FIGURE 11.19
2D and 3D figures of Eq. (11.43).

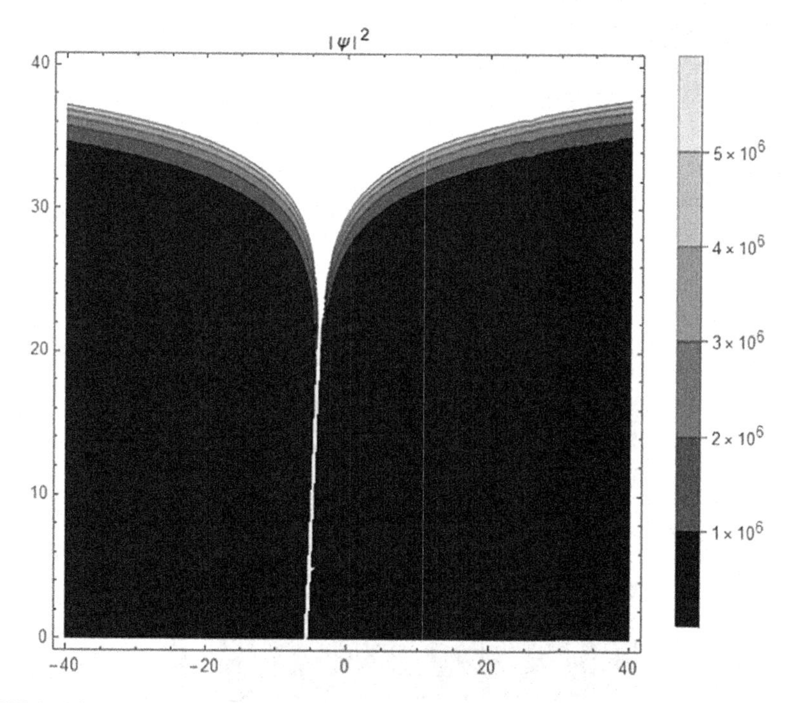

FIGURE 11.20
Contour figures of Eq. (11.43).

Case 2

If we take other coefficients for Eq. (11.40) given by $A_0 = \frac{\alpha\mu B_0}{2\sqrt{\Upsilon}}, A_1 = \frac{\alpha(2B_0 + \lambda B_1)}{2\sqrt{\Upsilon}}$, $A_2 = \frac{\alpha B_1}{\sqrt{\Upsilon}}, \kappa = \frac{1}{4}\left(2p^2 + \alpha^2(\lambda^2 - 4\mu)\right)$, we can obtain the following.

Family 1: With the help of Family 1 conditions, we can write the following solution

$$\Psi_{2,1}(x,t) = \frac{e^{i\left(px - \theta t^{\theta} - \lambda x^2/4\right) + \lambda t/2}\alpha\left(\lambda^2 - 4\mu + \lambda\sqrt{\lambda^2 - 4\mu}\,\mathrm{Tanh}\left[\frac{1}{2}(EE + \zeta)\sqrt{\lambda^2 - 4\mu}\right]\right)}{2\sqrt{\Upsilon}\left(\lambda + \sqrt{\lambda^2 - 4\mu}\,\mathrm{Tanh}\left[\frac{1}{2}(EE + \zeta)\sqrt{\lambda^2 - 4\mu}\right]\right)}, \tag{11.44}$$

in which is $\lambda^2 - 4\mu > 0, \theta = \frac{1}{4\theta}\left(2p^2 + \alpha^2(\lambda^2 - 4\mu)\right)$.

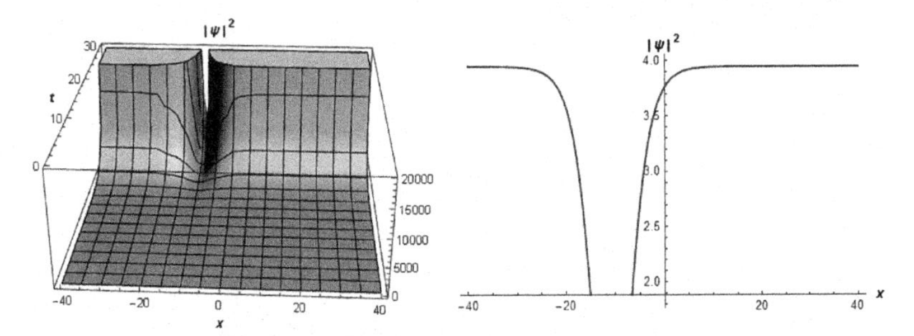

FIGURE 11.21
2D and 3D figures of Eq. (11.44).

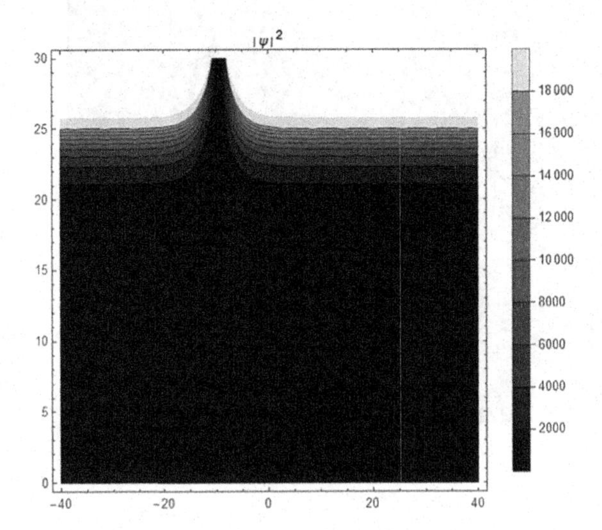

FIGURE 11.22
Contour figures of Eq. (11.44).

Family 2: Under Family 2 conditions, we can write the following solution

$$\Psi_{2,2}(x,t)=\frac{e^{i\left(px-\rho t^{\theta}-\lambda x^2/4\right)+\lambda t/2}\alpha\left(\lambda^2-4\mu-\lambda\sqrt{-\lambda^2+4\mu}\tan\left[\frac{1}{2}(EE+\zeta)\sqrt{-\lambda^2+4\mu}\right]\right)}{2\sqrt{\Upsilon}\left(\lambda-\sqrt{-\lambda^2+4\mu}\tan\left[\frac{1}{2}(EE+\zeta)\sqrt{-\lambda^2+4\mu}\right]\right)}, \tag{11.45}$$

when $\lambda^2-4\mu<0, \rho=\frac{1}{4\theta}\left(2p^2+\alpha^2\left(\lambda^2-4\mu\right)\right)$.

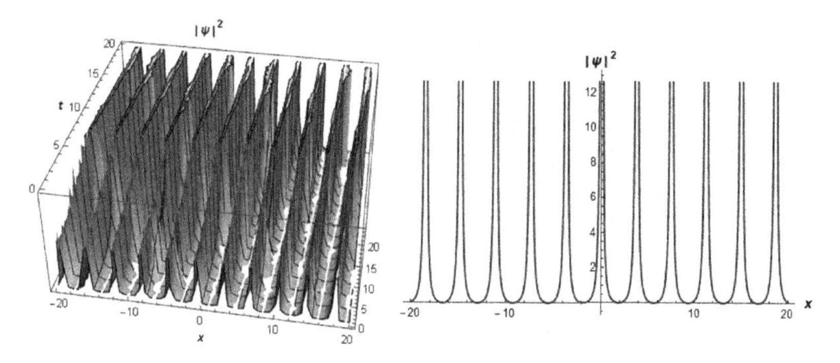

FIGURE 11.23
2D and 3D figures of Eq. (11.45).

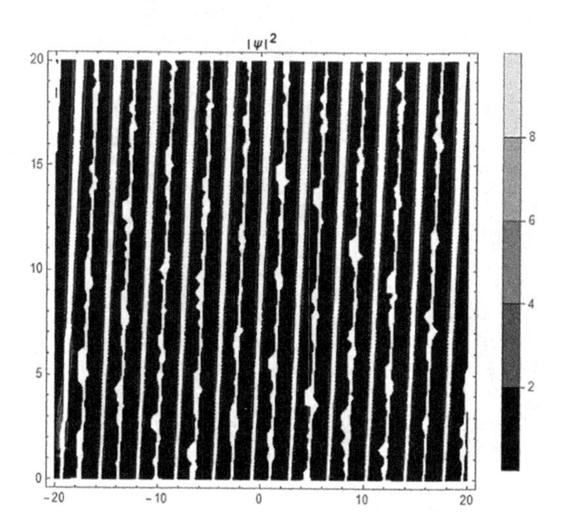

FIGURE 11.24
Contour figures of Eq. (11.45).

Family 3: With Family 3 conditions, the following may be found:

$$\Psi_{2,3}(x,t)=\frac{\alpha\lambda}{2\sqrt{\Upsilon}}e^{i\left(px-\frac{\left(2p^2+\alpha^2\left(\lambda^2-4\mu\right)\right)t^{\theta}}{4\theta}-\lambda x^2/4\right)+\lambda t/2}\coth\left[\frac{\left(-\alpha pt^{\theta}+EE\theta+\alpha\theta x\right)\lambda}{2\theta}\right], \tag{11.46}$$

where $\mu=0, \lambda\neq 0$.

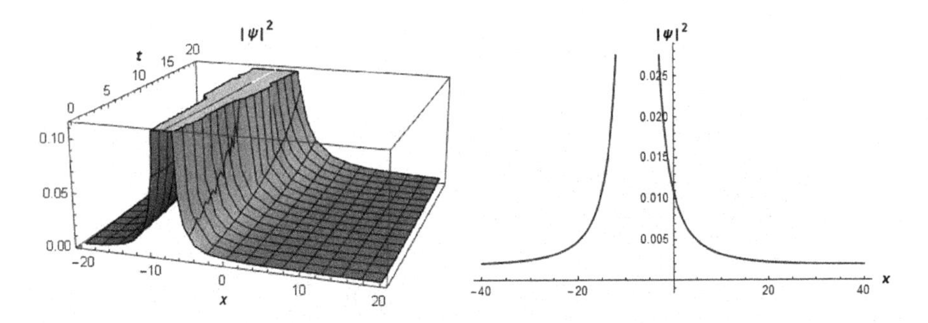

FIGURE 11.25
2D and 3D figures of Eq. (11.46).

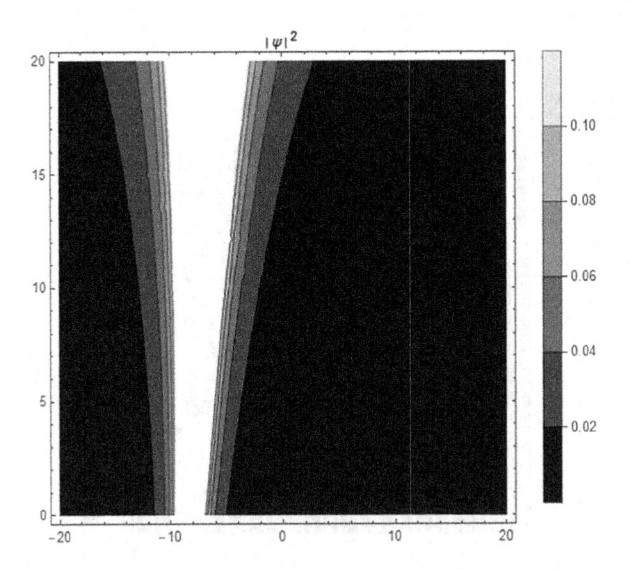

FIGURE 11.26
Contour figures of Eq. (11.46).

Family 4 The Family 4 conditions produce

$$\Psi_{2,4}(x,t)=\frac{-e^{i\left(px-\frac{p^2t^\theta}{2\theta}-\sqrt{\mu}x^2/2\right)+t\sqrt{\mu}}\alpha\theta\sqrt{\Upsilon}}{\sqrt{\Upsilon}\left(\theta\left(-1+\sqrt{\mu}\,(EE+\zeta)\right)-\alpha p\sqrt{\mu}t^\theta\right)} \tag{11.47}$$

when $\mu \neq 0$, $\lambda \neq 0$, and $\lambda^2 - 4\mu = 0$.

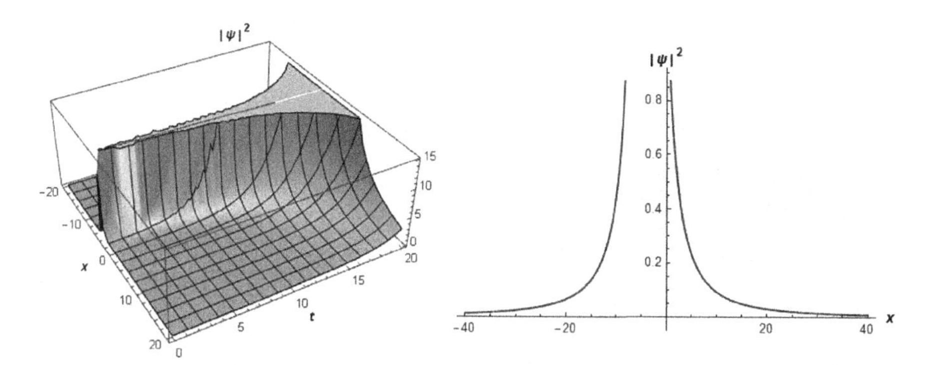

FIGURE 11.27
2D and 3D figures of Eq. (11.47).

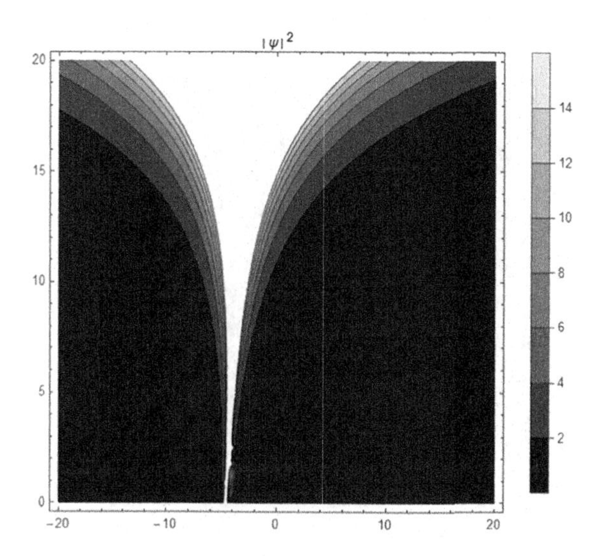

FIGURE 11.28
Contour figures of Eq. (11.47).

Family 5: With Family 5 conditions, we can obtain the following solution:

$$\Psi_{2,5}(x,t) = \frac{e^{i\left(px - \frac{p^2 t^\theta}{2\theta}\right)} \alpha\theta}{\sqrt{\Upsilon}\left(-\alpha p t^\theta + EE\theta + \alpha\theta x\right)}, \tag{11.48}$$

when $\mu = 0$, $\lambda = 0$, and $\lambda^2 - 4\mu = 0$.

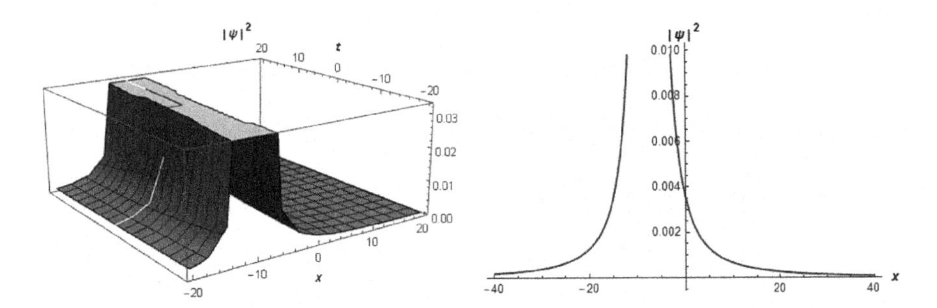

FIGURE 11.29
2D and 3D figures of Eq. (11.48).

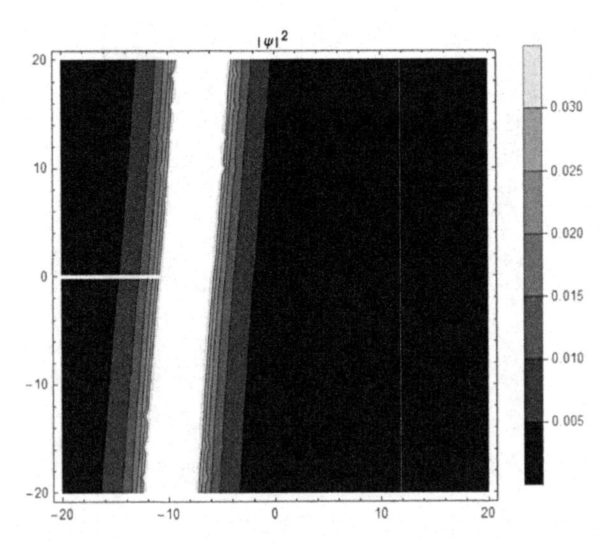

FIGURE 11.30
Contour figures of Eq. (11.48).

Case 3:

Selecting another coefficient given as follows $A_1 = -\frac{A_0}{\lambda}, A_2 = -\frac{6A_0}{\lambda^2}, B_1 = -\frac{3B_0}{\lambda}$, $\Upsilon = \frac{\alpha^2\lambda^2B_0^2}{4A_0^2}, \kappa = \frac{1}{4}\left(2p^2 + \alpha^2\left(\lambda^2 - 4\mu\right)\right)$, we obtain the following solution families.

Family 1: With

$$\Psi_{3,1}(x,t) = \frac{e^{i\left(px - \frac{\left(2p^2+\alpha^2\left(\lambda^2-4\mu\right)\right)t^\theta}{4\theta} - \lambda x^2/4\right) + \lambda t/2} A_0\left(\lambda^2 - 4\mu + \lambda\sqrt{\lambda^2-4\mu}\,\mathrm{Tanh}\left[\frac{1}{2}(EE+\zeta)\sqrt{\lambda^2-4\mu}\right]\right)}{\lambda B_0\left(\lambda + \sqrt{\lambda^2-4\mu}\,\mathrm{Tanh}\left[\frac{1}{2}(EE+\zeta)\sqrt{\lambda^2-4\mu}\right]\right)}, \tag{11.49}$$

in which $\lambda^2 - 4\mu > 0$.

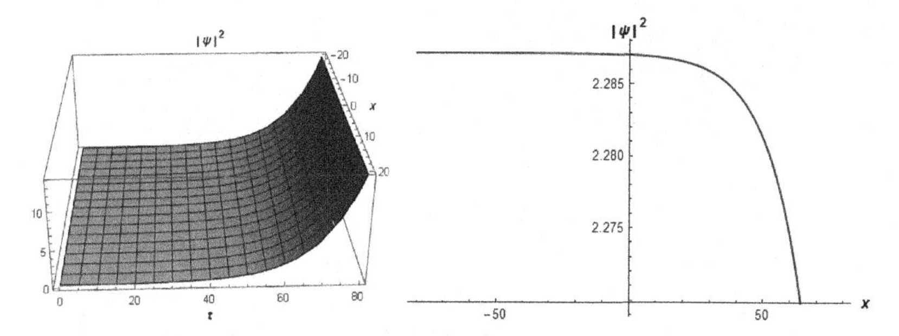

FIGURE 11.31
2D and 3D figures of Eq. (11.49).

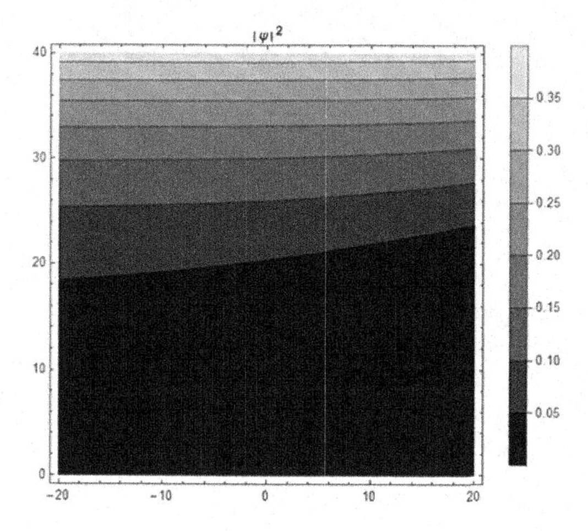

FIGURE 11.32
Contour figures of Eq. (11.49).

Family 2: If we consider this family condition, we reach the following results

$$\Psi_{3,2}(x,t)=\frac{A_0\left(\lambda^2-4\mu-\lambda\sqrt{-\lambda^2+4\mu}\,\mathrm{Tanh}\left[\frac{1}{2}(EE+\zeta)\sqrt{-\lambda^2+4\mu}\right]\right)}{\lambda B_0\left(\lambda-\sqrt{-\lambda^2+4\mu}\,\mathrm{Tanh}\left[\frac{1}{2}(EE+\zeta)\sqrt{-\lambda^2+4\mu}\right]\right)}e^{i\left(px-\frac{\left(2p^2+\alpha^2\left(\lambda^2-4\mu\right)\right)t^{\theta}}{4\theta}-\lambda x^2/4\right)+\lambda t/2}, \tag{11.50}$$

where $\lambda^2 - 4\mu < 0$.

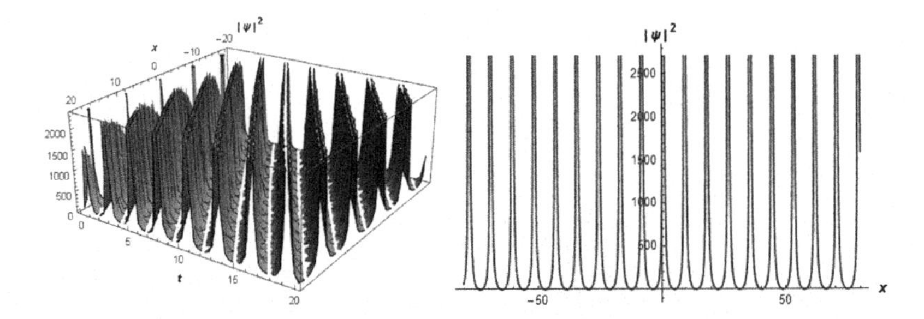

FIGURE 11.33
2D and 3D figures of Eq. (11.50).

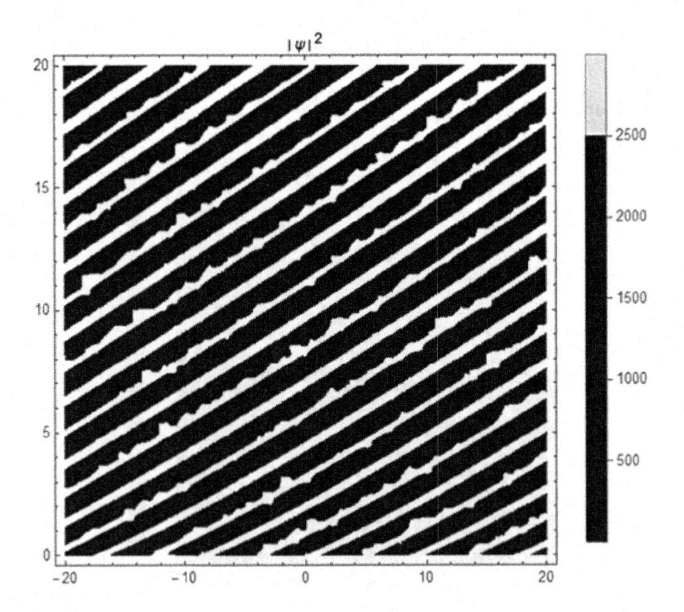

FIGURE 11.34
Contour figures of Eq. (11.50).

Family 3: If this family condition is taken, it may be extracted as

$$\Psi_{3,3}(x,t)=\frac{A_0}{B_0}\coth\left[\frac{\lambda}{2\theta}\left(-\alpha p t^{\theta}+EE\theta+\alpha\theta x\right)\right]e^{i\left(px-\frac{\left(2p^2+\alpha^2\left(\lambda^2-4\mu\right)\right)t^{\theta}}{4\theta}-\lambda x^2/4\right)+\lambda t/2}, \tag{11.51}$$

being $\mu = 0$, $\lambda \neq 0$, and $\lambda^2 - 4\mu > 0$.

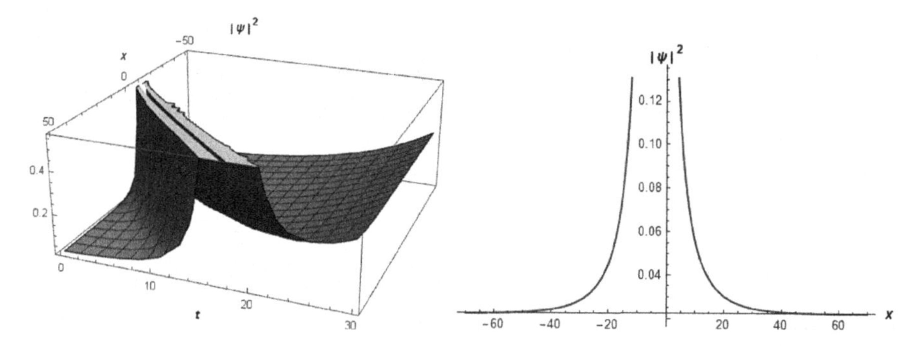

FIGURE 11.35
2D and 3D figures of Eq. (11.51).

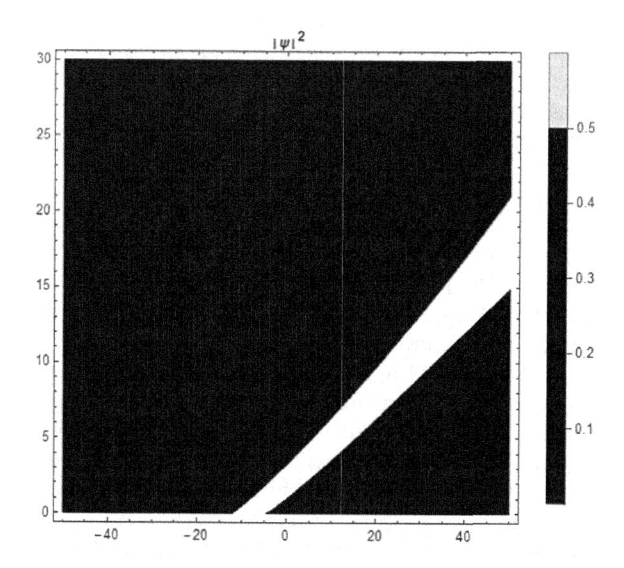

FIGURE 11.36
Contour figures of Eq. (11.51).

11.5 Conclusion

In this chapter, we successfully applied the sine-Gordon expansion method and modified exponential function method to the conformable Gross–Pitaevskii equation to extract some new traveling wave solutions. Many new analytical solutions to the governing model were reported. With suitably chosen solution parameters, various simulations were also plotted with the help of a computational program. From the figures and the results obtained in this chapter, it can be seen that the solutions symbolize the estimated wave distributions via surfaces. Future soliton theory research may include development of the general structure of the method used and the properties of soliton theory, which can also be applied to other nonlinear partial differential equations. Such models may be useful to explain deeper physical properties of real-world problem arising in science and engineering [12–32].

References

[1] H.G. Sun, Y. Zhang, D. Baleanu, W. Chen, Y.Q. Chen, A new collection of real world applications of fractional calculus in science and engineering, *CommunicationsinNonlinear Scienceand Numerical Simulation*, 64, 213–231, 2018.

[2] R. Khalil, M.A. Horani, A. Yousef, M. Sababheh, A new definition of fractional derivative, *Journal of Computational and Applied Mathematics*, 264, 65–70, 2014.

[3] A. Atangana, D. Baleanu, A. Alsaedi, New properties of conformable derivative, *Open Mathematics*, 13(1), 889–898, 2015.

[4] C. Yan, A simple transformation for nonlinear waves, *Physics Letters A*, 224(1–2), 77–84, 1996.

[5] Z. Yan, H. Zhang, New explicit and exact traveling wave solutions for a system of variant Boussinesq equations in mathematical physics, *Physics Letters A*, 252(6), 291–296, 1999.

[6] Y.D. Chong, MH2801: Complex Methods for the Sciences, Nanyang Technological University, 2016. Available from: http://www1.spms.ntu.edu.sg/~ydchong/teaching.html.

[7] M.G. Hafez, M.N. Alam, M.A. Akbar, Application of the exp (-∨(÷))-expansion Method to find exact solutions for the solitary wave equation in an unmagnatized dusty plasma, *World Applied Sciences Journal*, 32(10), 2150–2155, 2014.

[8] H.O. Roshid, M.A. Rahman, The exp($-\Phi(\eta)$)-expansion method with application in the (1+1)-dimensional classical Boussinesq equations, *Results in Physics*, 4, 150–155, 2014.

[9] Q.Y. Li et al. Nonautonomous bright and dark solitons of Bose–Einstein condensates with Feshbach-managed time-dependent scattering length, *Optics Communications*, 283, 3361–3366, 2010.

[10] Z.X. Liang, Z.D. Zhang, W.M. Liu, Dynamics of a bright soliton in Bose-Einstein condensates with time-dependent atomic scattering length in an expulsive parabolic potential, *Physical Review Letters*, 94, 050402, 2005.
[11] V.M. Pérez-García, V.V. Konotop, V.A. Brazhnyi, Feshbach resonance induced shock waves in Bose-Einstein condensates, *Physical Review Letters*, 92, 220403, 2004.
[12] H. Singh, A.M. Wazwaz, Computational method for reaction diffusion-model arising in a spherical catalyst, *International Journal of Applied and Computational Mathematics*, 7(3), 65, 2021.
[13] H. Singh, Analysis for fractional dynamics of Ebola virus model, *Chaos Solitons & Fractals*, 138, 109992, 2020.
[14] H.F. Ismael, H. Bulut, H.M. Baskonus, W shaped surfaces to the nematic liquid crystals with three nonlinearity laws, *Soft Computing*, (25), 4513–4524, 2021.
[15] H. Singh, Jacobi collocation method for the fractional advection-dispersion equation arising in porous media, *Numerical Methods for Partial Differential Equations*, 2021. DOI: 10.1002/num.22674.
[16] H. Singh, Numerical simulation for fractional delay differential equations, *International Journal of Dynamic and Control*, 2020.
[17] P. Veeresha, H.M. Baskonus, W. Gao, Strong interacting internal waves inrotating ocean: Novel fractional approach, *Axioms*, 10(2), 123, 2021.
[18] H. Singh, H.M. Srivastava, D. Kumar, A reliable algorithm for the approximate solution of the nonlinear Lane-Emden type equations arising in astrophysics, *Numerical Methods for Partial Differential Equations*, 34(5), 1524–1555, 2018.
[19] H. Singh, H.M. Srivastava, Numerical investigation of the fractional-order Liénard and duffing equations arising in oscillating circuit theory, *Frontiers in Physics*, 8, 120, 2020.
[20] H.F. Ismael, H. Bulut, H.M. Baskonus, W. Gao, Dynamicalbehaviors to the coupled Schrodinger-Boussinesq system with the beta derivative,*AIMS Mathematics*, 6(7), 7909–7928, 2021.
[21] H. Singh, J. Singh, S.D. Purohit, D. Kumar, *Advanced Numerical Methods for Differential Equations: Applications in Science and Engineering*, CRC Press Taylor and Francis, 2021.
[22] H. Singh, D. Kumar, D. Baleanu, *Methods of Mathematical Modelling: Fractional Differential Equations*, CRC Press Taylor and Francis, 2019.
[23] W. Gao, H.M. Baskonus, L. Shi, New investigation of Bats-Hosts-Reservoir-People coronavirus model and apply to 2019-nCoV system, *Advances inDifference Equations*, 2020(391), 1–11, 2020.
[24] H. Singh, Analysis of drug treatment of the fractional HIV infection model of CD4+ T-cells, *Chaos Solitons & Fractals* 146, 110868, 2021.
[25] K.K. Ali, R. Yilmazer, H.M. Baskonus, H. Bulut, Modulation instabilityanalysis and analytical solutions to the system of equations for the ion sound and Langmuirwaves, *Physica Scripta*, 95(065602), 1–10, 2020.
[26] W. Gao, H.F. Ismael, A.M. Husien, H. Bulut, H.M. Baskonus, Optical Soliton solutions of the Nonlinear Schrödinger and Resonant Nonlinear Schrödinger Equation with Parabolic Law, *Applied Science*, 10(1), 219, 1–20, 2020.
[27] W. Gao, G. Yel, H.M. Baskonus, C. Cattani, Complex solitons inthe conformable (2+1)-dimensional Ablowitz-Kaup-Newell-Segur equation, *AIMS Mathematics*, 5(1), 507–521, 2020.

[28] E.I. Eskitascioglu, M.B. Aktas, H.M. Baskonus, New complex and hyperbolic forms for Ablowitz-Kaup-Newell-Segur wave equation with fourth order, *Applied Mathematics and Nonlinear Sciences*, 4(1), 105–112, 2019.

[29] F. Dusunceli, New exact solutions for generalized (3+1) Shallow Water-Like (SWL) equation, *Applied Mathematics and Nonlinear Sciences*, 4(2), 365–370, 2019.

[30] D. Ziane, M.H. Cherif, C. Cattani, K. Belghaba, Yang-Laplace decomposition method for nonlinear system of local fractional partial differential equations, *Applied Mathematics and Nonlinear Sciences*, 4(2), 489–502, 2019.

[31] D. Arslan, The numerical study of a hybrid method for solving telegraph equation, *Applied Mathematics and Nonlinear Sciences*, 5(1), 293–302, 2020.

[32] G. Yel, T. Akturk, A new approach to (3+1) dimensional Boiti–Leon–Manna–Pempinelli equation, *Applied Mathematics and Nonlinear Sciences*, 5(1), 309–316, 2020.

[33] A. Yokus, H.M. Baskonus, T.A. Sulaiman, H. Bulut, Numerical simulation and solutions of the two-component second order KdV evolutionarysystem, *Numerical Methods for Partial Differential Equations*, 34(1), 211–227, 2018.

12

*New Fractional Integrals and Derivatives Results for the Generalized Mathieu-Type and Alternating Mathieu-Type Series**

Rakesh K. Parmar
University College of Engineering and Technology, Bikaner, India

Arjun K. Rathie
Vedant College of Engineering and Technology (Rajasthan Technical University), Bundi, India

S. D. Purohit
Rajasthan Technical University, Kota, India

CONTENTS

12.1 Introduction

Several authors have recently proved the utility of fractional calculus operators in a variety of contexts. For more than four decades, the subject of fractional calculus has grown in prominence and appeal, owing to its

* 2010 Mathematics Subject Classification: Primary 26A33, 33B20, 33C20; Secondary 26A09, 33B15, 33C05.

DOI: 10.1201/9781003263517-12

demonstrated applications in a wide range of seemingly unrelated domains of science and engineering. Fractional calculus operators of any arbitrary real or complex order, such as Marichev–Saigo–Maeda (MSM), Saigo's, Riemann–Liouville (RL), and Erdélyi–Kober (EK), are the most often exploited tools in the theory and applications of fractional integrals and derivatives [1, 2]. Various recent developments in this area can be found in [3–11].

Let $\mathbb{C}$ denote the sets of complex numbers, $\mathbb{R}^+$ be the set of positive real numbers and $\mathbb{N}$ be the set of positive integers and let $\mathbb{N}_0 := \mathbb{N} \cup \{0\}$. Let μ, μ', ξ, ξ', $\varpi \in \mathbb{C}$ and $x > 0$, then for $\Re(\varpi) > 0$, the left- and right-hand-sided Marichev–Saigo–Maeda (MSM) fractional integration operators are defined by

$$\begin{aligned}\left(\mathcal{I}_{0+}^{\mu,\mu',\xi,\xi',\varpi} f\right)(x) = \frac{x^{-\mu}}{\Gamma(\varpi)} \int_0^x (x-t)^{\varpi-1} t^{-\mu'} \\ \times F_3\left(\mu,\mu',\xi,\xi';\varpi;1-\frac{t}{x},1-\frac{x}{t}\right) f(t)\,\mathrm{d}t\end{aligned} \tag{12.1}$$

and

$$\begin{aligned}\left(\mathcal{I}_{-}^{\mu,\mu',\xi,\xi',\varpi} f\right)(x) = \frac{x^{-\mu'}}{\Gamma(\varpi)} \int_x^\infty (t-x)^{\varpi-1} t^{-\mu} \\ \times F_3\left(\mu,\mu',\xi,\xi';\varpi;1-\frac{x}{t},1-\frac{t}{x}\right) f(t)\,\mathrm{d}t,\end{aligned} \tag{12.2}$$

respectively, $\Gamma(.)$ being the Euler Gamma function [12]. The corresponding left- and right-hand-sided Marichev–Saigo–Maeda (MSM) fractional differentiation operators have the respective forms

$$\begin{aligned}\left(D_{0+}^{\mu,\mu',\xi,\xi',\varpi} f\right)(x) &= \left(\mathcal{I}_{0+}^{-\mu',-\mu,-\xi',-\xi,-\varpi} f\right)(x) \\ &= \left(\frac{d}{dx}\right)^n \left(\mathcal{I}_{0+}^{-\mu',-\mu,-\xi'+n,-\xi,-\varpi+n} f\right)(x) \quad (n=[Re(\varpi)]+1) \\ &= \frac{1}{\Gamma(n-\varpi)}\left(\frac{d}{dx}\right)^n x^{\mu'} \int_0^x (x-t)^{n-\eta-1} t^{\varrho} \\ &\quad \times F_3\left(-\mu',-\mu,n-\xi',-\xi;n-\varpi;1-\frac{t}{x},1-\frac{x}{t}\right) f(t)\,\mathrm{d}t\end{aligned} \tag{12.3}$$

and

$$\begin{aligned}\left(\mathcal{D}_{-}^{\mu,\mu',\xi,\xi',\varpi} f\right)(x) &= \left(\mathcal{I}_{-}^{-\mu',-\mu,-\xi',-\xi,-\varpi} f\right)(x) \\ &= \left(-\frac{d}{dx}\right)^n \left(\mathcal{I}_{-}^{-\mu',-\mu,-\xi',-\xi'+n,-\varpi+n} f\right)(x) \quad (n=[Re(\varpi)]+1) \\ &= \frac{1}{\Gamma(n-\varpi)}\left(-\frac{d}{dx}\right)^n x^{\mu'} \int_x^\infty (t-x)^{n-\varpi-1} t^{\varrho'} \\ &\quad \times F_3\left(-\mu',-\mu,\xi',n-\xi;n-\varpi;1-\frac{x}{t},1-\frac{t}{x}\right) f(t)\,\mathrm{d}t,\end{aligned} \tag{12.4}$$

where $[\Re(\varpi)]$ is the integral part of the $\Re(\varpi)$ and $F_3(.)$ denotes the third Appell's hypergeometric function of two variables [13]:

$$F_3[\mu_1,\mu_2,\xi_1,\xi_2;\varpi;x_1,x_2]=\sum_{m_1,m_2=0}^{\infty}\frac{(\mu)_{m_1}(\mu)_{m_2}(\xi_1)_{m_1}(\xi_2)_{m_2}}{(\varpi)_{m_1+m_2}}\frac{x_1^{m_1}}{m_1!}\frac{x_2^{m_2}}{m_2!} \quad (\max\{\Re(x_1),\Re(x_2)\}<1), \tag{12.5}$$

which reduces to Gauss' hypergeometric function [13]

$$\begin{aligned}{}_2F_1[\mu_1,\xi_1;\varpi;x_1]&=F_3[\mu_1,\mu_2,\xi_1,\xi_2;\varpi;x_1,0]\\&=F_3[\mu_1,0,\xi_1,\xi_2;\varpi;x_1,x_2]\\&=F_3[\mu_1,\mu_2,\xi_1,0;\varpi;x_1,x_2].\end{aligned} \tag{12.6}$$

These left- and right-hand-sided Marichev–Saigo–Maeda (MSM) fractional integral (12.1) and (12.2), and differentiation operators (12.3) and (12.4), reduce to the left- and right-hand-sided Saigo fractional integral and differential operators involving the hypergeometric function ${}_2F_1$, respectively, by substituting the specific values to the parameters as $\mu=\mu+\xi$, $\mu'=\xi'=0$, $\xi=-\eta$ and $\varpi=\mu$:

$$\left(\mathcal{I}_{0+}^{\mu,\xi,\eta}f\right)(x)=\frac{x^{-\mu-\xi}}{\Gamma(\mu)}\int_0^x(x-t)^{\mu-1}\,{}_2F_1\left(\mu+\xi,-\eta;\mu;1-\frac{t}{x}\right)f(t)\,dt, \tag{12.7}$$

$$\left(\mathcal{I}_{-}^{\mu,\xi,\eta}f\right)(x)=\frac{1}{\Gamma(\mu)}\int_x^\infty(t-x)^{\mu-1}t^{-\mu-\xi}\,{}_2F_1\left(\mu+\xi,-\eta;\mu;1-\frac{x}{t}\right)f(t)\,dt, \tag{12.8}$$

$$\begin{aligned}\left(\mathcal{D}_{0+}^{\mu,\xi,\eta}f\right)(x)&=\left(\mathcal{I}_{0+}^{-\mu,-\xi,\mu+\eta}f\right)(x)\\&=\left(\frac{d}{dx}\right)^n\left(\mathcal{I}_{0+}^{-\mu+n,-\xi-n,\mu+\eta-n}f\right)(x)\quad(n=[\Re(\mu)]+1),\end{aligned} \tag{12.9}$$

and

$$\begin{aligned}\left(\mathcal{D}_{-}^{\mu,\xi,\eta}f\right)(x)&=\left(\mathcal{I}_{-}^{-\mu,-\xi,\mu+\eta}f\right)(x)\\&=(-1)^n\left(\frac{d}{dx}\right)^n\left(\mathcal{I}_{-}^{-\mu+n,-\xi-n,\mu+\eta}f\right)(x)\quad(n=[\Re(\mu)]+1).\end{aligned} \tag{12.10}$$

Again these left- and right-hand-sided Saigo fractional integral (12.7) and (12.8), and differentiation operators (12.9) and (12.10) reduce to the left- and right-hand-sided Riemann–Liouville (RL) fractional integral and differential operators by setting $\xi = -\mu$ (see, e.g., [14–17]):

$$\left(\mathcal{I}_{0+}^{\mu} f\right)(x) = \frac{1}{\Gamma(\mu)} \int_0^x (x-t)^{\mu-1} f(t)\,\mathrm{d}t, \tag{12.11}$$

$$\left(\mathcal{I}_{-}^{\mu} f\right)(x) = \frac{1}{\Gamma(\mu)} \int_x^\infty (t-x)^{\mu-1} f(t)\,\mathrm{d}t, \tag{12.12}$$

and

$$\begin{aligned}\left(\mathcal{D}_{0+}^{\mu} f\right)(x) &= \left(\frac{d}{dx}\right)^n \left(\mathcal{I}_{0+}^{n-\mu} f\right)(x) \qquad (n = [\Re(\mu)]+1)\\ &= \left(\frac{d}{dx}\right)^n \frac{1}{\Gamma(n-\mu)} \int_0^x (x-t)^{n-\mu-1} f(t)\,dt\end{aligned} \tag{12.13}$$

$$\begin{aligned}\left(\mathcal{D}_{-}^{\mu} f\right)(x) &= (-1)^n \left(\frac{d}{dx}\right)^n \left(\mathcal{I}_{-}^{n-\mu} f\right)(x) \qquad (n = [\Re(\mu)]+1)\\ &= (-1)^n \left(\frac{d}{dx}\right)^n \frac{1}{\Gamma(n-\mu)} \int_x^\infty (t-x)^{n-\mu-1} f(t)\,\mathrm{d}t.\end{aligned} \tag{12.14}$$

Further, these left- and right-hand-sided Saigo fractional integral (12.7) and (12.8), and differentiation operators (12.9) and (12.10), reduce to the left- and right-hand-sided Erdélyi–Kober (EK) fractional integral and differential operators by setting $\xi = 0$ (see, e.g., [14, 16–18]):

$$\left(\mathcal{I}_{\eta,\mu}^{+} f\right)(x) = \frac{x^{-\mu-\eta}}{\Gamma(\mu)} \int_0^x (x-t)^{\mu-1} t^{\eta} f(t)\,\mathrm{d}t, \tag{12.15}$$

$$\left(\mathcal{K}_{\eta,\mu}^{-} f\right)(x) \equiv \frac{x^{\eta}}{\Gamma(\mu)} \int_x^\infty (t-x)^{\mu-1} t^{-\mu-\eta} f(t)\,\mathrm{d}t, \tag{12.16}$$

$$\begin{aligned}\left(\mathcal{D}_{\eta,\mu}^{+} f\right)(x) &= \left(\frac{d}{dx}\right)^n \left(\mathcal{I}_{0+}^{-\mu+n,-\mu,\mu+\eta-n} f\right)(x) \quad (n = [\Re(\mu)]+1)\\ &= x^{-\eta} \left(\frac{d}{dx}\right)^n \frac{1}{\Gamma(n-\mu)} \int_0^x t^{\mu+\eta} (x-t)^{n-\mu-1} f(t)\,\mathrm{d}t,\end{aligned} \tag{12.17}$$

and

$$\begin{aligned}\left(\mathcal{D}_{\eta,\mu}^{-}f\right)(x)&=(-1)^{n}\left(\frac{d}{dx}\right)^{n}\left(\mathcal{I}_{-}^{-\mu+n,-\mu,\mu+\eta}f\right)(x)\quad (n=[\Re(\mu)]+1)\\&=x^{\eta+\mu}\left(\frac{d}{dx}\right)^{n}\frac{1}{\Gamma(n-\mu)}\int_{x}^{\infty}t^{-\eta}(t-x)^{n-\mu-1}f(t)\,dt.\end{aligned}\tag{12.18}$$

In 1890, Émile Leonard Mathieu [19] investigated the functional series $S(r)$ of the form

$$S(r)=\sum_{m\geq 1}\frac{2m}{(m^{2}+r^{2})^{2}},\qquad r>0\tag{12.19}$$

popularly known as the Mathieu series. Further, Pogány et al. [20] introduced the alternating Mathieu functional series $\tilde{S}(r)$:

$$\tilde{S}(r)=\sum_{m\geq 1}(-1)^{m-1}\frac{2m}{(m^{2}+r^{2})^{2}},\qquad r>0.\tag{12.20}$$

Closed integral forms for $S(r)$ and $\tilde{S}(r)$ are presented as (see, e.g., [20, 21]):

$$S(r)=\frac{1}{r}\int_{0}^{\infty}\frac{x\sin(rx)}{e^{x}-1}dx\tag{12.21}$$

and

$$\tilde{S}(r)=\frac{1}{r}\int_{0}^{\infty}\frac{x\sin(rx)}{e^{x}+1}dx.\tag{12.22}$$

Moreover by making use of fractional power λ in (12.19) and (12.20), the generalized Mathieu-type functional series $S_{\lambda}(r)$ and its alternating version $\tilde{S}_{\lambda}(r)$ studied in [22, p. 2, Eq. (12.6)] (see also [23, p. 181]) are defined by:

$$S_{\lambda}(r)=\sum_{m\geq 1}\frac{2m}{(m^{2}+r^{2})^{\lambda}}\qquad (r>0,\lambda>1),\tag{12.23}$$

and

$$\tilde{S}_{\lambda}(r)=\sum_{m\geq 1}(-1)^{m-1}\frac{2m}{(m^{2}+r^{2})^{\lambda}}\qquad (r>0,\lambda>1).\tag{12.24}$$

In the mathematical literature, these series have received a good deal of attention (see [20, 22, 24]).

Recently, Saxena et al. [25] studied several integral transforms, in particular Mellin, Laplace, Euler, and Hankel transforms for the generalized fractional-order Mathieu-type functional series

$$S_\lambda(r;z) = \sum_{m\geq 1} \frac{2mz^{m-1}}{(m^2+r^2)^\lambda}, \qquad (\lambda > 1, r \in \mathbb{R}, |z| < 1), \tag{12.25}$$

and written inform of a Mellin–Barnes-type contour integral for $|\arg(-z)| < \pi$ as follows [25]:

$$S_\lambda(r;z) = \frac{1}{\pi i}\int_C \frac{\Gamma(s)\Gamma(2-s)\{\Gamma(1\pm ir-s)\}^\lambda}{\{\Gamma(2\pm ir-s)\}^\lambda}(-z)^{-s}, \tag{12.26}$$

where $C = C_{(i\kappa;\infty)}$ is the contour of integration (loop located in a horizontal strip) beginning from the point $\kappa - i\infty$ and ending at the point $\kappa + i\infty$, where $\kappa \in \mathbb{R} = (+\infty, -\infty)$ such that all the poles of the Gamma function $\Gamma(2 - s)$ are separated from the poles of the Gamma function $\Gamma(s)$ with the usual indentations and assuming that the poles of the integrand are simple and integral converges.

They also studied alternating generalized Mathieu-type functional series as:

$$\tilde{S}_\lambda(r;z) = \sum_{m\geq 1}(-1)^{m-1} \frac{2mz^{m-1}}{(m^2+r^2)^\lambda}, \qquad (\lambda > 1, r \in \mathbb{R}, |z| < 1), \tag{12.27}$$

and in the form of Mellin–Barnes-type contour integral

$$\tilde{S}_\lambda(r;z) = \frac{1}{\pi i}\int_C \frac{\Gamma(s)\Gamma(2-s)\{\Gamma(1\pm ir-s)\}^\lambda}{\{\Gamma(2\pm ir-s)\}^\lambda} z^{-s} \quad |\arg(z)| < \pi. \tag{12.28}$$

Several other investigations, extensions and generalizations of the Mathieu series with its alternating variants can be found in [23, 26–33]. More recently a new theory for multi-parameter Mathieu-type and alternating Mathieu-type functional series has been studied by Parmar et al. [34].

Our aim in this chapter is to derive fractional calculus results for the generalized Mathieu-type and alternating Mathieu-type functional series by making extensive use of Marichev–Saigo–Maeda operator tools (12.1), (12.2), (12.3), and (12.4) in terms of $\bar{H}$-function. Also, particular cases for Saigo's,

Riemann–Liouville (RL) and Erdélyi–Kober (EK) fractional integral and differentiation operators are established. Moreover, from the special case and application point of view, all the results are also deduced in terms of Fox's H-function [35]. Furthermore, we also observe that all the results derived here can also be represented in terms of I-function, and could be potentially useful in the areas of mathematics for engineering and mathematical physics.

12.2 Results of Fractional Integration of the Mathieu Series in Terms of $\bar{H}$-Function

In 1987, Inayat-Hussain [36] investigated a number of interesting properties and characteristics of the hypergeometric functions of several variables in an attempt to evaluate in two different ways some Feynman-type integrals that occur in perturbation calculations of the equilibrium properties of a magnetic model of phase transitions. More importantly, while demonstrating the usage of these Feynman-type integrals, Inayat-Hussain [37] was led to a unique generalization of Charles Fox's well-known H-function (1897–1977) [35]. The polylogarithm of a complex order and the precise partition function of the Gaussian model in statistical mechanics are examples of particular cases of this novel $\bar{H}$-function of Inayat-Hussain [37]. Indeed, it is defined as follows in terms of a Mellin–Barnes-type contour integral (see for details [38, 39]):

$$\overline{H}_{p,q}^{m,n}[z] = \overline{H}_{p,q}^{m,n}\left[z \left| \begin{matrix} (e_r, E_r; \mu_j)_{\overline{1,n}}, \cdots, (e_r, E_r)_{\overline{n+1,p}} \\ (f_r, F_r)_{\overline{1,m}}, \cdots, (f_r, F_r; \xi_j)_{\overline{m+1,q}} \end{matrix} \right. \right] = \frac{1}{2\pi i} \int_{£} \chi_{p,q}^{m,n}(s)\, z^s\, ds \qquad (\forall z, z \neq 0) \tag{12.29}$$

where the Mellin–Barnes-type integral, taken over the path $£ = £_{(i\kappa;\infty)}$ beginning from $\kappa - i\infty$ and ending at the point $\kappa + i\infty$ ($\kappa \in \mathbb{R}$), with suitable indentations to avoid poles of the factors $\Gamma(f_r - F_r s)$ from poles of the factors $\{\Gamma(1 - e_r + E_r s)\}\mu^r$ and $\chi_{p,q}^{m,n}(s)$ is given by

$$\chi_{p,q}^{m,n}(s) = \frac{E(s)\bar{F}(s)}{E'(s)\bar{F}'(s)}, \tag{12.30}$$

$$E(s) = \prod_{r=1}^{m} \Gamma(f_r - F_r s), \quad \bar{F}(s) = \prod_{r=1}^{n} \{\Gamma(1 - e_r + E_r s)\}^{\mu_r}, \tag{12.31}$$

$$\overline{E}'(s)=\prod_{r=m+1}^{q}\{\Gamma(1-f_r+F_rs)\}^{\xi_r},\ \ F'(s)=\prod_{r=n+1}^{p}\Gamma(e_r-E_rs), \tag{12.32}$$

with $e_r \in \mathbb{C}$ $(r=1,\cdots,p)$, $f_r \in \mathbb{C}$ $(r=1,\cdots,q)$, $E_r \in \mathbb{R}^+$ $(r=1,\cdots,p)$ and $F_r \in \mathbb{R}^+$ $(r=1,\cdots,q)$, and the exponents μ_r $(r=1,\cdots,n)$ and ξ_r $(r=m+1,\cdots,q)$ can assume non-integer values. The conditions for the absolute convergence of the Mellin–Barnes-type contour integral which defines analytic function for $|\arg(z)|<\frac{\pi}{2}\wedge$ in (12.29) is given by Buschman and Srivastava [38, p. 4708], where

$$\wedge=\sum_{r=1}^{m}F_r+\sum_{r=1}^{n}|\mu_j|E_r-\sum_{r=m+1}^{q}|\xi_r|F_r-\sum_{r=n+1}^{p}E_r>0.$$

Before starting our main results, we provide image formulas of the power function $t^{\varrho-1}$ for the fractional calculus operators

$$\mathcal{I}_{0+}^{\mu,\mu',\xi,\xi',\varpi},\ \mathcal{I}_{-}^{\mu,\mu',\xi,\xi',\varpi},$$
$$\mathcal{I}_{0+}^{\mu,\xi,\eta},\ \mathcal{I}_{-}^{\mu,\xi,\eta},$$
$$I_{0+}^{\mu},\ I_{-}^{\mu},\ \mathcal{I}_{\eta,\mu}^{+},\ \mathcal{K}_{\eta,\mu}^{-},$$

which will be useful in deriving our main results.

Lemma 2.1 *Let $\varrho, \mu, \mu', \xi, \xi', \varpi, \varrho \in \mathbb{C}$* and $x>0$. Then there is the following relationship:

(a) *If ¾$(\varpi)>0$ and* $\Re(\varrho)>\max\{0,\Re(\mu+\mu'+\xi-\varpi),\Re(\mu'-\xi')\}$, then

$$\left(\mathcal{I}_{0+}^{\mu,\mu',\xi,\xi',\varpi}t^{\varrho-1}\right)(x)=\frac{\Gamma(\varrho)\Gamma(\varrho+\varpi-\mu-\mu'-\xi)\Gamma(\varrho+\xi'-\mu')}{\Gamma(\varrho+\xi')\Gamma(\varrho+\varpi-\mu-\mu')\Gamma(\varrho+\varpi-\mu'-\xi)}x^{\varrho+\varpi-\mu-\mu'-1} \tag{12.33}$$

(b) If $\Re(\varpi)>0$ and $\Re(\varrho)<1+\min\{\Re(-\xi),\Re(\mu+\mu'-\varpi),\Re(\mu+\xi'-\varpi)\}$, then

$$\left(\mathcal{I}_{-}^{\mu,\mu',\xi,\xi',\varpi}t^{\varrho-1}\right)(x)=\frac{\Gamma(1-\varrho-\xi)\Gamma(1-\varrho-\varpi+\mu+\mu')\Gamma(1-\varrho-\varpi+\mu+\xi')}{\Gamma(1-\varrho)\Gamma(1-\varrho-\varpi+\mu+\mu'+\xi')\Gamma(1-\varrho+\mu-\xi)}x^{\varrho+\varpi-\mu-\mu'-1}. \tag{12.34}$$

Lemma 2.2 Let ϱ, μ, ξ, η, $\varrho \in \mathbb{C}$ and $x > 0$. Then there is the following relationship:

(a) If $\Re(\mu) > 0$ and $\Re(\varrho) > \max[0, \Re(\xi - \eta)]$, then

$$(\mathcal{I}_{0+}^{\mu,\xi,\eta} t^{\varrho-1})(x) = \frac{\Gamma(\varrho)\Gamma(\varrho+\eta-\xi)}{\Gamma(\varrho-\xi)\Gamma(\varrho+\mu+\eta)} x^{\varrho-\xi-1} \tag{12.35}$$

(b) If $\Re(\mu) > 0$ and $\Re(\varrho) < 1 + \min[\Re(\xi), \Re(\eta)]$, then

$$(\mathcal{I}_{-}^{\mu,\xi,\eta} t^{\varrho-1})(x) = \frac{\Gamma(1-\varrho+\xi)\Gamma(1-\varrho+\eta)}{\Gamma(1-\varrho)\Gamma(1-\varrho+\mu+\xi+\eta)} x^{\varrho-\xi-1}. \tag{12.36}$$

Lemma 2.3 Let $\varrho, \mu, \xi, \varrho \in \mathbb{C}$ and $x > 0$. Then there is the following relationship:

(a) If $\Re(\mu) > 0$ and $\Re(\varrho) > 0$, then

$$(\mathcal{I}_{0+}^{\mu} t^{\varrho-1})(x) = \frac{\Gamma(\varrho)}{\Gamma(\varrho+\mu)} x^{\varrho+\mu-1}. \tag{12.37}$$

(b) If $0 < \Re(\mu) < 1 - \Re(\varrho)$, then

$$(\mathcal{I}_{-}^{\mu} t^{\varrho-1})(x) = \frac{\Gamma(1-\mu-\varrho)}{\Gamma(1-\varrho)} x^{\varrho+\mu-1}. \tag{12.38}$$

Lemma 2.4 Let $\varrho, \mu, \eta, \varrho \in \mathbb{C}$ and $x > 0$. Then there is the following relationship:

(a) If $\Re(\mu) > 0$, $\Re(\varrho) > -\Re(\eta)$, then

$$(\mathcal{I}_{\eta,\mu}^{+} t^{\rho-1})(x) = \frac{\Gamma(\varrho+\eta)}{\Gamma(\varrho+\mu+\eta)} x^{\varrho-1}. \tag{12.39}$$

(b) If $\Re(\varrho) < 1 + \Re(\varrho)$, then

$$(\mathcal{K}_{\eta,\mu}^{-} t^{\varrho-1})(x) = \frac{\Gamma(1-\varrho+\eta)}{\Gamma(1-\varrho+\mu+\eta)} x^{\varrho-1}. \tag{12.40}$$

In this section, we present two composition formulas for Marichev–Saigo–Maeda (MSM) fractional integration operators (12.1) and (12.2) involving the generalized Mathieu-type functional series $S_\lambda(r,z)$ (12.26) in terms of the $\bar{H}$-function (12.29).

Theorem 2.1 Let $\lambda - 1$, $r \in \mathbb{R}^+$, $\varrho > 0$ and let the contour $\mathcal{C}$ be taken as in (12.28). Let $\varrho, \mu, \mu', \xi, \xi', \eta, \varpi \in \mathbb{C}$ and satisfy the conditions $\Re(\varpi) > 0$ and $\Re(\varrho) + \min\{0, \Re(\varpi - \mu - \mu' - \xi), \Re(-\xi' - \mu')\} > 0$. Then the following assertion for the MSM fractional operator $\mathcal{I}_{0+}^{\mu,\mu',\xi,\xi',\varpi}$ of the $S_\lambda(r,t^\varrho)$ (12.26) exists and holds true:

$$\left(\mathcal{I}_{0+}^{\mu,\mu',\xi,\xi',\varpi}\left\{t^{\varrho-1}S_\lambda(r,t^\rho)\right\}\right)(x) = 2x^{\varrho+\varpi-\mu-\mu'-1}$$
$$\times \bar{H}_{6,6}^{1,6}\left[-x^\rho \left| \begin{matrix} (1-\varrho,\rho;1),(1-\varrho-\varpi+\mu+\mu'+\xi,\rho;1),(1-\varrho-\xi'+\mu',\rho;1),(-1,1;1),(\pm ir,1;\lambda) \\ (0,1),(\pm ir-1,1;\lambda),(1-\varrho-\xi',\rho;1),(1-\rho-\varpi+\mu+\mu',\rho;1),(1-\varrho-\varpi+\mu'+\xi,\rho;1) \end{matrix} \right.\right]. \tag{12.41}$$

Proof. By making use of the operator (12.1) and expressing the definition of the generalized Mathieu-type functional series $S_\lambda(r,z)$ in the form of the contour integral representation (12.26) and then interchanging the order of integration and applying the relation (12.33), we have

$$\left(\mathcal{I}_{0+}^{\mu,\mu',\xi,\xi',\varpi}\left\{t^{\varrho-1}S_\lambda(r,t^\rho)\right\}\right)(x)$$
$$= \frac{1}{\pi i}\int_{\mathcal{C}} \frac{\Gamma(s)\Gamma(2-s)\{\Gamma(1\pm ir-s)\}^\lambda(-1)^{-\rho s}}{\{\Gamma(2\pm ir-s)\}^\lambda}\left(\mathcal{I}_{0+}^{\mu,\mu',\xi,\xi',\varpi}\, t^{\varrho-\rho s-1}\right)(x)\, ds$$
$$= x^{\varrho+\varpi-\mu-\mu'-1}\frac{1}{\pi i}\int_{\mathcal{C}} \frac{\Gamma(s)\Gamma(2-s)\{\Gamma(1\pm ir-s)\}^\lambda}{\{\Gamma(2\pm ir-s)\}^\lambda}$$
$$\times \frac{\Gamma(\varrho-\rho s)\Gamma(\varrho+\varpi-\mu-\mu'-\xi-\rho s)\Gamma(\varrho+\xi'-\mu'-\rho s)}{\Gamma(\varrho+\xi'-\rho s)\Gamma(\varrho+\varpi-\mu-\mu'-\rho s)\Gamma(\varrho+\varpi-\mu'-\xi-\rho s)}(-x)^{-\rho s}\, ds$$

which, interpreted with the help of the definition (12.29), yields the required result (12.41).

Theorem 2.2 Let $\lambda - 1$, $r \in \mathbb{R}^+$, $\varrho > 0$ and let the contour $\mathcal{C}$ be taken as in (12.28). Let $\varrho, \mu, \mu', \xi, \xi', \eta, \varpi \in \mathbb{C}$ and satisfy the conditions $\Re(\varpi) > 0$ and $\Re(\varrho) < 1 + \min\{\Re(-\xi), \Re(\mu + \mu' - \varpi), \Re(\mu + \xi' - \varpi)\}$. Then the following assertion for the MSM fractional operator $\mathcal{I}_{-}^{\mu,\mu',\xi,\xi',\varpi}$ of the $S_\lambda(r,t^\varrho)$ (12.26) exists and holds true:

$$\left(\mathcal{I}_{-}^{\mu,\mu',\xi,\xi',\varpi}\left\{t^{\varrho-1}S_\lambda(r,t^\rho)\right\}\right)(x) = 2x^{\varrho+\varpi-\mu-\mu'-1}$$
$$\times \bar{H}_{6,6}^{4,3}\left[-x^\rho \left| \begin{matrix} (-1,1;1),(\pm ir,1;\lambda),(1-\varrho,\rho),(1-\varrho-\varpi+\mu+\mu'+\xi',\rho),(1-\varrho+\mu-\xi,\rho) \\ (0,1),(1-\varrho-\xi,\rho),(1-\varrho-\varpi+\mu+\mu',\rho),(1-\varrho-\varpi+\mu+\xi',\rho),(\pm ir-1,1;\lambda) \end{matrix} \right.\right]. \tag{12.42}$$

Proof. By making use of the operator (12.2) and expressing the definition of the generalized Mathieu-type functional series $S_\lambda(r,z)$ in the form of the contour integral representation (12.26) and then changing the order of integrals and applying the relation (12.34), we have

$$\begin{aligned}
&\left(\mathcal{I}_{-}^{\mu,\mu',\xi,\xi',\varpi}\left\{t^{\rho-1}S_\lambda(r,t^\rho)\right\}\right)(x)\\
&=\frac{1}{\pi i}\int_C \frac{\Gamma(s)\Gamma(2-s)\{\Gamma(1\pm ir-s)\}^\lambda(-1)^{-\rho s}}{\{\Gamma(2\pm ir-s)\}^\lambda}\left(\mathcal{I}_{-}^{\mu,\mu',\xi,\xi',\varpi}t^{\varrho-\rho s-1}\right)(x)\,ds\\
&=x^{\varrho+\varpi-\mu-\mu'-1}\frac{1}{\pi i}\int_C \frac{\Gamma(s)\Gamma(2-s)\{\Gamma(1\pm ir-s)\}^\lambda}{\{\Gamma(2\pm ir-s)\}^\lambda}\\
&\times\frac{\Gamma(1-\varrho-\xi+\rho s)\Gamma(1-\varrho-\varpi+\mu+\mu'+\rho s)\Gamma(1-\varrho-\varpi+\mu+\xi'+\rho s)}{\Gamma(1-\varrho+\rho s)\Gamma(1-\varrho-\varpi+\mu+\mu'+\xi'+\rho s)\Gamma(1-\varrho+\mu-\xi+\rho s)}(-x)^{-\rho s}\,ds
\end{aligned}$$

which, interpreted with the help of the definition (12.29), yields the required result (12.42).

We also deduce the fractional integral formulas for the Saigo's by substituting the specific values to the parameters as $\mu = \mu + \xi$, $\mu' = \xi' = 0$, $\xi = -\eta$ and $\varpi = \mu$, which are asserted by Corollary 2.1 and Corollary 2.2 below.

Corollary 2.1.

Let $\lambda - 1$, $r \in \mathbb{R}^+$, $\varrho > 0$ and let the contour C be taken as in (12.28). Let ϱ, μ, ξ, η, $\varpi \in \mathbb{C}$ and satisfy the conditions $\Re(\mu) > 0$ and $\Re(\varrho) + \min[0, \Re(\eta - \xi)] > 0$. Then the following assertion for the Saigo's fractional operator $\mathcal{I}_{0+}^{\mu,\xi,\eta}$ of the $S_\lambda(r,t^\rho)$ (12.26) exists and holds true:

$$\begin{aligned}
&\left(\mathcal{I}_{0+}^{\mu,\xi,\eta}\left\{t^{\varrho-1}S_\lambda(r,t^\rho)\right\}\right)(x)\\
&=2x^{\varrho-\xi-1}\overline{H}_{5,5}^{1,5}\left[-x^\rho\left|\begin{matrix}(1-\varrho,\rho;1),(1-\varrho+\xi-\eta,\rho;1),(-1,1;1),(\pm ir,1;\lambda)\\(0,1),(\pm ir-1,1;\lambda),(1-\varrho+\xi,\rho;1),(1-\varrho-\mu-\eta,\rho;1)\end{matrix}\right.\right].
\end{aligned}\tag{12.43}$$

Corollary 2.2. Let $\lambda - 1$, $r \in \mathbb{R}^+$, $\varrho > 0$ and let the contour C be taken as in (12.28). Let ϱ, μ, ξ, η, $\varpi \in \mathbb{C}$ and satisfy the conditions $\Re(\mu) > 0$ and $\Re(\varrho) < 1 + \min[\Re(\xi), \Re(\eta)]$. Then the following assertion for the Saigo's fractional operator $\mathcal{I}_{-}^{\mu,\xi,\eta}$ of the $S_\lambda(r,t^\rho)$ (12.26) exists and holds true:

$$\begin{aligned}
&\left(\mathcal{I}_{-}^{\mu,\xi,\eta}\left\{t^{\varrho-1}S_\lambda(r,t^\rho)\right\}\right)(x)\\
&=2x^{\varrho-\xi-1}\overline{H}_{5,5}^{3,3}\left[-x^\rho\left|\begin{matrix}(-1,1;1),(\pm ir,1;\lambda),(1-\varrho,\rho),(1-\varrho+\mu+\xi+\eta,\rho)\\(0,1),(1-\varrho+\xi,\rho),(1-\varrho+\eta,\rho),(\pm ir-1,1;\lambda)\end{matrix}\right.\right].
\end{aligned}\tag{12.44}$$

If we let $\xi = -\mu$ and $\xi = 0$ respectively, in Corollary 2.1 and Corollary 2.2, we deduce the following fractional integral formulas for the Riemann–Liouville (RL) and Erdélyi–Kober (EK), which are asserted by Corollary 2.3, Corollary 2.4, Corollary 2.5, and Corollary 2.6.

Corollary 2.3. Let $\lambda - 1$, $r \in \mathbb{R}^+$, $\rho > 0$ and let the contour C be taken as in (12.28). Let $\varrho, \mu, \in \mathbb{C}$ and satisfy the conditions $\Re(\mu) > 0$ and $\Re(\varrho) + \min[0, \Re(\eta)] > 0$. Then the following assertion for the RL fractional operator $\mathcal{I}_{0+}^{\mu}$ of the $S_\lambda(r, t^\rho)$ (12.26) exists and holds true:

$$\left(\mathcal{I}_{0+}^{\mu}\left\{t^{\varrho-1}S_\lambda(r,t^\rho)\right\}\right)(x) = 2x^{\varrho+\mu-1}\overline{H}_{4,4}^{1,4}\left[-x^\rho \left| \begin{matrix} (1-\varrho,\rho;1),(-1,1;1),(\pm ir,1;\lambda) \\ (0,1),(\pm ir-1,1;\lambda),(1-\varrho-\mu,\rho;1) \end{matrix} \right.\right]. \tag{12.45}$$

Corollary 2.4. Let $\lambda - 1$, $r \in \mathbb{R}^+$, $\rho > 0$ and let the contour C be taken as in (12.28). Let $\varrho, \mu, \in \mathbb{C}$ and satisfy the conditions $0 < \Re(\mu) < 1 - \Re(\varrho)$. Then the following assertion for the RL fractional operator $\mathcal{I}_{-}^{\mu}$ of the $S_\lambda(r, t^\rho)$ (12.26) exists and holds true:

$$\left(\mathcal{I}_{-}^{\mu}\left\{t^{\varrho-1}S_\lambda(r,t^\rho)\right\}\right)(x) = 2x^{\varrho+\mu-1}\overline{H}_{4,4}^{2,3}\left[-x^\rho \left| \begin{matrix} (-1,1;1),(\pm ir,1;\lambda),(1-\varrho,\rho) \\ (0,1),(1-\varrho-\mu,\rho),(\pm ir-1,1;\lambda) \end{matrix} \right.\right]. \tag{12.46}$$

Corollary 2.5. Let $\lambda - 1$, $r \in \mathbb{R}^+$, $\rho > 0$ and let the contour C be taken as in (12.28). Let $\varrho, \mu, \eta, \varpi \in \mathbb{C}$ and satisfy the conditions $\Re(\mu) > 0$ and $\Re(\varrho) + \min[0, \Re(\eta)] > 0$. Then the following assertion for the EK fractional operator $\mathcal{I}_{\eta,\mu}^{+}$ of the $S_\lambda(r, t^\rho)$ (12.26) exists and holds true:

$$\left(\mathcal{I}_{\eta,\mu}^{+}\left\{t^{\varrho-1}S_\lambda(r,t^\rho)\right\}\right)(x) = 2x^{\varrho-1}\overline{H}_{4,4}^{1,4}\left[-x^\rho \left| \begin{matrix} (1-\varrho-\eta,\rho;1),(-1,1;1),(\pm ir,1;\lambda) \\ (0,1),(\pm ir-1,1;\lambda),(1-\varrho-\mu-\eta,\rho;1) \end{matrix} \right.\right]. \tag{12.47}$$

Corollary 2.6. Let $\lambda - 1$, $r \in \mathbb{R}^+$, $\rho > 0$ and let the contour C be taken as in (12.28). Let $\varrho, \mu, \eta, \varpi \in \mathbb{C}$ and satisfy the conditions $\Re(\mu) > 0$ and $\Re(\varrho) < 1 + \Re(\eta)$. Then the following assertion for the EK fractional operator $\mathcal{K}_{\eta,\mu}^{-}$ of the $S_\lambda(r, t^\rho)$ (12.26) exists and holds true:

$$\left(\mathcal{K}_{\eta,\mu}^{-}\left\{t^{\varrho-1}S_\lambda(r,t^\rho)\right\}\right)(x) = 2x^{\varrho-1}\overline{H}_{4,4}^{2,3}\left[-x^\rho \left| \begin{matrix} (-1,1;1),(\pm ir,1;\lambda),(1-\varrho+\mu+\eta,\rho) \\ (0,1),(1-\varrho+\eta,\rho),(\pm ir-1,1;\lambda) \end{matrix} \right.\right]. \tag{12.48}$$

12.3 Results of Fractional Differentiation of the Mathieu Series in Terms of $\bar{H}$-Function

In this section, we provide image formulas of the power function $t^{\rho-1}$ for the fractional calculus operators

$$\mathcal{D}_{0+}^{\mu,\mu',\xi,\xi',\varpi},\ \mathcal{D}_{-}^{\mu,\mu',\xi,\xi',\varpi},$$
$$\mathcal{D}_{0+}^{\mu,\xi,\eta},\ \mathcal{D}_{-}^{\mu,\xi,\eta},$$
$$\mathcal{D}_{0+}^{\mu},\ \mathcal{D}_{-}^{\mu},\ \mathcal{D}_{\eta,\mu}^{+},\ D_{\eta,\mu}^{-},$$

which will be useful in deriving our main results.

Lemma 3.1 Let $\varrho, \mu, \mu', \xi, \xi', \varpi, \varrho \in \mathbb{C}$ and $x > 0$. Then there is the following relationship:

(a) If $\Re(\varpi) > 0$ and $\Re(\varrho) > \max\{0, \Re(\varpi - \mu - \mu' + \xi'), \Re(\xi - \mu)\}$, then

$$\left(\mathcal{D}_{0+}^{\mu,\mu',\xi,\xi',\varpi} t^{\varrho-1}\right)(x) = \frac{\Gamma(\varrho)\Gamma(\varrho-\varpi+\mu+\mu'+\xi')\Gamma(\varrho-\xi+\mu)}{\Gamma(\varrho-\xi)\Gamma(\varrho-\varpi+\mu+\mu')\Gamma(\varrho-\varpi+\mu+\xi')} x^{\varrho-\varpi+\mu+\mu'-1} \quad (12.49)$$

(b) If $\Re(\varpi) > 0$ and $\Re(\varrho) < 1 + \min\{\Re(\xi'), \Re(\varpi - \mu - \mu', \Re(\eta - \mu' - \xi)\}$, then

$$\left(\mathcal{D}_{-}^{\mu,\mu',\xi,\xi',\varpi} t^{\varrho-1}\right)(x) = \frac{\Gamma(1-\varrho-\xi')\Gamma(1-\varrho+\varpi-\mu-\mu')\Gamma(1-\varrho+\varpi-\mu'-\xi)}{\Gamma(1-\varrho)\Gamma(1-\varrho+\varpi-\mu-\mu'-v)\Gamma(1-\varrho-\mu'-\xi')} x^{\varrho-\varpi+\mu+\mu'-1}. \quad (12.50)$$

Lemma 3.2 Let $\varrho, \mu, \xi, \eta \in \mathbb{C}$ and $x > 0$. Then there is the following relationship:

(a) If $\Re(\mu) > 0$ and $\Re(\varrho) > -\min[0, \Re(\mu + \xi + \eta)]$, then

$$(\mathcal{D}_{0+}^{\mu,\xi,\eta} t^{\varrho-1})(x) = \frac{\Gamma(\varrho)\Gamma(\varrho+\mu+\xi+\eta)}{\Gamma(\varrho+\xi)\Gamma(\varrho+\eta)} x^{\varrho+\xi-1} \quad (12.51)$$

(b) If $\Re(\mu) > 0$, $\Re(\varrho) < 1 + \min[\Re(-\xi - n), \Re(\mu + \eta)]$ and $n = [\Re(\mu)] + 1$, then

$$(\mathcal{D}_{-}^{\mu,\xi,\eta} t^{\varrho-1})(x) = \frac{\Gamma(1-\varrho-\xi)\Gamma(1-\varrho+\mu+\eta)}{\Gamma(1-\varrho)\Gamma(1-\varrho+\eta-\xi)} x^{\varrho+\xi-1}. \quad (12.52)$$

Lemma 3.3 Let $\varrho, \mu \in \mathbb{C}$ and $x > 0$. Then there is the following relationship:

(a) If $\mathfrak{R}(\mu) > 0$ and $\mathfrak{R}(\varrho) > 0$, then

$$(\mathcal{D}_{0+}^{\mu} t^{\varrho-1})(x) = \frac{\Gamma(\varrho)}{\Gamma(\varrho-\mu)} x^{\varrho-\mu-1}, \tag{12.53}$$

(b) If $\mathfrak{R}(\mu) > 0$, $\mathfrak{R}(\varrho) < 1 + \mathfrak{R}(\mu) - n)$ and $n = [\mathfrak{R}(\mu)] + 1$, then

$$(\mathcal{D}_{-}^{\mu} t^{\varrho-1})(x) = \frac{\Gamma(1-\varrho+\mu)}{\Gamma(1-\varrho)} x^{\varrho-\mu-1}, \tag{12.54}$$

Lemma 3.4 Let $\varrho, \mu, \eta \in \mathbb{C}$ and $x > 0$. Then there is the following relationship:

(a) If $\mathfrak{R}(\mu) > 0$ and $\mathfrak{R}(\varrho) > -\mathfrak{R}(\mu + \eta)$, then

$$(\mathcal{D}_{\eta,\mu}^{+} t^{\varrho-1})(x) = \frac{\Gamma(\varrho+\mu+\eta)}{\Gamma(\varrho+\eta)} x^{\varrho-1}. \tag{12.55}$$

(b) If $\mathfrak{R}(\mu) > 0$, $\mathfrak{R}(\varrho) < 1 + \mathfrak{R}(\mu + \eta) - n$ and $n = [\mathfrak{R}(\mu)] + 1$, then

$$(\mathcal{D}_{\eta,\mu}^{-} t^{\varrho-1})(x) = \frac{\Gamma(1-\varrho+\mu+\eta)}{\Gamma(1-\varrho-\eta)} x^{\varrho-1}. \tag{12.56}$$

In this section, we present two composition formulas for Marichev–Saigo–Maeda (MSM) fractional differentiation operators (12.3) and (12.4) involving the generalized Mathieu-type functional series $S_\lambda(r,z)$ (12.26) in terms of the $\overline{H}$-function (12.29).

Theorem 3.1 Let $\lambda - 1$, $r \in \mathbb{R}^+$, $\varrho > 0$ and let the contour C be taken as in (12.28). Let $\varrho, \mu, \mu', \xi, \xi', \eta, \varpi \in \mathbb{C}$ and satisfy the conditions $\mathfrak{R}(\varpi) > 0$ and $\mathfrak{R}(\varrho) > \max\{0, \mathfrak{R}(\varpi - \mu - \mu' + \xi'), \mathfrak{R}(\xi - \mu)\}$. Then the following assertion for the MSM fractional operator $\mathcal{D}_{0+}^{\mu,\mu',\xi,\xi',\varpi}$ of the $S_\lambda(r,t^\rho)$ (12.26) exists and holds true:

$$\left(\mathcal{D}_{0+}^{\mu,\mu',\xi,\xi',\varpi}\left\{t^{\varrho-1} S_\lambda(r,t^\rho)\right\}\right)(x) = 2x^{\varrho-\xi-1}$$
$$\times \overline{H}_{6,6}^{1,6}\left[-x^\rho \left| \begin{matrix} (1-\varrho,\rho;1),(1-\varrho+\varpi-\mu-\mu'-\xi',\rho;1),(1-\varrho+\xi-\mu,\rho;1),(-1,1;1),(\pm ir,1;\lambda) \\ (0,1),(\pm ir-1,1;\lambda),(1-\varrho+\xi,\rho;1),(1-\varrho+\varpi-\mu-\mu',\rho;1),(1-\varrho+\varpi-\mu-\xi',\rho;1) \end{matrix} \right.\right]. \tag{12.57}$$

Proof. By making use of the operator (12.3) and expressing the definition of the generalized Mathieu-type functional series $S_\lambda(r,z)$ in the form of the contour integral representation (12.26) and then changing the order of integrals and applying the relation (12.49), we have for $x > 0$

$$\begin{aligned}
&\left(\mathcal{D}_{0+}^{\mu,\mu',\xi,\xi',\varpi}\left\{t^{\varrho-1}S_\lambda(r,t^\rho)\right\}\right)(x) \\
&= \frac{1}{\pi i}\int_C \frac{\Gamma(s)\Gamma(2-s)\{\Gamma(1\pm ir-s)\}^\lambda(-1)^{-\rho s}}{\{\Gamma(2\pm ir-s)\}^\lambda}\left(D_{0+}^{\mu,\mu',\xi,\xi',\varpi}t^{\varrho-\rho s-1}\right)(x)\,ds \\
&= x^{\varrho-\xi-1}\frac{1}{\pi i}\int_C \frac{\Gamma(s)\Gamma(2-s)\{\Gamma(1\pm ir-s)\}^\lambda}{\{\Gamma(2\pm ir-s)\}^\lambda} \\
&\quad\times\frac{\Gamma(\varrho-\rho s)\Gamma(\varrho-\varpi+\mu+\mu'+\xi'-\rho s)\Gamma(\varrho-\xi+\mu-\rho s)}{\Gamma(\varrho-\xi-\rho s)\Gamma(\varrho-\varpi+\mu+\mu'-\rho s)\Gamma(\varrho-\varpi+\mu+\xi'-\rho s)}(-x)^{-\rho s}\,ds
\end{aligned}$$

which, interpreted with the help of the definition (12.29), yields the required derivative formula (12.57).

Theorem 3.2 Let $\lambda - 1$, $r \in \mathbb{R}^+$, $\varrho > 0$ and let the contour $\mathcal{C}$ be taken as in (12.28). Let $\varrho, \mu, \mu', \xi, \xi', \eta, \varpi \in \mathbb{C}$ and satisfy the conditions $\Re(\varpi) > 0$ and $\Re(\varrho) < 1 + \min\{\Re(\xi'), \Re(\varpi-\mu-\mu', \Re(\eta-\mu'-\xi)\}$. Then the following assertion for the MSM fractional operator $\mathcal{D}_{-}^{\mu,\mu',\xi,\xi',\varpi}$ of the $S_\lambda(r,t^\rho)$ (12.26) exists and holds true:

$$\begin{aligned}
&\left(\mathcal{D}_{-}^{\mu,\xi,\eta}\left\{t^{\varrho-1}S_\lambda(r,t^\rho)\right\}\right)(x) = 2x^{\varrho+\xi-1} \\
&\times\overline{H}_{6,6}^{4,3}\left[-x^\rho\left|\begin{matrix}(-1,1;1),(\pm ir,1;\lambda),(1-\varrho,\rho),(1-\varrho+\varpi-\mu-\mu'-v,\rho),(1-\varrho-\mu'-\xi',\rho)\\(0,1),(1-\varrho-\xi',\rho),(1-\varrho+\varpi-\mu-\mu',\rho),(1-\varrho+\varpi-\mu'-\xi,\rho),(\pm ir-1,1;\lambda)\end{matrix}\right.\right].
\end{aligned}\tag{12.58}$$

Proof. By making use of the operator (12.4) and expressing the definition of the generalized Mathieu-type functional series $S_\lambda(r,z)$ in the form of the contour integral representation (12.26) and then changing the order of integrals and applying the relation (12.50), we have for $x > 0$

$$\begin{aligned}
&\left(\mathcal{D}_{-}^{\mu,\mu',\xi,\xi',\varpi}\left\{t^{\varrho-1}S_\lambda(r,t^\rho)\right\}\right)(x) \\
&= \frac{1}{\pi i}\int_C \frac{\Gamma(s)\Gamma(2-s)\{\Gamma(1\pm ir-s)\}^\lambda(-1)^{-\rho s}}{\{\Gamma(2\pm ir-s)\}^\lambda}\left(\mathcal{D}_{-}^{\mu,\mu',\xi,\xi',\varpi}t^{\varrho-\rho s-1}\right)(x)\,ds \\
&= x^{\varrho+\xi-1}\frac{1}{\pi i}\int_C \frac{\Gamma(s)\Gamma(2-s)\{\Gamma(1\pm ir-s)\}^\lambda}{\{\Gamma(2\pm ir-s)\}^\lambda} \\
&\quad\times\frac{\Gamma(1-\varrho-\xi'+\rho s)\Gamma(1-\varrho+\varpi-\mu-\mu'+\rho s)\Gamma(1-\varrho+\varpi-\mu'-\xi+\rho s)}{\Gamma(1-\varrho+\rho s)\Gamma(1-\varrho+\varpi-\mu-\mu'-v+\rho s)\Gamma(1-\varrho-\mu'-\xi'+\rho s)}(-x)^{-\rho s}\,ds
\end{aligned}$$

which, interpreted with the help of the definition (12.29), yields the required formula (12.58).

We also establish the fractional integral formulas for the Saigo's by substituting the specific values to the parameters as $\mu = \mu + \xi, \mu' = \xi' = 0, \xi = -\eta$ and $\varpi = \mu$, which are asserted by Corollary 3.1 and Corollary 3.2 below.

Corollary 3.1. Let $\lambda - 1$, $r \in \mathbb{R}^+$, $\varrho > 0$ and let the contour $\mathcal{C}$ be taken as in (12.28). Let $\varrho, \mu, \xi, \eta, \varpi \in \mathbb{C}$ with $\Re(\mu) > 0$, $\Re(\mu + \xi + \eta) \neq 0$ and satisfy the conditions $\Re(\varrho) + \min[0, \Re(\mu + \xi + \eta)] > 0$. Then the following assertion for the Saigo's fractional operator $\mathcal{D}_{0+}^{\mu,\xi,\eta}$ of the $S_\lambda(r, t^\rho)$ (12.26) exists and holds true:

$$\left(\mathcal{D}_{0+}^{\mu,\xi,\eta}\left\{t^{\varrho-1}S_\lambda(r,t^\rho)\right\}\right)(x)$$
$$= 2x^{\varrho-\xi-1}\overline{H}_{5,5}^{1,5}\left[-x^\rho\left|\begin{matrix}(1-\varrho,\rho;1),(1-\varrho-\mu-\xi-\eta,\rho;1),(-1,1;1),(\pm ir,1;\lambda)\\(0,1),(\pm ir-1,1;\lambda),(1-\varrho-\xi,\rho;1),(1-\varrho-\eta,\rho;1)\end{matrix}\right.\right] \tag{12.59}$$

Corollary 3.2. Let $\lambda - 1$, $r \in \mathbb{R}^+$, $\varrho > 0$ and let the contour $\mathcal{C}$ be taken as in (12.28). Let $\varrho, \mu, \xi, \eta, \varpi \in \mathbb{C}$ and satisfy the conditions $\Re(\mu) \geq 0$ and $\Re(\varrho) < 1 + \min[\Re(-\xi - n), \Re(\mu + \eta)]$, $n = [\Re(\mu)] + 1$. Then the following assertion for the Saigo's fractional operator $\mathcal{D}_{-}^{\mu,\xi,\eta}$ of the $S_\lambda(r, t^\rho)$ (12.26) exists and holds true:

$$\left(\mathcal{D}_{-}^{\mu,\xi,\eta}\left\{t^{\varrho-1}S_\lambda(r,t^\rho)\right\}\right)(x)$$
$$= 2x^{\varrho+\xi-1}\overline{H}_{5,5}^{3,3}\left[-x^\rho\left|\begin{matrix}(-1,1;1),(\pm ir,1;\lambda),(1-\varrho,\rho),(1-\varrho+\eta-\xi,\rho)\\(0,1),(1-\varrho-\xi,\rho),(1-\varrho+\mu+\eta,\rho),(\pm ir-1,1;\lambda)\end{matrix}\right.\right] \tag{12.60}$$

If we let $\xi = -\mu$ and $\xi = 0$ respectively, in Corollary 3.1 and Corollary 3.2, we deduce the following fractional integral formulas for the Riemann–Liouville (RL) and Erdélyi–Kober (EK), which are asserted by Corollary 3.3, Corollary 3.4, Corollary 3.5, and Corollary 3.6 below.

Corollary 3.3. Let $\lambda - 1$, $r \in \mathbb{R}^+$, $\varrho > 0$ and let the contour $\mathcal{C}$ be taken as in (12.28). Let $\varrho, \mu, \in \mathbb{C}$ and satisfy the conditions $\Re(\mu) > 0$ and $\Re(\varrho) + \min[0, \Re(\eta)] > 0$. Then the following assertion for the RL fractional operator $\mathcal{D}_{0+}^{\mu}$ of the $S\lambda(r, t\varrho)$ (12.26) exists and holds true:

$$\left(\mathcal{D}_{0+}^{\mu}\left\{t^{\varrho-1}S_\lambda(r,t^\rho)\right\}\right)(x) = 2x^{\varrho+\mu-1}\overline{H}_{4,4}^{1,4}\left[-x^\varrho\left|\begin{matrix}(1-\varrho,\rho;1),(-1,1;1),(\pm ir,1;\lambda)\\(0,1),(\pm ir-1,1;\lambda),(1-\varrho+\mu,\rho;1)\end{matrix}\right.\right]. \tag{12.61}$$

Corollary 3.4. Let $\lambda - 1$, $r \in \mathbb{R}^{+}$, $\varrho > 0$ and let the contour $\mathcal{C}$ be taken as in (12.28). Let ϱ, μ, $\in \mathbb{C}$ and satisfy the conditions $\Re(\mu) \geq 0$ and $\Re(\varrho) < \Re(\mu) - [\Re(\mu)]$. Then the following assertion for the RL fractional operator D_{-}^{μ} of the $S\lambda(r, t\varrho)$ (12.26) exists and holds true:

$$\left(\mathcal{D}_{-}^{\mu}\left\{t^{\varrho-1}S_{\lambda}(r,t^{\rho})\right\}\right)(x) = x^{\varrho-\mu-1}\overline{H}_{4,4}^{2,3}\left[-x^{\varrho}\left|\begin{matrix}(-1,1;1),(\pm ir,1;\lambda),(1-\varrho,\rho)\\(0,1),(1-\varrho+\mu,\rho),(\pm ir-1,1;\lambda)\end{matrix}\right.\right]. \tag{12.62}$$

Corollary 3.5. Let $\lambda - 1$, $r \in \mathbb{R}^{+}$, $\varrho > 0$ and let the contour $\mathcal{C}$ be taken as in (12.28). Let $\varrho, \mu, \eta, \varpi \in \mathbb{C}$ and satisfy the conditions $\Re(\mu) \geq 0$ and $\Re(\varrho) > -\Re(\mu + \eta)$. Then the following assertion for the EK fractional operator $\mathcal{D}_{\eta,\mu}^{+}$ of the $S\lambda(r, t\varrho)$ (12.26) exists and holds true:

$$\left(\mathcal{D}_{\eta,\mu}^{+}\left\{t^{\varrho-1}S_{\lambda}(r,t^{\rho})\right\}\right)(x) = 2x^{\varrho-1}\overline{H}_{4,4}^{1,4}\left[-x^{\rho}\left|\begin{matrix}(1-\varrho-\mu-\eta,\rho;1),(-1,1;1),(\pm ir,1;\lambda)\\(0,1),(\pm ir-1,1;\lambda),(1-\varrho-\eta,\rho;1)\end{matrix}\right.\right]. \tag{12.63}$$

Corollary 3.6. Let $\lambda - 1$, $r \in \mathbb{R}^{+}$, $\varrho > 0$ and let the contour $\mathcal{C}$ be taken as in (12.28). Let ϱ, μ, η, $\varpi \in \mathbb{C}$ and satisfy the conditions $\Re(\mu) > 0$ and $\Re(\varrho) < 1 + \Re(\eta)$. Then the following assertion for the EK fractional operator $D_{\eta,\mu}^{-}$ of the $S\lambda(r, t\varrho)$ (12.26) exists and holds true:

$$\left(\mathcal{D}_{\eta,\mu}^{-}\left\{t^{\varrho-1}S_{\lambda}(r,t^{\rho})\right\}\right)(x) = x^{\varrho-1}\overline{H}_{4,4}^{2,3}\left[-x^{\rho}\left|\begin{matrix}(-1,1;1),(\pm ir,1;\lambda),(1-\varrho+\eta,\rho)\\(0,1),(1-\varrho+\mu+\eta,\rho),(\pm ir-1,1;\lambda)\end{matrix}\right.\right]. \tag{12.64}$$

12.4 Special Cases in Terms of Fox's H-Function

The H-function is defined by C. Fox [35] in his studies of symmetrical Fourier kernels as the Mellin–Barnes-type path integral (see, for details, [40]):

$$\begin{aligned}H_{p,q}^{m,n}(z) &= H_{p,q}^{m,n}\left[z\left|\begin{matrix}(e_r,E_r)\\(f_r,F_r)\end{matrix}\right.\right] = H_{p,q}^{m,n}\left[z\left|\begin{matrix}(e_1,E_1),\cdots,(e_r,E_r)\\(f_1,F_1),\cdots,(f_r,F_r)\end{matrix}\right.\right]\\&= \frac{1}{2\pi i}\int_{\pounds}\psi_{p,q}^{m,n}(s)\,z^{s}\,ds \qquad (\forall z, z \neq 0)\end{aligned} \tag{12.65}$$

where

$$\psi_{p,q}^{m,n}(s) = \frac{E(s)F(s)}{E'(s)F'(s)}, \tag{12.66}$$

$$E(s) = \prod_{r=1}^{m} \Gamma(f_r - F_r s), \;\; B(s) = \prod_{rr=1}^{n} \Gamma(1 - e_r + E_r s), \tag{12.67}$$

$$E'(s) = \prod_{r=m+1}^{q} \Gamma(1 - f_r + F_r s), \;\; F'(s) = \prod_{r=n+1}^{p} \Gamma(e_r - E_r s), \tag{12.68}$$

with $e_r \in \mathbb{C}$ $(r = 1, \cdots, p)$, $f_r \in \mathbb{C}$ $(r = 1, \cdots, q)$, $E_r \in \mathbb{R}^+$ $(r = 1, \cdots, p)$ and $F_r \in \mathbb{R}^+$ $(r = 1, \cdots, q)$. and £ is a suitable contour of the Mellin–Barnes type separating the poles of $\Gamma(f_r + F_r s)$ $(r = 1, \cdots, m)$ from those of $\Gamma(1 - e_r - E_r s)$ $(r = 1, \cdots, n)$ with the usual indentations. An empty product, when it occurs, is taken to be interpreted as 1, the integers m, n, p, q satisfy the inequalities $0 \le m \le q$ and $0 \le n \le p$. The Mellin–Barnes-type contour integration representing in the H-function converges absolutely and defines an analytic function for $|\arg(z)| < \frac{\pi}{2}\Omega$, where

$$\Omega = \sum_{r=1}^{m} F_r - \sum_{r=m+1}^{q} F_r + \sum_{r=1}^{n} E_r - \sum_{r=n+1}^{p} E_r > 0.$$

Our results derived in previous sections are in terms of $\bar{H}$-function (12.29). An important special case of $\bar{H}$-function when integral powers assume positive integer values yields Fox's H-function (12.65). Also, if we put $\lambda = 2$ in all the results in Sections 12.2 and 12.3, we obtain the results for the Mathieu-type functional series

$$S(r;z) = \sum_{m \ge 1} \frac{2mz^{m-1}}{(m^2 + r^2)^2}, \qquad (r \in \mathbb{R}, |z| < 1). \tag{12.69}$$

We present only the main results for the Marichev–Saigo–Maeda (MSM) fractional operators as Corollaries 4.1 to 4.4 without proof.

Corollary 4.1. Let $\lambda - 1$, $r \in \mathbb{R}^+$, $\varrho > 0$ and let the contour $\mathcal{C}$ be taken as in (12.28). Let $\varrho, \mu, \mu', \xi, \xi', \eta, \varpi \in \mathbb{C}$ and satisfy the conditions $\Re(\varpi) > 0$ and $\Re(\varrho) + \min\{0, \Re(\varpi - \mu - \mu' - \xi), \Re(-\xi' - \mu')\} > 0$. Then the following assertion for the MSM fractional operator $\mathcal{I}_{0+}^{\mu,\mu',\xi,\xi',\varpi}$ of the $S(r,t^{\rho})$ (12.69) exists and holds true:

$$\left(\mathcal{I}_{0+}^{\mu,\mu',\xi,\xi',\varpi}\left\{t^{\varrho-1}S(r,t^{\rho})\right\}\right)(x)=2x^{\varrho+\varpi-\mu-\mu'-1}$$
$$\times H_{8,8}^{1,8}\left[-x^{\rho}\left|\begin{matrix}(1-\varrho,\rho),(1-\varrho-\varpi+\mu+\mu'+\xi,\rho),(1-\varrho-\xi'+\mu',\rho),(-1,1),(\pm ir,1),(\pm ir,1)\\(0,1),(\pm ir-1,1),(\pm ir-1,1),(1-\varrho-\xi',\rho),(1-\varrho-\varpi+\mu+\mu',\rho),(1-\varrho-\varpi+\mu'+\xi,\rho)\end{matrix}\right.\right].\tag{12.70}$$

Corollary 4.2. Let $\lambda-1$, $r\in\mathbb{R}^{+}$, $\rho>0$ and let the contour C be taken as in (12.28). Let ϱ, μ, μ', ξ, ξ', η, $\varpi\in\mathbb{C}$ and satisfy the conditions $\Re(\varpi)>0$ and $\Re(\varrho)<1+\min\{\Re(-\xi),\Re(\mu+\mu'-\varpi),\Re(\mu+\xi'-\varpi)\}$. Then the following assertion for the MSM fractional operator $\mathcal{I}_{-}^{\mu,\mu',\xi,\xi',\varpi}$ of the $S(r,t^{\rho})$ (12.69) exists and holds true:

$$\left(\mathcal{I}_{-}^{\mu,\mu',\xi,\xi',\varpi}\left\{t^{\varrho-1}S(r,t^{\rho})\right\}\right)(x)=2x^{\varrho+\varpi-\mu-\mu'-1}$$
$$\times H_{8,8}^{4,5}\left[-x^{\rho}\left|\begin{matrix}(-1,1),(\pm ir,1),(\pm ir,1),(1-\varrho,\rho),(1-\varrho-\varpi+\mu+\mu'+\xi',\rho),(1-\varrho+\mu-\xi,\rho)\\(0,1),(1-\varrho-\xi,\rho),(1-\varrho-\varpi+\mu+\mu',\rho),(1-\varrho-\varpi+\mu+\xi',\rho),(\pm ir-1,1),(\pm ir-1,1)\end{matrix}\right.\right].\tag{12.71}$$

Corollary 4.3. Let $\lambda-1$, $r\in\mathbb{R}^{+}$, $\rho>0$ and let the contour C be taken as in (12.28). Let $\varrho,\mu,\mu',\xi,\xi',\eta,\varpi\in\mathbb{C}$ and satisfy the conditions $\Re(\varpi)>0$ and $\Re(\varrho)>\max\{0,\Re(\varpi-\mu-\mu'+\xi'),\Re(\xi-\mu)\}$. Then the following assertion for the MSM fractional operator $\mathcal{D}_{0+}^{\mu,\mu',\xi,\xi',\varpi}$ of the $S(r,t^{\rho})$ (12.69) exists and holds true:

$$\left(\mathcal{D}_{0+}^{\mu,\mu',\xi,\xi',\varpi}\left\{t^{\varrho-1}S(r,t^{\rho})\right\}\right)(x)=2x^{\varrho-\xi-1}$$
$$\times H_{8,8}^{1,8}\left[-x^{\rho}\left|\begin{matrix}(1-\varrho,\rho),(1-\varrho+\varpi-\mu-\mu'-\xi',\rho),(1-\varrho+\xi-\mu,\rho),(-1,1),(\pm ir,1),(\pm ir,1)\\(0,1),(\pm ir-1,1),(\pm ir-1,1),(1-\varrho+\xi,\rho),(1-\varrho+\varpi-\mu-\mu',\rho),(1-\varrho+\varpi-\mu-\xi',\rho)\end{matrix}\right.\right].\tag{12.72}$$

Corollary 4.4. Let $\lambda-1$, $r\in\mathbb{R}^{+}$, $\rho>0$ and let the contour C be taken as in (12.28). Let $\varrho,\mu,\mu',\xi,\xi',\eta,\varpi\in\mathbb{C}$ and satisfy the conditions $\Re(\varpi)>0$ and $\Re(\varrho)<1+\min\{\Re(\xi'),\Re(\varpi-\mu-\mu',\Re(\eta-\mu'-\xi)\}$. Then the following assertion for the MSM fractional operator $\mathcal{D}_{-}^{\mu,\mu',\xi,\xi',\varpi}$ of the $S(r,t^{\rho})$ (12.69) exists and holds true:

$$\left(\mathcal{D}_{-}^{\mu,\xi,\eta}\left\{t^{\varrho-1}S(r,t^{\rho})\right\}\right)(x)=2x^{\varrho+\xi-1}$$
$$\times H_{8,8}^{4,5}\left[-x^{\rho}\left|\begin{matrix}(-1,1),(\pm ir,1),(\pm ir,1),(1-\varrho,\rho),(1-\varrho+\varpi-\mu-\mu'-v,\rho),(1-\varrho-\mu'-\xi',\rho)\\(0,1),(1-\varrho-\xi',\rho),(1-\varrho+\varpi-\mu-\mu',\rho),(1-\varrho+\varpi-\mu'-\xi,\rho),(\pm ir-1,1),(\pm ir-1,1)\end{matrix}\right.\right].\tag{12.73}$$

Interested readers can derive the other results for Saigo's, Riemann–Liouville (RL) and Erdélyi–Kober (EK) operators.

12.5 Further Observations and Applications

In the investigations of the previous section, we established various fractional calculus results for the generalized Mathieu-type and alternating Mathieu-type functional series by extensive use of the Marichev–Saigo–Maeda, Saigo's, Riemann–Liouville and Erdélyi–Kober tools. The (presumably) new and (potentially) useful results are expressed in terms of the generalized H-function, that is the $\bar{H}$-function. Moreover, as a special case and from the applications point of view, all the results are obtained in terms of the Fox H-function. From the point of view of other applications, all the results obtained in the previous sections can be written in terms of the I-function introduced by Rathie [41], which contains several special cases as various special functions, for example, Inayat-Hussein $\bar{H}$-function, H-function, and G-function in the following form:

$$I_{p,q}^{m,n}(z) = I_{p,q}^{m,n}\left[z \left| \begin{matrix} (e_r, E_r, \mu_r)_{\overline{1,p}} \\ (f_r, F_r, \xi_r)_{\overline{1,q}} \end{matrix} \right. \right] = I_{p,q}^{m,n}\left[z \left| \begin{matrix} (e_1, E_1, \mu_1), \cdots, (e_r, E_r, \mu_p) \\ (f_1, F_1, \xi_1), \cdots, (f_r, F_r, \xi_q) \end{matrix} \right. \right] \tag{12.74}$$

$$= \frac{1}{2\pi i} \int_{£} \phi_{p,q}^{m,n}(s)\, z^s\, ds \qquad (\forall z, z \neq 0)$$

for all $z \neq 0$, where

$$\phi_{p,q}^{m,n}(s) = \frac{\bar{E}(s)\bar{F}(s)}{\bar{E}'(s)\bar{F}'(s)}, \tag{12.75}$$

$$\bar{E}(s) = \prod_{r=1}^{m} \{\Gamma(f_r - F_r s)\}^{\xi_r},\ \bar{F}(s) = \prod_{r=1}^{n} \{\Gamma(1 - e_r + E_r s)\}^{\mu_r}, \tag{12.76}$$

$$\bar{E}'(s) = \prod_{j=m+1}^{q} \{\Gamma(1 - f_r + F_r s)\}^{\xi_j},\ \bar{F}'(s) = \prod_{j=n+1}^{p} \{\Gamma(e_r - E_r s)\}^{\mu_j}, \tag{12.77}$$

with $e_r \in \mathbb{C}$ $(j = 1, \cdots, p)$, $f_r \in \mathbb{C}$ $(j = 1, \cdots, q)$, $E_r \in \mathbb{R}^+$ $(j = 1, \cdots, p)$ and $F_r \in \mathbb{R}^+$ $(j = 1, \cdots, q)$. and the exponents μ_j $(j = 1, \cdots, n)$ and ξ_j $(j = m + 1, \cdots, q)$ can take non-integer values. £ is a suitable contour of the Mellin–Barnes type separating the poles of $\{\Gamma(f_r - F_r s)\}_{\xi j}$ $(j = 1, \cdots, m)$ from those of $\Gamma(1 - e_r + E_r s)\}_{\mu j}$ $(j = 1, \cdots, n)$ with the usual indentations. An empty product is interpreted as 1, the integers $m,\ n,\ p,\ q$ satisfy the inequalities $0 \leq m \leq q$ and $0 \leq n \leq p$. Also, the Mellin–Barnes contour integral representing the I-function converges absolutely and defines an analytic function for $|\arg(z)| < \frac{\pi}{2}\Delta$, where

$$\Delta = \sum_{j=1}^{m} |\xi_j| F_r + \sum_{j=1}^{n} |\mu_j| E_r - \sum_{j=m+1}^{q} |\xi_j| F_r - \sum_{j=n+1}^{p} |\mu_j| E_r > 0.$$

Example 5.1

Let $\lambda - 1$, $r \in \mathbb{R}^+$, $\varrho > 0$ and let the contour C be taken as in (12.28). Let ϱ, μ, μ', ξ, ξ', η, $\varpi \in \mathbb{C}$ and satisfy the conditions $\Re(\varpi) > 0$ and $\Re(\varrho) + \min\{0, \Re(\varpi - \mu - \mu' - \xi), \Re(-\xi' - \mu')\} > 0$. Then the following assertion for the MSM fractional operator $I_{0+}^{\mu,\mu',\xi,\xi',\varpi}$ of the $S_\lambda(r, t^\rho)$ (12.26) exists and holds true:

$$\left(I_{0+}^{\mu,\mu',\xi,\xi',\varpi}\left\{t^{\varrho-1}S_\lambda(r,t^\rho)\right\}\right)(x) = 2x^{\varrho+\varpi-\mu-\mu'-1}$$

$$\times I_{6,6}^{1,6}\left[-x^\rho \left| \begin{matrix} (1-\varrho,\rho,1),(1-\varrho-\varpi+\mu+\mu'+\xi,\rho,1),(1-\varrho-\xi'+\mu',\rho,1),(-1,1,1),(\pm ir,1,\lambda) \\ (0,1,1),(\pm ir-1,1,\lambda),(1-\varrho-\xi',\rho,1),(1-\varrho-\varpi+\mu+\mu',\rho,1),(1-\varrho-\varpi+\mu'+\xi,\rho,1) \end{matrix} \right.\right]. \tag{12.78}$$

Proof. By making use of the operator (12.1) and expressing the definition of the generalized Mathieu-type functional series $S_\lambda(r,z)$ in the form of the contour integral representation (12.26), and then changing the order of integrals and applying the relation (12.33), we have

$$\begin{aligned}
&\left(I_{0+}^{\mu,\mu',\xi,\xi',\varpi}\left\{t^{\varrho-1}S_\lambda(r,t^\rho)\right\}\right)(x) \\
&= \frac{1}{\pi i}\int_C \frac{\Gamma(s)\Gamma(2-s)\{\Gamma(1\pm ir-s)\}^\lambda(-1)^{-\rho s}}{\{\Gamma(2\pm ir-s)\}^\lambda}\left(I_{0+}^{\mu,\mu',\xi,\xi',\varpi}t^{\varrho-\rho s-1}\right)(x)\,ds \\
&= x^{\varrho+\varpi-\mu-\mu'-1}\frac{1}{\pi i}\int_C \frac{\Gamma(s)\Gamma(2-s)\{\Gamma(1\pm ir-s)\}^\lambda}{\{\Gamma(2\pm ir-s)\}^\lambda} \\
&\times \frac{\Gamma(\varrho-\rho s)\Gamma(\varrho+\varpi-\mu-\mu'-\xi-\rho s)\Gamma(\varrho+\xi'-\mu'-\rho s)}{\Gamma(\varrho+\xi'-\rho s)\Gamma(\varrho+\varpi-\mu-\mu'-\rho s)\Gamma(\varrho+\varpi-\mu'-\xi-\rho s)}(-x)^{-\rho s}\,ds
\end{aligned}$$

which, interpreted with the help of the definition (12.74), yields the required result (12.78).

Example 5.2

Let $\lambda - 1$, $r \in \mathbb{R}^+$, $\varrho > 0$ and let the contour C be taken as in (12.28). Let ϱ, μ, μ', ξ, ξ', η, $\varpi \in \mathbb{C}$ and satisfy the conditions $\Re(\varpi) > 0$ and $\Re(\varrho) < 1 + \min\{\Re(-\xi), \Re(\mu + \mu' - \varpi), \Re(\mu + \xi' - \varpi)\}$. Then the following assertion for the MSM fractional operator $I_{-}^{\mu,\mu',\xi,\xi',\varpi}$ of the $S_\lambda(r, t^\rho)$ (12.26) exists and holds true:

$$\left(I_{-}^{\mu,\mu',\xi,\xi',\varpi}\left\{t^{\varrho-1}S_\lambda(r,t^\rho)\right\}\right)(x) = 2x^{\varrho+\varpi-\mu-\mu'-1}$$

$$\times I_{6,6}^{4,3}\left[-x^\rho \left| \begin{matrix} (-1,1,1),(\pm ir,1,\lambda),(1-\varrho,\rho,1),(1-\varrho-\varpi+\mu+\mu'+\xi',\rho,1),(1-\varrho+\mu-\xi,\rho,1) \\ (0,1,1),(1-\varrho-\xi,\rho,1),(1-\varrho-\varpi+\mu+\mu',\rho,1),(1-\varrho-\varpi+\mu+\xi',\rho,1),(\pm ir-1,1,\lambda) \end{matrix} \right.\right]. \tag{12.79}$$

Proof. By making use of the operator (12.2) and expressing the definition of the generalized Mathieu-type functional series $S_\lambda(r,z)$ in the form of the contour integral representation (12.26), and then changing the order of integrals and applying the relation (12.34), we have

$$\left(I_{-}^{\mu,\mu',\xi,\xi',\varpi}\left\{t^{\varrho-1}S_{\lambda}(r,t^{\rho})\right\}\right)(x)$$
$$=\frac{1}{\pi i}\int_{C}\frac{\Gamma(s)\Gamma(2-s)\{\Gamma(1\pm ir-s)\}^{\lambda}(-1)^{-\rho s}}{\{\Gamma(2\pm ir-s)\}^{\lambda}}\left(I_{-}^{\mu,\mu',\xi,\xi',\varpi}t^{\varrho-\rho s-1}\right)(x)\,ds$$
$$=x^{\varrho+\varpi-\mu-\mu'-1}\frac{1}{\pi i}\int_{C}\frac{\Gamma(s)\Gamma(2-s)\{\Gamma(1\pm ir-s)\}^{\lambda}}{\{\Gamma(2\pm ir-s)\}^{\lambda}}$$
$$\times\frac{\Gamma(1-\varrho-\xi+\rho s)\Gamma(1-\varrho-\varpi+\mu+\mu'+\rho s)\Gamma(1-\varrho-\varpi+\mu+\xi'+\rho s)}{\Gamma(1-\varrho+\rho s)\Gamma(1-\varrho-\varpi+\mu+\mu'+\xi'+\rho s)\Gamma(1-\varrho+\mu-\xi+\rho s)}(-x)^{-\rho s}\,ds$$

which, interpreted with the help of the definition (12.74), yields the required result (12.79).

Example 5.3

Let $\lambda - 1$, $r \in \mathbb{R}^{+}$, $\varrho > 0$ and let the contour $\mathcal{C}$ be taken as in (12.28). Let ϱ, μ, μ', ξ, ξ', η, $\varpi \in \mathbb{C}$ and satisfy the conditions $\Re(\varpi) > 0$ and $\Re(\varrho) > \max\{0, \Re(\varpi - \mu - \mu' + \xi'), \Re(\xi - \mu)\}$. Then the following assertion for the MSM fractional operator $D_{0+}^{\mu,\mu',\xi,\xi',\varpi}$ of the $S_{\lambda}(r, t^{\rho})$ (12.26) exists and holds true:

$$\left(D_{0+}^{\mu,\mu',\xi,\xi',\varpi}\left\{t^{\varrho-1}S_{\lambda}(r,t^{\rho})\right\}\right)(x)=2x^{\varrho-\xi-1}$$
$$\times I_{6,6}^{1,6}\left[-x^{\rho}\left|\begin{matrix}(1-\varrho,\rho,1),(1-\varrho+\varpi-\mu-\mu'-\xi',\rho,1),(1-\varrho+\xi-\mu,\rho,1),(-1,1,1),(\pm ir,1,\lambda)\\(0,1,1),(\pm ir-1,1,\lambda),(1-\varrho+\xi,\rho,1),(1-\varrho+\varpi-\mu-\mu',\rho,1),(1-\varrho+\varpi-\mu-\xi',\rho,1)\end{matrix}\right.\right]. \tag{12.80}$$

Proof. By making use of the operator (12.3) and expressing the definition of the generalized Mathieu series $S_{\lambda}(r,z)$ in the form of the contour integral representation (12.26), and then changing the order of integrals and applying the relation (12.49), we have for $x > 0$

$$\left(D_{0+}^{\mu,\mu',\xi,\xi',\varpi}\left\{t^{\varrho-1}S_{\lambda}(r,t^{\rho})\right\}\right)(x)$$
$$=\frac{1}{\pi i}\int_{C}\frac{\Gamma(s)\Gamma(2-s)\{\Gamma(1\pm ir-s)\}^{\lambda}(-1)^{-\rho s}}{\{\Gamma(2\pm ir-s)\}^{\lambda}}\left(D_{0+}^{\mu,\mu',\xi,\xi',\varpi}t^{\varrho-\rho s-1}\right)(x)\,ds$$
$$=x^{\varrho-\xi-1}\frac{1}{\pi i}\int_{C}\frac{\Gamma(s)\Gamma(2-s)\{\Gamma(1\pm ir-s)\}^{\lambda}}{\{\Gamma(2\pm ir-s)\}^{\lambda}}$$
$$\times\frac{\Gamma(\varrho-\rho s)\Gamma(\varrho-\varpi+\mu+\mu'+\xi'-\rho s)\Gamma(\varrho-\xi+\mu-\rho s)}{\Gamma(\varrho-\xi-\rho s)\Gamma(\varrho-\varpi+\mu+\mu'-\rho s)\Gamma(\varrho-\varpi+\mu+\xi'-\rho s)}(-x)^{-\rho s}\,ds$$

which, interpreted with the help of the definition (12.74), yields the required derivative formula (12.80).

Example 5.4

Let $\lambda - 1$, $r \in \mathbb{R}^{+}$, $\varrho > 0$ and let the contour $\mathcal{C}$ be taken as in (12.28). Let ϱ, μ, μ', ξ, ξ', η, $\varpi \in \mathbb{C}$ and satisfy the conditions $\mathfrak{R}(\varpi) > 0$ and $\mathfrak{R}(\varrho) < 1 + \min\{\mathfrak{R}(\xi'), \mathfrak{R}(\varpi - \mu - \mu', \mathfrak{R}(\eta - \mu' - \xi)\}$. Then the following assertion for the MSM fractional operator $D_{-}^{\mu,\mu',\xi,\xi',\varpi}$ of the $S_{\lambda}(r, t^{\rho})$ (12.26) exists and holds true:

$$\left(D_{-}^{\mu,\xi,\eta}\left\{t^{\varrho-1} S_{\lambda}(r,t^{\rho})\right\}\right)(x) = 2x^{\varrho+\xi-1}$$

$$\times I_{6,6}^{4,3}\left[-x^{\rho}\left|\begin{matrix}(-1,1,1),(\pm ir,1,\lambda),(1-\varrho,\rho,1),(1-\varrho+\varpi-\mu-\mu'-v,\rho,1),(1-\varrho-\mu'-\xi',\rho,1)\\(0,1,1),(1-\varrho-\xi',\rho.1),(1-\varrho+\varpi-\mu-\mu',\rho,1),(1-\varrho+\varpi-\mu'-\xi,\rho,1),(\pm ir-1,1,\lambda)\end{matrix}\right.\right]. \tag{12.81}$$

Proof. By making use of the operator (12.4) and expressing the definition of the generalized Mathieu-type functional series $S_{\lambda}(r, z)$ in the form of the contour integral representation (12.26), and then changing the order of integrals and applying the relation (12.50), we have for $x > 0$

$$\left(D_{-}^{\mu,\mu',\xi,\xi',\varpi}\left\{t^{\varrho-1}S_{\lambda}(r,t^{\rho})\right\}\right)(x)$$

$$= \frac{1}{\pi i}\int_{C}\frac{\Gamma(s)\Gamma(2-s)\{\Gamma(1\pm ir-s)\}^{\lambda}(-1)^{-\rho s}}{\{\Gamma(2\pm ir-s)\}^{\lambda}}\left(D_{-}^{\mu,\mu',\xi,\xi',\varpi}t^{\varrho-\rho s-1}\right)(x)\,ds$$

$$= x^{\varrho+\xi-1}\frac{1}{\pi i}\int_{C}\frac{\Gamma(s)\Gamma(2-s)\{\Gamma(1\pm ir-s)\}^{\lambda}}{\{\Gamma(2\pm ir-s)\}^{\lambda}}$$

$$\times\frac{\Gamma(1-\varrho-\xi'+\rho s)\Gamma(1-\varrho+\varpi-\mu-\mu'+\rho s)\Gamma(1-\varrho+\varpi-\mu'-\xi+\rho s)}{\Gamma(1-\varrho+\rho s)\Gamma(1-\varrho+\varpi-\mu-\mu'-v+\rho s)\Gamma(1-\varrho-\mu'-\xi'+\rho s)}(-x)^{-\rho s}\,ds$$

which, interpreted with the help of the definition (12.74), yields the required formula (12.81).

Here, we have mentioned only the results for Marichev–Saigo–Maeda (MSM) fractional operators. Other results can be derived for Saigo's, Riemann–Liouville(RL) and Erdélyi–Kober (EK) operators, and these are left as an exercise for interested readers.

To conclude this section, we present some results for generalized alternating Mathieu-type functional series $\tilde{S}_{\lambda}(r;z)$ (12.28).

$$\tilde{S}_{\lambda}(r;z) = \sum_{n\geq 1}(-1)^{n-1}\frac{2nz^{n-1}}{(n^2+r^2)^{\lambda}}, \qquad (\lambda > 1, r \in \mathbb{R}, |z| < 1), \tag{12.82}$$

Theorem 5.1 Let $\lambda - 1$, $r \in \mathbb{R}^{+}$, $\varrho > 0$ and let the contour $\mathcal{C}$ be taken as in (12.28). Let $\varrho, \mu, \mu', \xi, \xi', \eta, \varpi \in \mathbb{C}$ and satisfy the conditions $\Re(\varpi) > 0$ and $\Re(\varrho) + \min\{0, \Re(\varpi - \mu - \mu' - \xi), \Re(-\xi' - \mu')\} > 0$. Then the following assertion for the MSM fractional operator $\mathcal{I}_{0+}^{\mu,\mu',\xi,\xi',\varpi}$ of the $S_\lambda\left(r,t^\rho\right)$ (12.28) exists and holds true:

$$\left(\mathcal{I}_{0+}^{\mu,\mu',\xi,\xi',\varpi}\left\{t^{\varrho-1}\tilde{S}_\lambda(r,t^\rho)\right\}\right)(x) = 2x^{\varrho+\varpi-\mu-\mu'-1}$$
$$\times \overline{H}_{6,6}^{1,6}\left[x^\rho \left| \begin{matrix} (1-\varrho,\rho;1),(1-\varrho-\varpi+\mu+\mu'+\xi,\rho;1),(1-\varrho-\xi'+\mu',\rho;1),(-1,1;1),(\pm ir,1;\lambda) \\ (0,1),(\pm ir-1,1;\lambda),(1-\varrho-\xi',\rho;1),(1-\varrho-\varpi+\mu+\mu',\rho;1),(1-\varrho-\varpi+\mu'+\xi,\rho;1) \end{matrix}\right.\right]. \tag{12.83}$$

Theorem 5.2 Let $\lambda - 1$, $r \in \mathbb{R}^{+}$, $\varrho > 0$ and let the contour $\mathcal{C}$ be taken as in (12.28). Let $\varrho, \mu, \mu', \xi, \xi', \eta, \varpi \in \mathbb{C}$ and satisfy the conditions $\Re(\varpi) > 0$ and $\Re(\varrho) < 1 + \min\{\Re(-\xi), \Re(\mu + \mu' - \varpi), \Re(\mu + \xi' - \varpi)\}$. Then the following assertion for the MSM fractional operator $\mathcal{I}_{-}^{\mu,\mu',\xi,\xi',\varpi}$ of the $S_\lambda\left(r,t^\rho\right)$ (12.28) exists and holds true:

$$\left(\mathcal{I}_{-}^{\mu,\mu',\xi,\xi',\varpi}\left\{t^{\varrho-1}\tilde{S}_\lambda(r,t^\rho)\right\}\right)(x) = 2x^{\varrho+\varpi-\mu-\mu'-1}$$
$$\times \overline{H}_{6,6}^{4,3}\left[x^\rho \left| \begin{matrix} (-1,1;1),(\pm ir,1;\lambda),(1-\varrho,\rho),(1-\varrho-\varpi+\mu+\mu'+\xi',\rho),(1-\varrho+\mu-\xi,\rho) \\ (0,1),(1-\varrho-\xi,\rho),(1-\varrho-\varpi+\mu+\mu',\rho),(1-\varrho-\varpi+\mu+\xi',\rho),(\pm ir-1,1;\lambda) \end{matrix}\right.\right]. \tag{12.84}$$

Theorem 5.3 Let $\lambda - 1$, $r \in \mathbb{R}^{+}$, $\varrho > 0$ and let the contour $\mathcal{C}$ be taken as in (12.28). Let $\varrho, \mu, \mu', \xi, \xi', \eta, \varpi \in \mathbb{C}$ and satisfy the conditions $\Re(\varpi) > 0$ and $\Re(\varrho) > \max\{0, \Re(\varpi - \mu - \mu' + \xi'), \Re(\xi - \mu)\}$. Then the following assertion for the MSM fractional operator $\mathcal{D}_{0+}^{\mu,\mu',\xi,\xi',\varpi}$ of the $S_\lambda\left(r,t^\rho\right)$ (12.28) exists and holds true:

$$\left(\mathcal{D}_{0+}^{\mu,\mu',\xi,\xi',\varpi}\left\{t^{\varrho-1}\tilde{S}_\lambda(r,t^\rho)\right\}\right)(x) = 2x^{\varrho-\xi-1}$$
$$\times \overline{H}_{6,6}^{1,6}\left[x^\rho \left| \begin{matrix} (1-\varrho,\rho;1),(1-\varrho+\varpi-\mu-\mu'-\xi',\rho;1),(1-\varrho+\xi-\mu,\rho;1),(-1,1;1),(\pm ir,1;\lambda) \\ (0,1),(\pm ir-1,1;\lambda),(1-\varrho+\xi,\rho;1),(1-\varrho+\varpi-\mu-\mu',\rho;1),(1-\varrho+\varpi-\mu-\xi',\rho;1) \end{matrix}\right.\right]. \tag{12.85}$$

Theorem 5.4 Let $\lambda - 1$, $r \in \mathbb{R}^{+}$, $\varrho > 0$ and let the contour $\mathcal{C}$ be taken as in (12.28). Let $\varrho, \mu, \mu', \xi, \xi', \eta, \varpi \in \mathbb{C}$ and satisfy the conditions $\Re(\varpi) > 0$ and $\Re(\varrho) < 1 + \min\{\Re(\xi'), \Re(\varpi - \mu - \mu', \Re(\eta - \mu' - \xi)\}$. Then the following assertion for the MSM fractional operator $\mathcal{D}_{-}^{\mu,\mu',\xi,\xi',\varpi}$ of the $S_{\lambda}\left(r,t^{\rho}\right)$ (12.28) exists and holds true:

$$\left(\mathcal{D}_{-}^{\mu,\xi,\eta}\left\{t^{\varrho-1}\tilde{S}_{\lambda}(r,t^{\rho})\right\}\right)(x) = 2x^{\varrho+\xi-1}$$
$$\times \overline{H}_{6,6}^{4,3}\left[x^{\rho}\left|\begin{matrix}(-1,1;1),(\pm ir,1;\lambda),(1-\varrho,\rho),(1-\varrho+\varpi-\mu-\mu'-v,\rho),(1-\varrho-\mu'-\xi',\rho)\\(0,1),(1-\varrho-\xi',\rho),(1-\varrho+\varpi-\mu-\mu',\rho),(1-\varrho+\varpi-\mu'-\xi,\rho),(\pm ir-1,1;\lambda)\end{matrix}\right.\right]. \tag{12.86}$$

12.6 Concluding Remarks

In this chapter, we have established some new and interesting fractional calculus results for the generalized Mathieu-type functional series and alternating Mathieu-type functional series by extensive use of the Marichev–Saigo–Maeda, Saigo's, Riemann–Liouville and Erdélyi–Kober tools. The results have been expressed in terms of generalized hypergeometric functions such as Rathie's I-function, Inayat-Hussain's $\overline{H}$-function, and Fox's H-function.

It is hoped that our results will have potential uses in the areas of mathematics for engineering and mathematical physics. Applications related to engineering are under investigation and will form part of a subsequent paper on this subject.

A similar process can be applied to the result obtained when $\lambda = 2$ for alternating Mathieu-type functional series $\tilde{S}(r;z)$:

$$\tilde{S}(r;z) = \sum_{m\geq 1}(-1)^{m-1}\frac{2mz^{m-1}}{(m^2+r^2)^2}, \qquad (\lambda > 1, r \in \mathbb{R}, |z| < 1), \tag{12.87}$$

Interested readers may wish to further research these results.

Acknowledgments

The work of Rakesh K. Parmar is supported by the research project MATRICS, Department of Science and Technology (DST), India (File No. MTR/2019/001328).

References

[1] R. P. Agarwal, A. Kılıçman, R. K. Parmar and A. K. Rathie, Certain generalized fractional calculus formulas and integral transforms involving (p,q)–Mathieu-type series, *Adv. Differ. Equ.* **221**, (2019), 1–11, https://doi.org/10.1186/s13662-019-2142-0.

[2] R.K. Saxena and R.K. Parmar, Fractional Integration and Differentiation of the Generalized Mathieu Series, *Axioms*, **6(3)**, (2017), 1–11, doi:10.3390/axioms6030018.

[3] H. Singh, D. Kumar and D. Baleanu, *Methods of Mathematical Modelling: Fractional Differential Equations*, CRC Press, Boca Raton, 2019.

[4] H. Singh, Analysis for fractional dynamics of Ebola virus model, *Chaos, Solit. Fract.*, **138** 109992 (2020).

[5] H. Singh, Analysis of drug treatment of the fractional HIV infection model of $CD4^+$ T-cells, *Chaos, Solitons and Fractals*, **146** 110868 (2021).

[6] H. Singh, Jacobi collocation method for the fractional advection-dispersion equation arising in porous media, *Numer. Methods Partial Differ. Equ.*, 2020, https://doi.org/10.1002/num.22674.

[7] H. Singh, Numerical simulation for fractional delay differential equations, *Int. J. Dynam. Control*, **9** (2021), 463–474.

[8] H. Singh, H.M. Srivastava and D. Kumar, A reliable algorithm for the approximate solution of the nonlinear Lane-Emden type equations arising in astrophysics, *Numer. Methods Partial Differ. Equ.*, (2017).

[9] H. Singh, J. Singh, S. D. Purohit and D. Kumar, *Advanced Numerical Methods for Differential Equations: Applications in Science and Engineering*, CRC Press, Boca Raton, 2021.

[10] H. Singh and H. M. Srivastava, Numerical investigation of the fractional-order Liénard and Duffing equations arising in oscillating circuit theory, *Front. Phys.* **8**: 120, 2020.

[11] H. Singh and A. M. Wazwaz, Computational method for reaction diffusion-model arising in a spherical catalyst, *Int. J. Appl. Comput. Math*, **7** 65 (2021).

[12] E. D. Rainville, *Special Functions*, Macmillan Company, New York, 1960; Reprinted by Chelsea Publishing Company, Bronx, New York, 1971.

[13] H. M. Srivastava and P. W. Karlsson, *Multiple Gaussian Hypergeometric Series*, Halsted Press (Ellis Horwood Limited, Chichester), John Wiley and Sons, New York, Chichester, Brisbane and Toronto, 1985.

[14] A. A. Kilbas, H. M. Srivastava and J.J. Trujillo, *Theory and Applications of Fractional Differential Equations, North-Holland Mathematical Studies*, Vol. **204**, Elsevier (North-Holland) Science Publishers, Amsterdam, London and New York, 2006.

[15] V. Kiryakova, *Generalized Fractional Calculus and Applications*, Pitman Research Notes in Mathematics Series, 301, Longman Scientific and Technical, Harlow; copublished in the United States with John Wiley and Sons, Inc., New York (1994).

[16] S. G. Samko, A. A. Kilbas and O. I. Marichev, *Fractional Integrals and Derivatives: Theory and Applications*, Translated from the Russian: *Integrals and Derivatives of Fractional Order and Some of Their Applications* (Nauka i Tekhnika, Minsk, 1987); Gordon and Breach Science Publishers: Reading, UK, 1993.

[17] H. M. Srivastava and R. K. Saxena, Operators of fractional integration and their applications, *Appl. Math. Comput.*, **118 (1)**, (2001), 1–52.

[18] M. Saigo, A remark on integral operators involving the Gauss hypergeometric functions, *Math. Rep. Kyushu Univ.* **11** (1977/78), 135–143.

[19] E. L. Mathieu, *Traité de Physique Mathématique, VI-VII: Théorie de l'élasticité des corps solides*, Gauthier–Villars, Paris, 1890.

[20] T. K. Pogány, H. M. Srivastava and Ž. Tomovski, Some families of Mathieu **a**–series and alternating Mathieu **a**–series, *Appl. Math. Comp.* **173** (2006), 69–108.

[21] O. Emersleben, Über die Reihe $\sum_{k=1}^{\infty} \frac{k}{\left(k^2+r^2\right)^2}$, *Math. Ann.* **125** (1952), 165–171.

[22] P. Cerone and C. T. Lenard, On integral forms of generalized Mathieu series, *JIPAM J. Inequal. Pure Appl. Math.* **4(5)** (2003), Art. No. 100, 1–11.

[23] Gradimir V. Milovanović and T. K. Pogány, New integral forms of generalized Mathieu series and related applications, *Appl. Anal. Discrete Math.* **7** (2013), 180–192.

[24] P. H. Diananda, Some inequalities related to an inequality of Mathieu, *Math. Ann.* **250** (1980), 95–98.

[25] R. K. Saxena, T. K. Pogány and R. Saxena, Integral transforms of the generalized Mathieu series, *J. Appl. Math. Stat. Inform.* **6** (2010), 5–16.

[26] Á. Baricz, P. L. Butzer and T. K. Pogány, Alternating Mathieu series, Hilbert - Eisenstein series and their generalized Omega functions, in T. Rassias, G. V. Milovanović (Eds), *Analytic Number Theory, Approximation Theory, and Special Functions - In Honor of Hari M.* Srivastava, Springer, New York, 2014, 775–808.

[27] J. Choi and H. M. Srivastava, Mathieu series and associated sums involving the Zeta functions, *Comput. Math. Appl.* **59** (2010), 861–867.

[28] N Elezović, H. M. Srivastava and Ž. Tomovski, Integral representations and integral transforms of some families of Mathieu type series, *Integral Transforms Spec. Funct.* **19(7)** (2008), 481–495.

[29] T. K. Pogány and Ž. Tomovski, Bounds improvement for alternating Mathieu type series, *J. Math. Inequal.* **4(3)** (2010), 315–324.

[30] Ž. Tomovski, Integral representations of generalized Mathieu series via Mittag-Leffler type functions, *Fract. Calc. Appl. Anal.* **10** (2007), 127–138.

[31] Ž. Tomovski and T. K. Pogány, New integral forms of generalized Mathieu series and related applications, *Appl. Anal. Discrete Math.* **7** (2013), 180–192.

[32] Ž. Tomovski and T. K. Pogány, New upper bounds for Mathieu-type series, *Banach J. Math. Anal.* **3(2)** (2009), 9–15.

[33] Ž. Tomovski and T. K. Pogány, Integral expressions for Mathieu-type power series and for the Butzer-Flocke-Hauss Ω–function, *Fract. Calc. Appl. Anal.* **14** (2011), 623–634.

[34] Rakesh K. Parmar, G. V. Milovanović and Tibor K. Pogány, Multi-parameter Mathieu, and alternating Mathieu series, *Appl. Math. Comp.* **400** (2021), 1–27. Article ID 126099. https://doi.org/10.1016/j.amc.2021.126099

[35] C. Fox, The G and H functions as symmetrical Fourier Kernels, *Trans. Amer. Math. Soc.* **98** (1961), 395–429.

[36] A. A. Inayat-Hussain, New properties of hypergeometric series derivable from Feynman integrals. I: Transformation and reduction formulae, *J. Phys. A: Math. Gen.* **20** (1987), 4109–4117.

[37] A. A. Inayat-Hussain, New properties of hypergeometric series derivable from Feynman integrals. II: A generalization of the *H*-function, *J. Phys. A: Math. Gen.* **20** (1987), 4119–4128.

[38] R. G. Buschman and H. M. Srivastava, The *H*–function associated with a certain class of Feynman integrals, *J. Phys. A: Math. Gen.* **23** (1990), 4707–4710.

[39] H. M. Srivastava, Shy-Der Lin and Pin-Yu Wang, Some Fractional-Calculus Results for the $\bar{H}$–Function Associated with a Class of Feynman Integrals, *Russ. J. Math. Phy.*, **13(1)** (2006), 94–100.

[40] A. M. Mathai, R. K. Saxena and H. J. Haubold, *The H-Functions: Theory and Applications*, Springer, New York, 2010.

[41] A. K. Rathie, A new generalization of generalized hypergeometric functions, *Le Matematiche* **52(2)** (1997), 297–310.

Index

Page numbers in *italics* refer to figures.

Printed in the United States
by Baker & Taylor Publisher Services